Catalyzed Mizoroki–Heck Reaction or C–H Activation

Catalyzed Mizoroki–Heck Reaction or C–H Activation

Special Issue Editor

Sabine Berteina-Raboin

MDPI • Basel • Beijing • Wuhan • Barcelona • Belgrade

Special Issue Editor
Sabine Berteina-Raboin
University of Orléans
France

Editorial Office
MDPI
St. Alban-Anlage 66
4052 Basel, Switzerland

This is a reprint of articles from the Special Issue published online in the open access journal *Catalysts* (ISSN 2073-4344) from 2017 to 2019 (available at: https://www.mdpi.com/journal/catalysts/special_issues/mizoroki_heck).

For citation purposes, cite each article independently as indicated on the article page online and as indicated below:

LastName, A.A.; LastName, B.B.; LastName, C.C. Article Title. *Journal Name* **Year**, *Article Number*, Page Range.

ISBN 978-3-03928-138-1 (Pbk)
ISBN 978-3-03928-139-8 (PDF)

© 2020 by the authors. Articles in this book are Open Access and distributed under the Creative Commons Attribution (CC BY) license, which allows users to download, copy and build upon published articles, as long as the author and publisher are properly credited, which ensures maximum dissemination and a wider impact of our publications.

The book as a whole is distributed by MDPI under the terms and conditions of the Creative Commons license CC BY-NC-ND.

Contents

About the Special Issue Editor vii

Sabine Berteina-Raboin
Catalyzed Mizoroki-Heck Reaction or C-H Activation
Reprinted from: *Catalysts* **2019**, *9*, 925, doi:10.3390/catal9110925 1

Jing Yang, Hua-Wen Zhao, Jian He and Cheng-Pan Zhang
Pd-Catalyzed Mizoroki-Heck Reactions Using Fluorine-Containing Agents as the
Cross-Coupling Partners
Reprinted from: *Catalysts* **2018**, *8*, 23, doi:10.3390/catal8010023 4

Sangeeta Jagtap
Heck Reaction—State of the Art
Reprinted from: *Catalysts* **2017**, *7*, 267, doi:10.3390/catal7090267 39

Benjamin Large, Flavien Bourdreux, Aurélie Damond, Anne Gaucher and Damien Prim
Palladium-Catalyzed Regioselective Alkoxylation via C–H Bond Activation in the
Dihydrobenzo[c]acridine Series
Reprinted from: *Catalysts* **2018**, *8*, 139, doi:10.3390/catal8040139 92

Joana F. Campos, Maria-João R. P. Queiroz and Sabine Berteina-Raboin
The First Catalytic Direct C–H Arylation on C2 and C3 of Thiophene Ring Applied to
Thieno-Pyridines, -Pyrimidines and -Pyrazines
Reprinted from: *Catalysts* **2018**, *8*, 137, doi:10.3390/catal8040137 101

Shuai Shi, Khan Shah Nawaz, Muhammad Kashif Zaman and Zhankui Sun
Advances in Enantioselective C–H Activation/Mizoroki-Heck Reaction and Suzuki Reaction
Reprinted from: *Catalysts* **2018**, *8*, 90, doi:10.3390/catal8020090 115

Geoffrey Dumonteil, Marie-Aude Hiebel and Sabine Berteina-Raboin
Solvent-Free Mizoroki-Heck Reaction Applied to the Synthesis of Abscisic Acid and
Some Derivatives
Reprinted from: *Catalysts* **2018**, *8*, 115, doi:10.3390/catal8030115 148

Magda H. Abdellattif and Mohamed Mokhtar
MgAl-Layered Double Hydroxide Solid Base Catalysts for Henry Reaction: A Green Protocol
Reprinted from: *Catalysts* **2018**, *8*, 133, doi:10.3390/catal8040133 157

About the Special Issue Editor

Sabine Berteina-Raboin received her Ph.D. in organic chemistry from the University of Paris-Sud XI in 1994 under the supervision of Prof. A. Lubineau in the field of oligosaccharide chemistry. After a year of research in the Analytical Research and Development Department of Bristol-Myers Squibb (Paris, France) and 2.5 years of postdoctoral work in the field of combinatorial chemistry in the group of Dr. A. De Mesmaeker, first in the Central Research Laboratories of Ciba-Geigy, then in the Novartis Crop Protection division in Basel, Switzerland, she joined the Institute of Organic and Analytical Chemistry of the University of Orléans (France) in 1998 as a lecturer in the team of Prof. G. Guillaumet. Her research work was dedicated to solid phase synthesis and heterocyclic chemistry. Since 2007, she has been the president of an association aiming to promote scientific collaboration and technology transfer between industrial corporations and academic research, at regional, national, and international levels. She became Professor of Organic Chemistry in 2009 at University of Orléans. For the last six years, her research activities have focused on green chemistry and natural products.

Editorial

Catalyzed Mizoroki-Heck Reaction or C-H Activation

Sabine Berteina-Raboin

Institut de Chimie Organique et Analytique (ICOA), Université d'Orléans UMR-CNRS 7311, BP 6759, rue de Chartres, 45067 Orléans CEDEX 2, France; sabine.berteina-raboin@univ-orleans.fr

Received: 29 October 2019; Accepted: 30 October 2019; Published: 6 November 2019

In the last few decade, research conducted on the development by catalytic processes of C-C bonds formation on the one hand and on the other hand on the activation of C-H bonds has grown considerably [1,2]. For their outstanding contribution to development of Palladium-Catalyzed cross-coupling reaction, Richard F. Heck with Akira Suzuki and Ei-ichi Negishi obtained the 2010 Nobel Prize in Chemistry. However, many improvements are still possible in terms of selectivity or even enantioselectivity via the development of new ligands or the study of the catalytic effect of other metals to carry out the same chemical transformations.

Zhang et al. emphasize, in their review [3], that the Mizoroki-Heck reaction is one of the most important catalytic methods to generate C-C bonds in organic synthesis. This reaction is highly efficient and has good chemo- and stereoselectivity. The authors discuss the interest of using fluorine-containing agents as cross-coupling partners, in this reaction, to introduce fluorine atom(s) into organic molecules because these compounds show advantageous physicochemical and biological properties. However, only few organo-fluorinated natural compounds exist. The Mizoroki-Heck Cross-coupling reaction between fluorinated alkenes with aryl halide or equivalent and/or some alkenes with fluorinated aryl or alkyl halide or equivalent is the best pathway to obtain these fluorinated organic compounds in a view to drug discovery and advanced materials. This review presents the various combination of reagents usable with this process.

Sangeeta Jagtap [4] gives an overview of the state of the art of the Heck reaction. This reaction has played an important role in the elaboration of numerous compounds in many fields and is probably one of the most studied cross-coupling reactions in organic and medicinal chemistry. Sangeeta Jagtap has summarized many reviews on this topic mainly about catalysts, ligands and various conditions used but also suggested mechanisms.

Damien Prim's group worked on acridines, aza-polycyclic compound having a broad range of properties and applications in therapeutic, pigments, dyes, imaging probes, sensor and some other materials activities [5]. The 5,6-dihydrobenzo[c]acridine, comprises of four fused cycles of which one is partially hydrogenated leading to not fully planar compounds usable in the preparation of helical-shaped molecules. The authors succeeded to functionalize this kind of tetracyclic molecule by regioselective alkoxylation via C-H bond activation avoiding the standard pre-halogenation.

Berteina-Raboin's group developed the first direct C-H arylation on C2 and C3 thiophene ring for a convenient one-pot synthesis of thienopyridine, thienopyrimidine, and thienopyrazine scaffolds [6].

The attention paid to environmentally friendly methods in terms of the quantities of catalysts, ligands and solvents is currently indispensable. In this context, Shi, Nawaz, Zaman and Sun [7] summarized recent advances in enantioselective C-H activation/functionalization via Mizoroki-Heck reaction or Suzuki reaction. These are methodologies that are used to generate synthetic, hemisynthetic or natural compounds with high added value whether in medicinal chemistry or agrochemistry. Conventional methods require pre-functionalization of the substrates which generates additional steps of synthesis and purification whereas the direct activation of the C-H bond allows an atom-economy and a more sustainable chemistry. However, the direct activation of inert C-H bonds still remains difficult because of the poor reactivity and selectivity. In their review the authors discuss the latest progress on enantioselective C-H activation via Mizoroki-Heck or Suzuki reaction with a particular interest on the origin of chirality and discussion on mechanisms with chiral ligands used.

Always in the interest of developing in a more environmentally sound manner, Berteina-Raboin's group [8] described the synthesis of the ABA phytohormone (abscisic acid) which exhibit some interesting biological activities and synthesis of new analogues performed solvent and ligand free Mizoroki-Heck reaction conditions. Some delicate dienes and trienes were obtained without isomerization in moderate to good yields (27–78%). The phytohormone ABA was synthesized with this process in four steps from commercially available diketone in 54% global yields.

Finally, Mokhtar's team has developed a series of heterogeneous catalysts, the MgAl-layered double hydroxide, its calcined form at 500 °C (MgALOx) and the rehydrated form (MgAl-HT-RH) for the Henry reaction between nitroalkanes and various aldehydes [9]. This is the first study for understanding the effect of mesoporous and basic nature of this kind of catalysts for the Henri reaction. These catalysts have been fully characterized and the large surface area of mesoporous catalysts as well as the strong basic sites of rehydrated catalyst allowed a very efficient catalytic activity. In addition, the catalyst is reusable without loss of activity after five catalytic cycles what makes this process a green protocol.

This Special Issue on "Catalyzed Mizoroki-Heck Reaction or C-H activation" was focussed on new advances in the formation of C-C bonds via the Mizoroki-heck reaction or new C-H activation methods. I would like sincerely thank all authors for their valuable contributions, original research papers and short reviews on synthesis of biologically active compounds using these catalytic processes, identification of new catalysts, of new conditions allowing selectivity or enantioselectivity, the activity and stability of catalysts under turnover conditions and all improvements in catalytic processes. I also sincerely thank the editorial team of Catalysts for their kind support and fast responses. Without you all, this special issue would not have been possible.

Excellent research is being performed worldwide on new processes to increase the efficiency of bioactive compounds elaboration and numerous efforts were made to develop sustainable chemistry but we only are at the beginning.

Conflicts of Interest: The author declares no conflicts of interest.

References

1. Kim, D.S.; Park, W.J.; Jun, C.H. Metal–Organic Cooperative Catalysis in C–H and C–C Bond Activation. *Chem. Rev.* **2017**, *117*, 8977–9015. [CrossRef] [PubMed]
2. Ritleng, V.; Sirlin, C.; Pfeffer, M. Ru-, Rh-, and Pd-Catalyzed C−C Bond Formation Involving C−H Activation and Addition on Unsaturated Substrates: Reactions and Mechanistic Aspects. *Chem. Rev.* **2002**, *102*, 1731–1770. [CrossRef] [PubMed]
3. Yang, J.; Zhao, H.W.; He, J.; Zhang, C.P. Pd-Catalyzed Mizoroki-Heck Reactions Using Fluorine-Containing Agents as the Cross-Coupling Partners. *Catalysts* **2018**, *8*, 23. [CrossRef]
4. Jagtap, S. Heck Reaction-State of the Art. *Catalysts* **2017**, *7*, 267. [CrossRef]
5. Campos, J.F.; Berteina-Raboin, S. The First Catalytic Direct C-H Arylation on C2 and C3 of Thiophene Ring Applied to Thieno-Pyridines, -Pyrimidines and –Pyrazines. *Catalysts* **2018**, *8*, 137. [CrossRef]
6. Large, B.; Bourdreux, F.; Damond, A.; Anne Gaucher, A.; Prim, D. Palladium-Catalyzed Regioselective Alkoxylation via C-H Bond Activation in the Dihydrobenzo[c]acridine Series. *Catalysts* **2018**, *8*, 139. [CrossRef]
7. Dumonteil, G.; Hiebel, M.-A.; Berteina-Raboin, S. Solvent-Free Mizoroki-Heck Reaction Applied to the Synthesis of Abscisic Acid and some Derivatives. *Catalysts* **2018**, *8*, 115. [CrossRef]

8. Shi, S.; Nawaz, K.S.; Zaman, M.K.; Sun, Z. Advances in Enantioselective C-H Activation/Mizoroki-Heck Reaction and Suzuki Reaction. *Catalysts* **2018**, *8*, 90. [CrossRef]
9. Abdellattif, M.H.; Mokhtar, M. MgAl-Layered Double Hydroxide Solid Base Catalysts for Henry Reaction: A Green Protocol. *Catalysts* **2018**, *8*, 133. [CrossRef]

© 2019 by the author. Licensee MDPI, Basel, Switzerland. This article is an open access article distributed under the terms and conditions of the Creative Commons Attribution (CC BY) license (http://creativecommons.org/licenses/by/4.0/).

Review

Pd-Catalyzed Mizoroki-Heck Reactions Using Fluorine-Containing Agents as the Cross-Coupling Partners

Jing Yang [1], Hua-Wen Zhao [1,*], Jian He [1] and Cheng-Pan Zhang [1,2,*]

[1] Department of Chemistry, College of Pharmacy, Army Medical University, Shapingba, Chongqing 400038, China; yangjing93@whut.edu.cn (J.Y.); freedpower@tmmu.edu.cn (J.H.)
[2] School of Chemistry, Chemical Engineering and Life Science, Wuhan University of Technology, 205 Luoshi Road, Wuhan 430070, China
* Correspondence: sydzhw@aliyun.com (H.-W.Z.); cpzhang@whut.edu.cn (C.-P.Z.); Tel.: +86-023-6877-2357 (H.-W.Z. & C.-P.Z.)

Received: 27 December 2017; Accepted: 10 January 2018; Published: 14 January 2018

Abstract: The Mizoroki-Heck reaction represents one of the most convenient methods for carbon-carbon double bond formation in the synthesis of small organic molecules, natural products, pharmaceuticals, agrochemicals, and functional materials. Fluorine-containing organic compounds have found wide applications in the research areas of materials and life sciences over the past several decades. The incorporation of fluorine-containing segments into the target molecules by the Mizoroki-Heck reactions is highly attractive, as these reactions efficiently construct carbon-carbon double bonds bearing fluorinated functional groups by simple procedures. This review summarizes the palladium-catalyzed Mizoroki-Heck reactions using various fluorine-containing reagents as the cross-coupling partners. The first part of the review describes the Pd-catalyzed Mizoroki-Heck reactions of aryl halides or pseudo-halides with the fluorinated alkenes, and the second part discusses the Pd-catalyzed Mizoroki-Heck reactions of the fluorinated halides or pseudo-halides with alkenes. Variants of the Pd-catalyzed Mizoroki-Heck reactions with fluorine-containing reagents are also briefly depicted. This work supplies an overview, as well as a guide, to both younger and more established researchers in order to attract more attention and contributions in the realm of Mizoroki-Heck reactions with fluorine-containing participants.

Keywords: Mizoroki-Heck reaction; Pd-catalyzed; fluorine; cross-coupling; alkenes; halides

1. Introduction

The palladium-catalyzed carbon-carbon cross-coupling of an aryl or vinyl halide and an alkene in the presence of a base is referred as the "Mizoroki-Heck reaction" [1–6]. The reaction was discovered independently by Heck and Mizoroki more than 45 years ago. Heck first reported the Li_2PdCl_4-mediated reactions of organomercury compounds with olefins in acetonitrile or methanol at room temperature [7]. Then, Mizoroki and co-workers disclosed the first cross-couplings of aryl iodide with alkenes catalyzed by $PdCl_2$ in methanol in the presence of potassium acetate at 120 °C [8]. In 1972, Heck and co-worker proposed a possible mechanism for the reactions of aryl, benzyl, or styryl halides (R–X) with alkenes and a catalytic amount of $Pd(OAc)_2$ under milder conditions (Scheme 1) [9]. In these conversions, an oxidative addition occurs between Pd(0) (formed in situ from reduction of $Pd(OAc)_2$ by olefin) and R–X (**1**), presumably generating a very reactive solvated organopalladium(II) halide ([R–Pd–X], **2**), which is probably the same intermediate produced previously in the exchange reactions between palladium halides and organomercury compounds [7]. [R–Pd–X] undergoes an addition reaction with olefin (**3**) to yield a palladium adduct (**4**), which decomposes by elimination of a hydridopalladium halide ([H–Pd–X], **6**) to form the substituted olefinic compound (**5**). Reductive elimination of HX from [H–Pd–X] in the presence of a certain base regenerates the Pd(0) species, maintaining the catalytic cycle. Formally speaking, the vinylic hydrogen atom of alkene is substituted by the organic residue (R) of R–X in the reactions (Scheme 1).

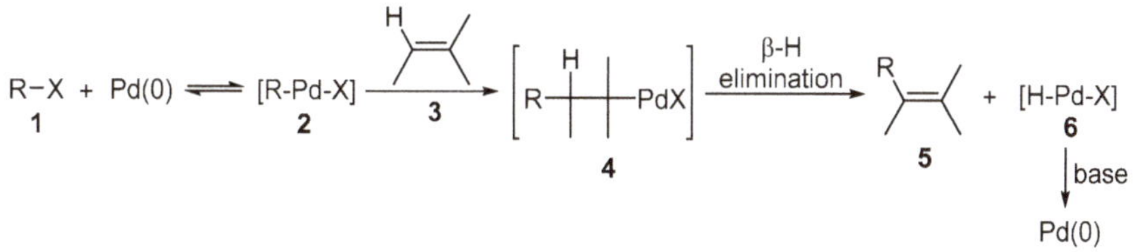

Scheme 1. A possible mechanism for the Mizoroki-Heck reactions.

Due to its high efficiency, easy operation, and good chemo- and stereoselectivity, the Mizoroki-Heck reaction has been extensively used for functionalization of various organic scaffolds since its discovery. Currently, the Mizoroki-Heck reaction has become one of the most important tools for the formation of carbon-carbon double bonds [1–6]. To manifest the importance of the Mizoroki-Heck reaction, Richard F. Heck, together with Ei-ichi Negishi and Akira Suzuki, was honored with the 2010 Nobel Prize in Chemistry for their great contribution in the development of Pd-catalyzed cross-coupling reactions in organic synthesis. The Mizoroki-Heck reactions have exhibited good functional group tolerance with a wide range of substrates under mild conditions. At present, not only aryl, vinyl, benzyl, and alkyl halides [1–6], but also the corresponding pseudo halides such as sulfonates [10–12], sulfonyl chlorides [13–15], carboxylic acid derivatives [16–18], diazonium salts [19–21], iodonium salts [22,23], phosphonium salts [24], and sulfonium salts [25], have been successfully employed as electrophiles in Heck-type cross-couplings. Both electron-poor and -rich alkenes (such as acrylic esters, enolethers, and ethylene) have proved to be viable cross-coupling partners in the reactions [1–6]. The Mizoroki-Heck reactions are originally catalyzed by palladium [1–6]. Other transition metals, such as nickel, cobalt, copper, gold, and iron, are also active catalysts for Heck-type reactions [26–33]. The visible light-induced Pd-catalyzed Mizoroki-Heck reactions between sterically hindered alkyl halides and vinyl arenes have been accomplished, as well [34,35].

On the other hand, fluorine is a very intriguing atom for its unique properties. Introduction of fluorine atom(s) into organic molecules usually brings about a dramatic impact on the physicochemical and biological properties of the molecules [36–38]. Fluorine-containing organic compounds have found wide application in the areas of chemistry, biology, and materials science over the past several decades [39–45]. There have been as many as 25% of pharmaceuticals and 30–40% of agrochemicals on the market containing at least a single fluorine atom [41]. Because only a few naturally-occurring organofluorides have been discovered, most of the fluorinated organic compounds have to be manually synthesized [36–51]. It is undoubted that the development of efficient methods to construct fluorine-containing molecules is of great importance [46–51]. In general, the fluorinated compounds can be synthesized by direct fluorination or fluoroalkylation, or through reactions with the fluorine-containing building blocks [39–51]. The incorporation of fluorine-containing fragments into organic frameworks by Pd-catalyzed Mizoroki-Heck reactions has proved to be the simplest and most convenient pathway to build diverse alkenes bearing fluorinated functionalities. This strategy includes the Pd-catalyzed cross-couplings of fluorinated alkenes with aryl halides or pseudo halides, and Pd-catalyzed reactions of alkenes with the fluorinated aryl or alkyl halides or pseudo halides. Fluorine-containing alkenes are versatile building blocks in the synthesis of bioactive molecules for drug discovery and advanced materials for specific applications (see Sections 2–4). The concise, straightforward, selective, and highly efficient preparation of the fluorinated alkenes by the Mizoroki-Heck reactions has made these compounds easy to access and diversify.

To our knowledge, there has been no review article systematically summarizing the Pd-catalyzed Mizoroki-Heck reactions using fluorine-containing agents as the cross-coupling partners. To fill the gap in this area, we present an overview of the recent advances in the Pd-catalyzed Mizoroki-Heck reactions with the fluorinated cross-coupling participants including the fluorinated alkenes and/or the fluorinated aryl or alkyl halides. This review offers as a guide to both younger and more established researchers, and is intended to attract more attention to and contributions in the development of the Mizoroki-Heck reactions with fluorine-containing cross-coupling reagents.

2. Mizoroki-Heck Reactions of Aryl Halides or Pseudo-Halides with Fluorine-Containing Alkenes

2.1. Fluoroalkenes as Cross-Coupling Participants

The Mizoroki-Heck reaction provides a convenient method for the arylation of olefins [1–6]. In most cases, the vinylic hydrogen atom is formally substituted by the organic residue of an organic halide under the Heck-type reaction conditions. High regioselectivities of the arylation at the less substituted site of the carbon-carbon double bond of the unsymmetrically substituted olefins are usually observed, which may be attributed to the steric factors [1–6]. One of the key steps of the reaction is β-hydride elimination. However, when aryl bromide or iodide (**7**) reacted with vinylidene difluoride (**8**) in the presence of Pd(OAc)$_2$, the expected β-H elimination product, β,β-difluorostyrene (**9**), was formed only in a very small amount (Scheme 2) [45,52]. The major product of the reaction was α-fluorostyrene (**10**). Moreover, the reaction of vinyl fluoride (**11**) with **7a** gave styrene (**12**) and stilbene (**13**), the ratio of which strongly depended on the reaction conditions. Treatment of **7a** with trifluoroethylene (**14**) produced an isomeric mixture of **15**, **16** and **17** (45%), with **15** being predominant (86% by GC). Unexpectedly, **15** again was the product when chlorotrifluoroethylene (**18**) was treated with **7a**. These results suggested the substitution of a vinylic fluorine atom in all cases, which were distinct from the known Heck-type reactions with non-fluorinated olefins [1–6]. The transformations represented the first examples of charge-controlled Heck-type reaction, which was only significant in the presence of fluoroolefins [52].

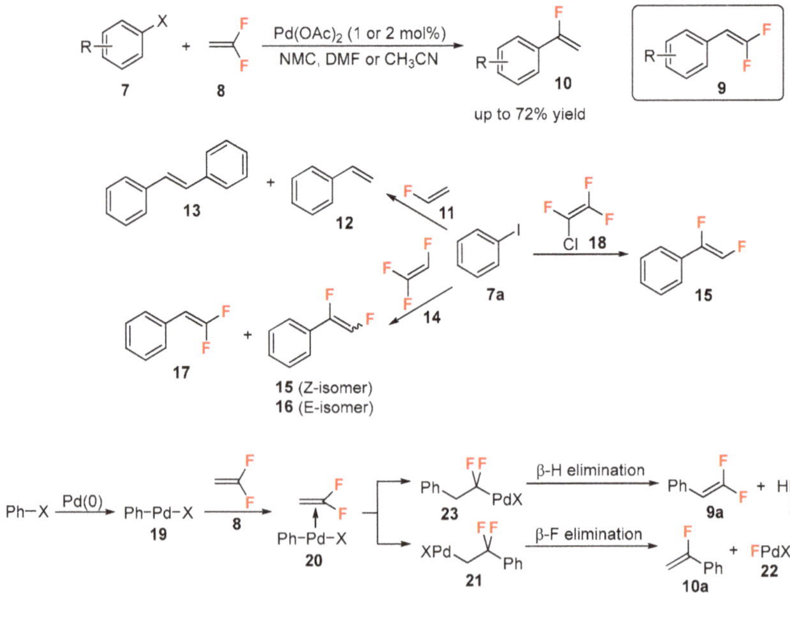

Scheme 2. Pd-catalyzed reactions of aromatic halides with fluoroolefins.

Mechanistically, the reaction starts with the formation of a palladium adduct (**19**) by the oxidative addition of the in situ generated Pd(0) species with **7a**, which undergoes olefin coordination to produce complex **20** (Scheme 2) [52]. Then, the phenyl group in **20** transfers from the Pd center to the CF_2 site of **8**, affording **21**. A β-fluorine elimination of **21** yields α-fluorostyrene (**10a**) as the final product. Compound **23** would be an intermediate, if the steric aspects were relevant, and a subsequent β-hydride elimination of **23** could form **9a**. However, the formation of **23** must be of low probability, as only trace amounts of **9a** were detected. The favorable formation of **21** in the reaction of **8** could be explained by the charge-controlled mechanism, which was verified by the MNDO-calculations [52]. Furthermore, the reactions of **7a** with **11** and **14** obeyed a mechanism similar to that of **7a** with **8**, and the reaction of **7a** with **18** underwent both the olefinic fluorine substitution and the C–Cl bond reduction. The β-fluorine elimination seemed to be the preferred type of elimination, even though the competitive β-hydride elimination was a possible pathway [45,52]. This procedure constituted a convenient method for the preparation of α-fluorostyrenes.

More than fifteen years later, Patrick and co-workers found that 3-fluoro-3-buten-2-one (**26**) reacted smoothly with aryl iodides (**25**) under the Heck-type cross-coupling conditions to give 3-fluorobenzalacetones (**27**) in good yields with only Z-stereoselectivity (Scheme 3) [53]. The reaction used $Pd(OAc)_2$ as a catalyst, triphenylphosphine as a ligand, and triethylamine as a base in DMF. The conjugate addition products and the fluoride elimination products were not observed. The preferable *trans* relationship between the aryl and acyl groups during the reaction was maintained in the configuration of the final product (**27**). The required *syn*-elimination of $HPdL_2$ in intermediate **28** sustained very small steric repulsion between the aryl group and the fluorine atom.

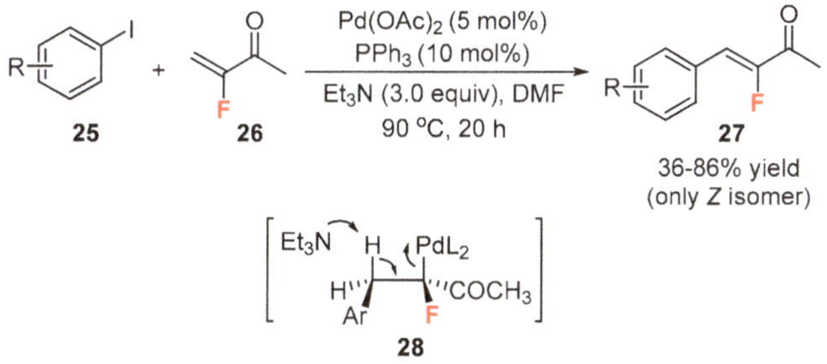

Scheme 3. Pd-catalyzed Mizoroki-Heck reactions of aryl iodides with 3-fluoro-3-buten-2-one.

In 2016, Couve-Bonnaire and co-workers reported the ligand-free palladium-catalyzed Mizoroki-Heck reactions of methyl α-fluoroacrylate (**30**) with aryl or heteroaryl iodides (**29**), leading to a cheap, efficient, and stereoselective synthesis of fluoroacrylate derivatives (**31**) in good to quantitative yields (Scheme 4) [54]. The transformation had good functional group tolerance and could be extended to more steric hindered trisubstituted alkenes, which were previously the reluctant substrates in the Mizoroki-Heck reactions. The reactions of trisubstituted (*E*)-3-alkyl-2-fluoroacrylate (**32**) with **29** under the standard conditions gave the corresponding tetrasubstituted fluoroacrylates (**33**) in fair to good yields [54]. These results constituted the first examples for the synthesis of tetrasubstituted alkenes by using the Mizoroki-Heck reaction. This methodology was also applicable to the preparation of a fluorinated analogue of a therapeutic agent against inflammation and cancers.

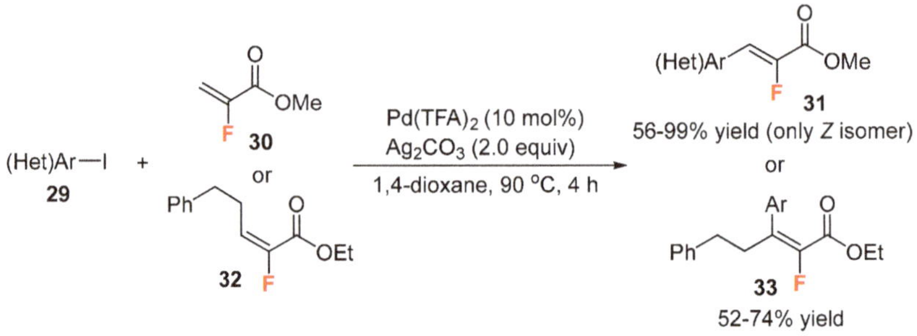

Scheme 4. Pd-catalyzed Mizoroki-Heck reactions of methyl α-fluoroacrylates with aryl or heteroaryl iodides.

Similarly, Hanamoto and co-worker disclosed the Mizoroki-Heck reactions of (1-fluorovinyl)methyldiphenylsilane (**35**) with aryl iodides (**34**) catalyzed by Pd(OAc)$_2$ (5 mol%) in the presence of Ag$_2$CO$_3$ (3 equiv) and 4 Å MS in 1,4-dioxane (Scheme 5) [55]. The reactions supplied a series of (E)-β-aryl-(α-fluorovinyl)methyldiphenylsilanes (**36**) in good yields with excellent stereoselectivity. Desilylation/protonation of the product gave the corresponding (E)-β-fluorostyrene derivative with complete retention of the configuration of the double bond, which illustrated the synthetic scope of this method.

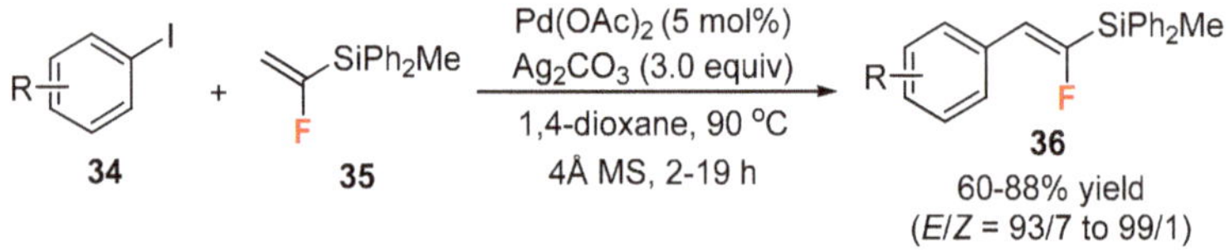

Scheme 5. Pd-catalyzed Mizoroki-Heck reactions of (1-fluorovinyl)methyldiphenylsilane with aryl iodides.

Moreover, the Mizoroki-Heck reaction of ethyl (Z)-3-fluoropropenoate (**Z-37**) or ethyl (E)-3-fluoropropenoate (**E-37**) with iodobenzene in the presence of Pd(OAc)$_2$ (5 mol%) produced ethyl 3-fluoropropenoate (**Z-39**) as a sole product (Scheme 6) [56]. The reaction proceeded smoothly at the β-position with specific stereoselectivity. The stereochemistry of **E-37** was completely inverted and only **Z-39** was produced. Meanwhile, compound **40** was formed as a side product via loss of a fluorine atom. It was possible that the catalyst system caused the isomerization. However, the exact mechanism for the high stereoselectivity of the reaction remained unclear.

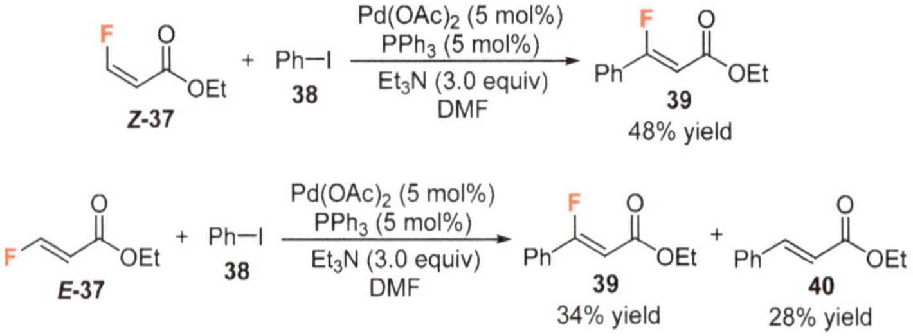

Scheme 6. Pd-catalyzed Mizoroki-Heck reactions of ethyl (E)- and (Z)-3-fluoropropenoate with iodobenzene.

Pd-Catalyzed intramolecular cyclization of O-(3,3-difluoroallyl)phenyl triflate (**41**) and 3,3-difluoroallyl ketone oximes (**46**) by the Mizoroki-Heck reactions of the polarized carbon-carbon double bonds of the 1,1-difluoro-1-alkene moieties was accomplished (Scheme 7) [45,57,58]. In the first step of the reactions, an arylpalladium or aminopalladium intermediate (**42** or **47**) bearing a 2,2-difluorovinyl group is formed from **41** or **46**, respectively. Then, intermediate **42** or **47** undergoes a 5-endo-trig alkene insertion and subsequent β-fluorine elimination to afford ring-fluorinated indene (**45**) or 3H-pyrroles (**49**). In both cases, the CF_2 unit was very essential for the cyclization as the corresponding monofluoroalkene, fluorine-free alkene, dichloroalkene, and dibromoalkene didn't give the cyclized products under the same reaction conditions [57,58].

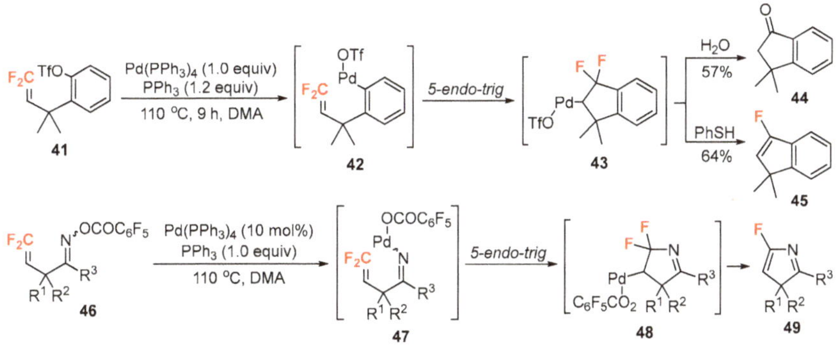

Scheme 7. Heck-type 5-endo-trig cyclization promoted by vinylic fluorines.

2.2. Fluorine-Containing Vinyl Sulfur Compounds as the Cross-Coupling Participants

Ethenesulfonyl fluoride (ESF) is a highly reactive and versatile reagent in the synthesis of a wide variety of organosulfur compounds, which behaves as a strong Michael acceptor or a Diels-Alder dienophile to conveniently introduce an SO_2F group [59]. In 2016, Wu and Sharpless described a Pd-catalyzed Heck-Matsuda process for the synthesis of the otherwise difficult to access β-arylethenesulfonyl fluorides (Scheme 8) [60]. In this reaction, ethenesulfonyl fluoride (**51**) underwent β-arylation with the stable and readily prepared arenediazonium tetrafluoroborates (**50**) in the presence of catalytic palladium(II) acetate to afford the E-isomer sulfonyl analogues of cinnamoyl fluoride (**52**) in 43–97% yield. The products **52** proved to be selectively addressable bis-electrophiles for sulfur(VI) fluoride exchange (SuFEx) click chemistry, in which either the alkenyl moiety or the sulfonyl fluoride group could be exclusively attacked by nucleophiles under defined conditions, making these simple cores attractive for covalent drug discovery [60].

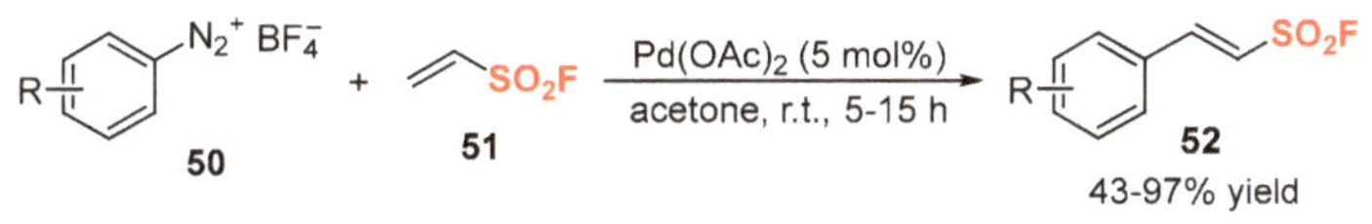

Scheme 8. Heck-Matsuda reactions of ethenesulfonyl fluoride (ESF) with aryldiazonium salts.

Later, Qin and Sharpless employed a similar strategy for the synthesis of 2-(hetero)arylethenesulfonylfluorides (**54**) and 1,3-dienylsulfonyl fluorides (**56**) (Scheme 9) [61]. They found that a combination of catalytic $Pd(OAc)_2$ with a stoichiometric amount of silver(I) trifluoroacetate enabled the coupling process between either an (hetero)aryl or alkenyl iodide (**53** or **55**) and ethenesulfonyl fluoride (ESF, **51**). The reaction was demonstrated in the successful synthesis of eighty-eight compounds in up to 99% yields, including the unprecedented 2-heteroarylethenesulfonyl fluorides and 1,3-dienylsulfonyl fluorides [61]. These substituted ethenesulfonyl fluorides are useful building blocks for consequent synthetic transformations [61].

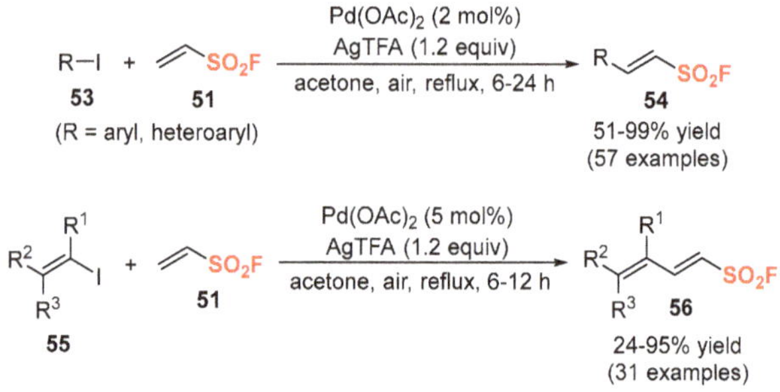

Scheme 9. Palladium-catalyzed fluorosulfonylvinylation of organic iodides.

Furthermore, the oxidative Heck cross-coupling reactions have become attractive for modern organic synthesis due to advantages such as efficiency, mild reaction conditions, good functional group tolerance, and widespread applications [62–64]. In 2017, Arvidsson and co-workers reported an operationally simple method for ligand- and additive-free oxidative Heck couplings of aryl boronic acids (**57**) with ESF (**51**) (Scheme 10) [65]. The reactions proceeded at room temperature with good chemoselectivity and E-selectivity and offered facile access to a wide range of β-aryl/heteroaryl ethenesulfonyl fluorides (**58**) from the commercially available boronic acids (**57**). The products (**58**) have a "dual warhead" with two electrophilic sites that have been used as covalent enzyme inhibitors and as synthetic reagents [65]. The authors also demonstrated that aryl-substituted β-sultams could be prepared through a one-pot procedure in which an excess of primary amine was added to the reaction mixture before workup [65].

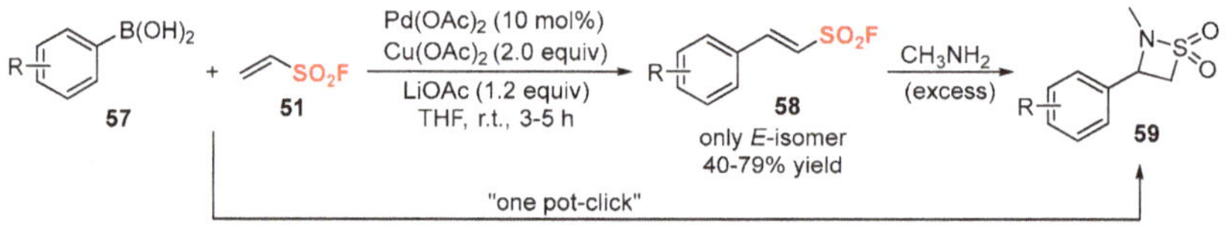

Scheme 10. The oxidative Heck couplings of boronic acids with ethenesulfonyl fluoride.

Likewise, Qin and co-workers disclosed the base-free palladium-catalyzed fluorosulfonylvinylation of (hetero)arylboronic acids (**60**) with ESF (**51**) under oxidative conditions (Scheme 11) [66]. Aryl- and heteroaryl-boronic acids (**60**) reacted with ESF in the presence of a catalytic amount of Pd(OAc)$_2$ and excess 2,3-dichloro-5,6-dicyano-p-benzoqui-none (DDQ) or AgNO$_3$ in AcOH to stereoselectively afford the corresponding E-isomer of β-arylethenesulfonyl fluoride products (**61**) in up to 99% yield. The utility of the reactions was exemplified by an expanded scope of 47 examples including N-, O-, and S-containing heteroaromatics, demonstrating chemoselectivity over aryl iodides [66].

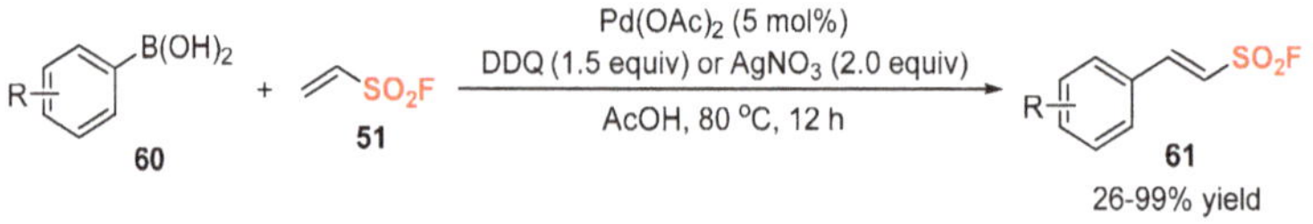

Scheme 11. The oxidative Heck reactions of arylboronic acids with ethenesulfonyl fluoride.

It should be noted that the sulfur-containing olefins (e.g., vinyl sulfides, sulfoxides) are the least investigated substrates in the Mizoroki-Heck reactions, as these substrates may poison the Pd-catalysts to form stable metal-sulfur complexes [67]. To date, only a handful of successful Heck cross-coupling reactions based on sulfoxides have been reported, even though these moieties have potential for many synthetic applications. Among this type of molecule, perfluoroalkyl vinyl sulfoxides possess a strongly polarized double bond and is highly reactive, which makes them interesting for investigation. In 2015, Sokolenko and co-workers explored the Heck-type reactions of trifluoromethyl or tridecafluorohexyl vinyl sulfoxides (**62**) with aryl iodides (**63**) (Scheme 12) [67]. Palladium(II) acetate was found to be the most suitable catalyst for the reactions. By this method, a series of E-1-aryl-2-perfluoroalkylsulfinylethylenes (**64**) were synthesized. Styrenes (**65**) without a perfluoroalkylsulfinyl group were also formed (as byproducts) in these cases. The perfluoroalkyl vinyl sulfoxides (**62**) may undergo both terminal and internal additions of the aryl group (Scheme 12). In the former case, β-elimination of the palladium species and hydride from **67** yields **64**. In the latter case, β-elimination of the palladium species and perfluoroalkylsulfinyl group from **68** leads to **65**. The formation of stable Pd–S bonds might facilitate the latter process.

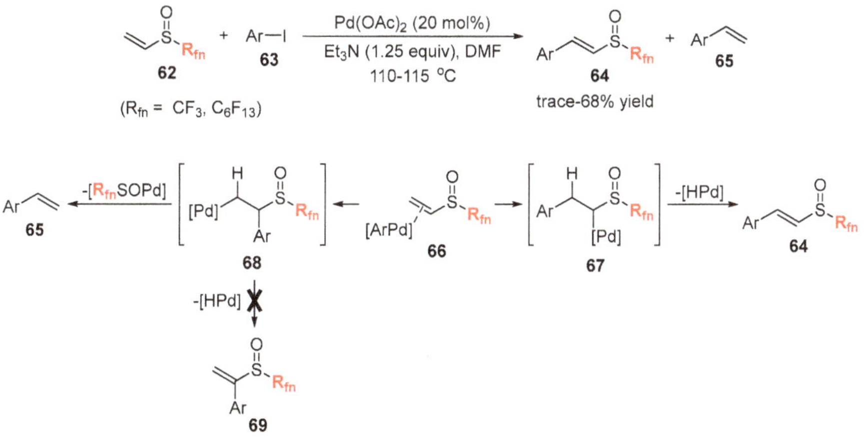

Scheme 12. The Mizoroki-Heck reactions of perfluoroalkyl vinyl sulfoxides with aryl iodides.

2.3. Fluoroalkylated Alkenes as the Cross-Coupling Partners

Perfluoroalkylated alkenes are important feedstocks for the synthesis of useful fluorine-containing molecules. In 1981, Ojima and co-workers reported the Pd-catalyzed Mizoroki-Heck reactions of 3,3,3-trifluoropropene or pentafluorostyrene (**70**) with aromatic halides (**71**) (Scheme 13) [68]. The cross-couplings proceeded smoothly by simply heating the mixtures of aryl iodides or bromides, trifluoropropene or pentafluorostyrene, a Pd-catalyst (1 mol%), and a base (such as Et$_3$N or KOAc), which gave a variety of trans-β-trifluoromethylstyrenes or trans-β-pentafluorophenylstyrenes (**72**) in good to high yields via a one-step procedure. The arylation was not sensitive to the electronic nature of the substituents on the aryl halides but was rather sensitive to the steric hindrance. However, aryl chlorides such as chlorobenzene and chlorotoluene were unreactive under the same reaction conditions.

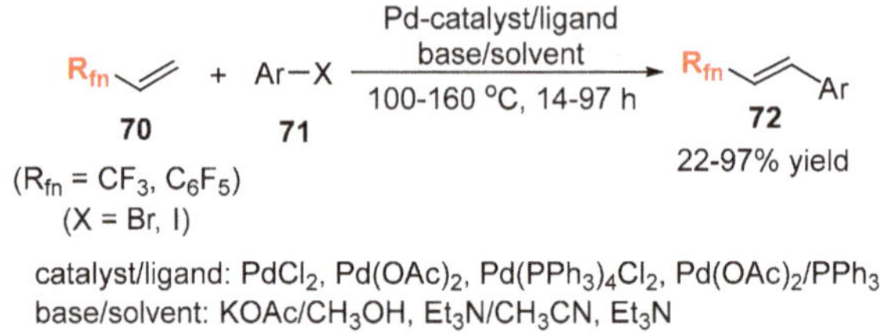

Scheme 13. The Heck-type reactions of aryl halides with 3,3,3-trifluoropropene or pentafluorostyrene.

In 2001, Xiao and co-workers found that Pd-catalyzed olefination of aryl halides (73) with 1H,1H,2H-perfluoro-1-alkenes (74) provided *trans*-fluorous ponytail-substituted aryl compounds (75) in good to excellent yields (Scheme 14) [69]. Typically, the reactions of 73 with 74 in the presence of NaOAc and the Herrmann-Beller palladacycle catalyst (0.5–1 mol%) in DMF supplied the *trans* olefins 75 in more than 90% isolated yields in most cases without optimization. Purification of 75 by flash chromatography allowed easy reduction of the C=C double bonds of 75 under the standard Pd/C- or Rh/C-catalyzed hydrogenation conditions [69]. This method was successfully applied to the synthesis of binaphathols bearing fluorinated ponytails, which are potential ligands for catalysis in scCO$_2$ and fluorous solvents [69,70]. The application of these ligands in the ruthenium-catalyzed hydrogenation of dimethyl itaconate revealed that the fluorous ponytails on the ligands imposed significant effects on the hydrogenation activity, but not on the enantioselectivity [70].

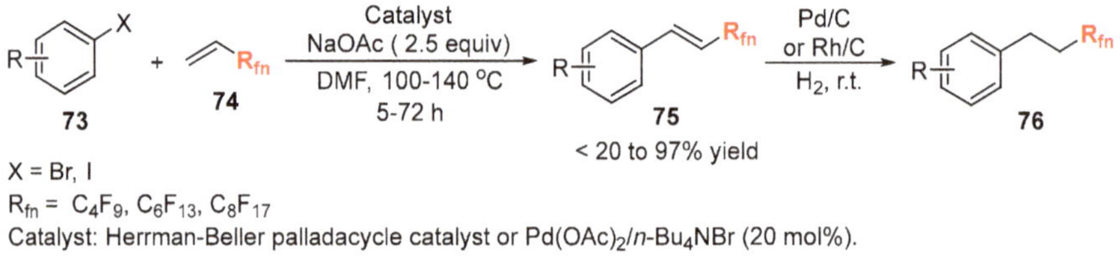

Scheme 14. The Mizoroki-Heck reactions between perfluoroalkyl alkenes and aryl halides.

Later, Cai and co-worker developed a PCP Pincer-Pd catalyst (77), which was favorably used in Mizoroki-Heck cross-couplings between perfluoroalkyl alkenes (74) and aryl halides (78) for the preparation of perfluoroalkenylated aryl compounds (79) (Scheme 15) [71]. Due to the unique tridentate coordination architecture, the Pincer complex was stable, selective, and highly reactive, and it permitted low catalyst loadings and gave the possibility of fine-tuning the catalytic properties of the metal center. Catalyst 77 showed high catalytic activity in this reaction and the corresponding arylated products were obtained in moderate to excellent yields. Moreover, 77 could be easily separated by F-SPE (Fluorous solid phase extraction) technique and reused three times without significant loss of activity.

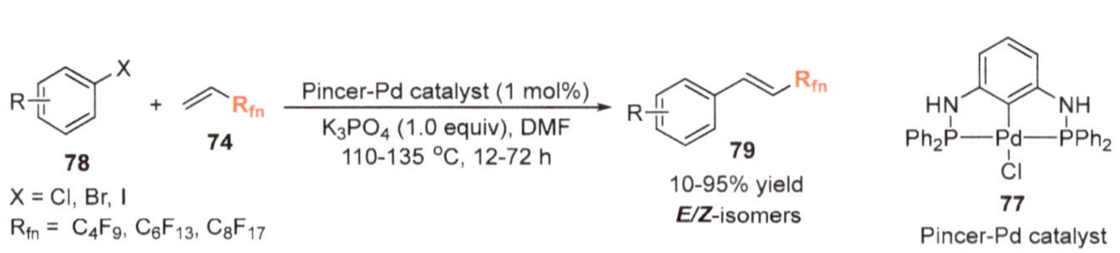

Scheme 15. The Mizoroki-Heck cross-couplings of aryl halides with perfluoroalkyl-substituted alkenes.

Gladysz and co-workers reported a convenient and scalable procedure for the cross-couplings of fluorous alkenes (**80**) with aryl bromides (**81**) using a modified Jeffery version of the Mizoroki-Heck reaction (Scheme 16) [72]. Fluorous alkenes (**80**) reacted with aryl monobromides and polybromides (**81**), such as 1,3- and 1,4-$C_6H_4Br_2$, 1,3,5-$C_6H_3Br_3$, 1,3,5-$C_6H_3Br_2Cl$, 1,4-XC_6H_4Br (X = CF_3, C_8F_{17}, $COCH_3$, CN, 1,4-OC_6H_4Br), 1,2-$O_2NC_6H_4Br$, 5-bromo-isoquinoline, 5-bromopyrimidine, 3-bromo-5-methoxypyridine, and 3,5-dibromopyridine, under the modified Mizoroki-Heck coupling conditions to afford the corresponding fluorophilic adducts (**82**) in good to high yields. Typically, 1.2–2.4 equiv. of alkene were employed per Ar-Br bond, together with $Pd(OAc)_2$ (4–5 mol%/Ar-Br bond), n-Bu_4NBr (0.8–1.0 equiv/Ar-Br bond), NaOAc (1.2–2.4 equiv/Ar-Br bond), and DMF/THF (3:1 w/w) as solvent (120 °C). It was not necessary to exclude air or moisture, and the reactions could be conducted on >10 g scales. Only E-isomers of the products **82** were detected. Hydrogenation of thirteen representative products with Pd/C and balloon pressure H_2 gave Ar($CH_2CH_2R_{fn}$)$_m$ in 92% average isolated yield.

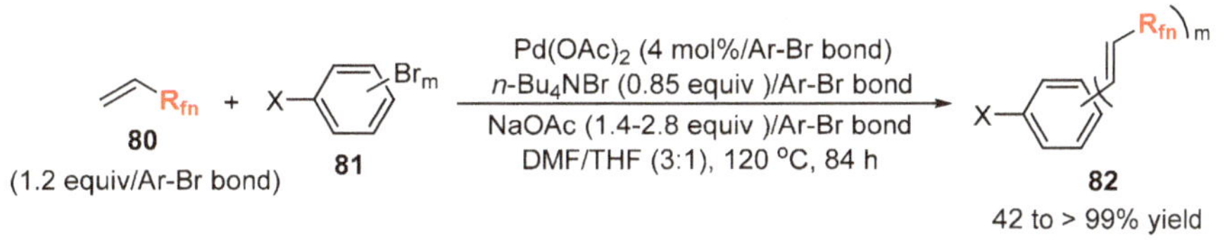

Scheme 16. The Mizoroki-Heck reactions of aromatic bromides and polybromides with fluorous alkenes.

In addition, a family of diazonium-functionalized oligo(phenylene vinylene)s (OPVs) tetramers were synthesized by alternating the Mizoroki-Heck cross-coupling and the Horner-Wadswoth-Emmons (HWE) reaction using fluorous mixture synthesis (FMS) technology (Scheme 17) [73]. The FMS technology was found to be superior to the solid-phase organic synthesis (SPOS) techniques. The Mizoroki-Heck couplings of bromide **83** with 1H,1H,2H-perfluoroalkenes (**84**) employed a palladacycle catalyst which proved to be a better catalyst than $Pd(OAc)_2$. Hydrogenation of the coupled alkenes (**85**) followed by removal of the Boc groups gave secondary amines as fluorous tags for diazonium-substituted aryl iodides (**87**). The final stage of OPV tetramer growth was a Heck-type reaction with an end-capping styrene derivative (**89**). At the end, the tagged products were detagged by cleaving the triazene linkage and generating a series of aryl diazonium compounds. The fluorous tags could be reused. The aryl diazonium functionalities in the products allowed them to be used as surface-grafting moieties in hybrid silicon/molecule assemblies [73].

Genêt and co-workers described the palladium-catalyzed Mizoroki-Heck reactions between aryldiazonium salts (**91**) and perfluoroalkyl alkenes (**92**) (Scheme 18) [74]. The reactions carried out in methanol at 40 °C in the presence of 0.5 mol% of $Pd(OAc)_2$ afforded a series of long-chain perfluoroalkyl-substituted aromatic compounds (**93**) in good to excellent yields. However, coupling of 4-iodotoluene (**91f**) under the same reaction conditions failed to give the desired product (**93bb**; R = CH_3, R_{fn} = C_8F_{17}). Coupling of aryl iodide in the presence of $Pd(PPh_3)_4$ and triethylamine in toluene didn't form the expected product neither. Interestingly, under Jeffery's conditions ($Pd(OAc)_2$, DMF, $NaHCO_3$, n-Bu_4NHSO_4), the reaction of 4-iodotoluene (**91f**) with **92b** (R_{fn} = C_8F_{17}) afforded **93bb** in 80% yield, but 4-bromotoluene (**91g**) remained much less reactive. Thanks to the difference in the reactivity of these functionalities, the Heck-type reaction of **92b** with 4-bromobenzenediazonium tetrafluoroborate (**91c**) produced **93cb** (R = Br, R_{fn} = C_8F_{17}) exclusively, showing good chemoselectivity. Hydrogenation of the carbon-carbon double bond of **93** afforded the fluorous aromatic compounds. Perfluoroalkyl-substituted aryl bromides and anilines were also accessed by this method. This protocol represented one of the most efficient methods to install a perfluoroalkyl chain onto an aromatic ring. The simple purification and the high purities of products allowed an easy scale up of this procedure.

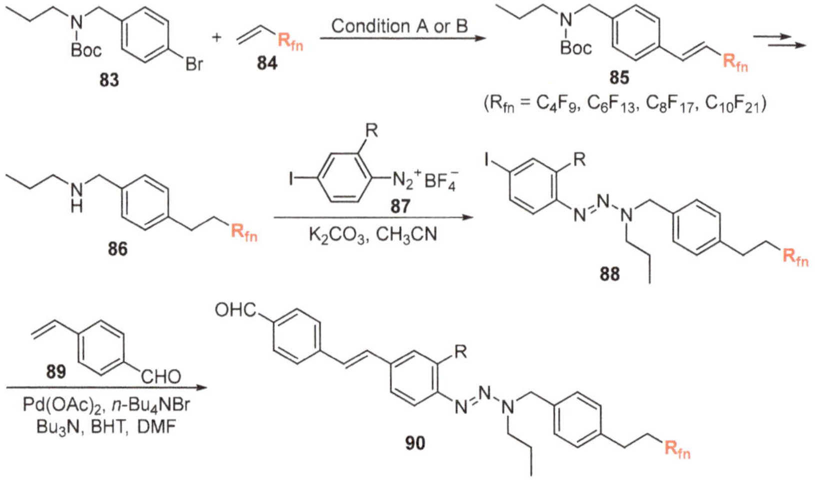

Condition A: $Pd(OAc)_2$, $NaHCO_3$, n-Bu_4NHSO_4, DMF, 90 °C
Condition B: Palladacycle, NaOAc, DMF, 125 °C

Scheme 17. Preparation of diazonium-functionalized oligo(phenylenevinylene)s by FMS technology.

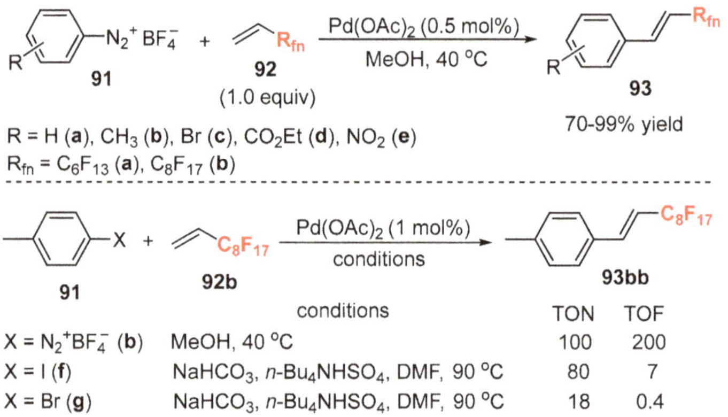

Scheme 18. The Heck-type reactions between perfluoroalkyl alkenes and aryldiazonium salts.

In 2015, Matsugi and co-workers synthesized f-Fmoc reagents (**97**) bearing C_3F_7, C_4F_9, or C_6F_{13} chains by the $Pd(OAc)_2$-catalyzed Mizoroki-Heck reaction of **94** with **95a**, **95b**, or **95c**, followed by hydroxymethylation and introduction of the *N*-hydroxysuccinimide group (Scheme 19) [75]. The method applied on a multi-gram scale after eight steps gave the desired products **97a**, **97b**, and **97c** with overall yields of 73%, 60%, and 90%, respectively. The Pd-catalyzed doubly tagging Heck reaction of **94** with a mixture of the fluorous alkenes (**95**) was also studied, leading to an encoded mixture synthesis of f-Fmoc reagents. Addition of **95** in an order of **95c**, **95b**, **95a** to the reaction mixture was found to be an effective condition in the mixed tagging Heck reaction. The target f-Fmoc reagents with C_3F_7, C_4F_9, and C_6F_{13} groups could be separated owing to the different fluorine content of the molecules. This method provided a reasonable solution to obtain various f-Fmoc reagents in a one-pot procedure. These f-Fmoc reagents would be useful protecting groups for the effective synthesis of peptide isomer libraries.

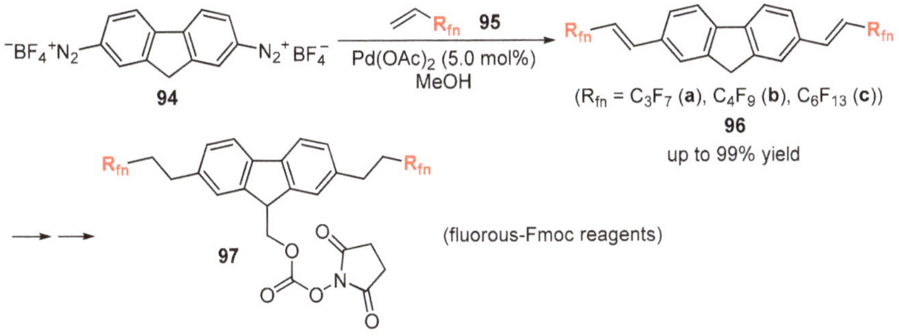

Scheme 19. Multi-gram scale and divergent preparation of fluorous-Fmoc reagents via the Mizoroki-Heck reactions.

In 2017, Konno and co-workers investigated the Pd-catalyzed Mizoroki-Heck reactions of aryldiazonium tetrafluoroborates with different types of CF_2CF_2-containing alkenes (Scheme 20) [76]. A survey of the optimized reaction conditions revealed that the reactions of 4-bromo-3,3,4,4-tetrafluoro-1-butene (**98**) with aryldiazonium salts (**99**) in the presence of 0.5 mol% Pd(OAc)$_2$ in MeOH at 40 °C for 1 h gave the best yields of the products. The *E*-configured products (**100**) were exclusively obtained, and no *Z*-isomers were detected. In the cases of 4-bromo-3,3,4,4-tetrafluoro-1-aryl-1-butenes (**101**), a bulky P(*o*-tolyl)$_3$-coordinated palladium catalyst was found to be a good catalytic system, leading to an increased yield of the product (**102**). High stereoselectivity of the reaction was achieved if **102** containing an electron-deficient aromatic ring, whereas a slight decrease in the stereoselectivity was observed if **102** bearing an electron-donating substituent on the aromatic ring. In the reactions of 4-bromo-3,3,4,4-tetrafluoro-2-aryl-1-butenes (**103**) with **99**, the geometry of the major isomers of **104** was determined as *E*-configuration on the basis of a proposed reaction mechanism [76]. This protocol provided a convenient and highly stereoselective method for the preparation of multi-substituted alkenes (**100**, **102**, and **104**) bearing a tetrafluoroethylene fragment, which are promising building blocks for the synthesis of novel CF_2CF_2-containing organic molecules [76].

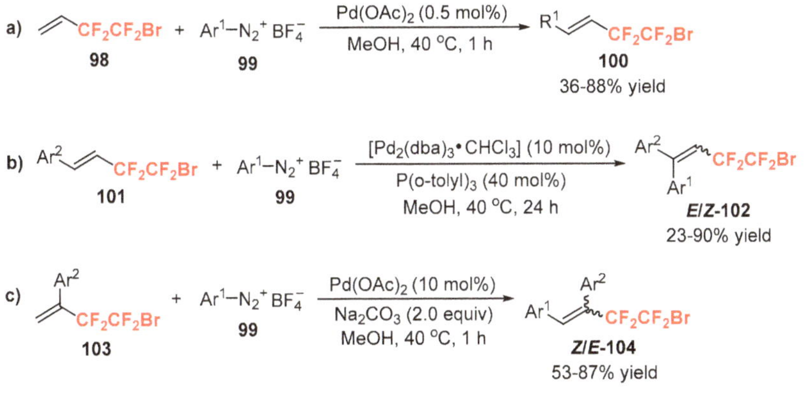

Scheme 20. The Mizoroki-Heck reactions of CF_2CF_2-containing alkenes with aryldiazonium tetrafluoroborates.

In a similar manner, Konno and co-workers examined the palladium-catalyzed Heck reactions of (*E*)-4,4,4-trifluoro-1-phenyl-2-buten-1-one (**105a**) with the readily available aryldiazonium salts

(**106**) (Scheme 21) [77]. The reactions allowed for easy and regio- and stereoselective access to a variety of 4,4,4-trifluoro-2-aryl-1-phenyl-2-buten-1-ones (**107**) in good yields. Aryldiazonium salts bearing electron-donating groups or halogens on the phenyl rings reacted nicely under the Heck-type cross-coupling conditions, preferentially forming the Z-isomers of the products (**Z-107**). However, substrates with electron-withdrawing groups or bulky groups on the aryldiazonium moieties didn't give the satisfactory results. The Michael addition adducts (**108**) were formed as byproducts in most cases [77]. Later, the same research group extended the scope of the heck reaction to other types of fluorine-containing electron-deficient olefins (Scheme 21) [78]. They found that the electron-withdrawing group (EWG) on alkenes had a big influence on the efficacy of the reaction. α,β-Unsaturated ester reacted with phenyldiazonium salt to provide the corresponding adduct in very low yield. Vinylphosphonate and vinylsulfone were found to be unreactive, with only the starting materials being recovered. In the case of nitroalkene, neither the starting material nor the desired Heck adduct was detected, and the reaction became very complicated. Additionally, changing the fluoroalkyl group from CF_3 to CF_2H group caused a significant decrease in the chemical yield, although the relatively high regioselectivity was obtained. A plausible reaction mechanism was suggested in Scheme 21 [78]. The reaction presumably proceeded via (1) oxidative addition of aryldiazonium salt (**106**) to Pd(0), leading to an arylpalladium complex (**110**), (2) coordination of the electron-deficient olefin (**105**) to the metal center of **110** and subsequent insertion into the Ar-Pd bond to generate **112** other than **113**, (3) carbon-carbon bond rotation of **112** to produce intermediate **114**, and (4) reductive elimination of [HPd]BF_4 from **114** to finally afford tri-substituted alkene (**107**) and to regenerate the Pd(0) catalyst.

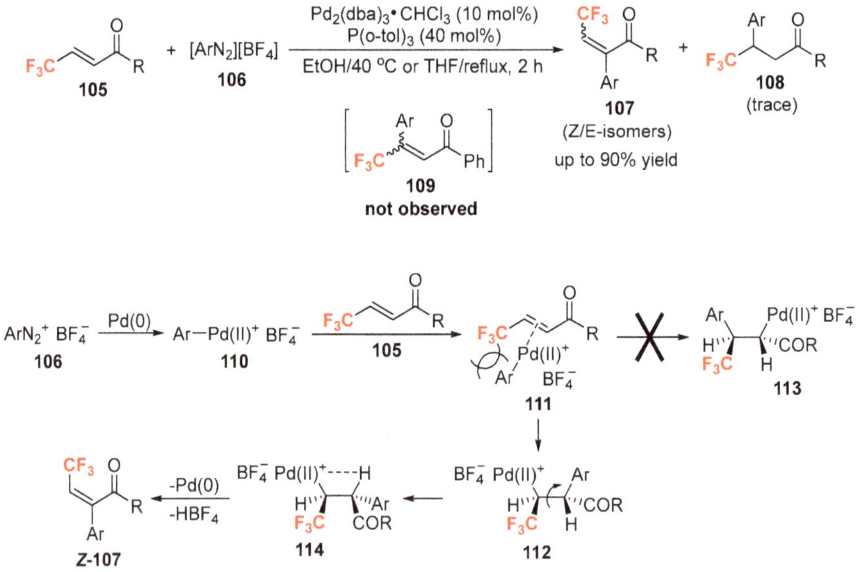

Scheme 21. The Heck-type reactions of fluorine-containing electron-deficient olefins with aryldiazonium salts.

Arylboronic acids are an important class of compounds for coupling reactions, which are stable in air and to moisture, and they are compatible with a broad range of common functional groups [62–64]. In 2015, Lu and co-workers presented the first oxidative Heck-type reactions of arylboronic acids with fluoroalkylated olefin [79]. The Pd-catalyzed cross-couplings of the commercially available 2,3,3,3-tetrafluoroprop-1-ene (**116**) with diverse arylboronic acids (**115**) provided a variety of

(Z)-β-fluoro-β-(trifluoromethyl)styrene derivatives (**117**) in good yields (Scheme 22) [79]. The wide range of substrates and the good functional group tolerance made this strategy facile and practical for the streamlined synthesis of functional styrenes.

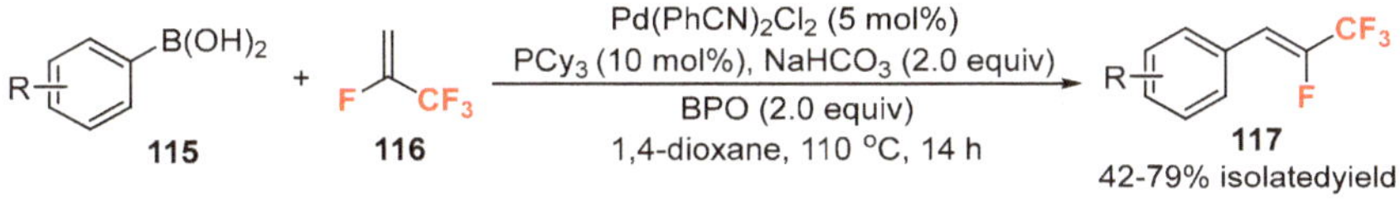

Scheme 22. The oxidative Heck reactions of fluorinated olefins with arylboronic acids by palladium catalysis.

Furthermore, Pd-catalyzed Heck-type cyclization of 2-trifluoromethyl-1-alkenes (**118**) bearing an O-acyloxime moiety was accomplished (Scheme 23) [45,80]. The reaction represented a rare example of 5-endo mode alkene insertion into transition-metal species via oxidative addition of the N–O bond (**119**). Although there were two possible pathways, namely β-fluorine elimination and β-hydrogen elimination of **120** after ring formation, the former exclusively took place to form an exo-difluoromethylene unit. This catalytic process provided facile access to 4-difluoromethylene-1-pyrrolines (**121**). To elucidate the role of the CF$_3$ group, the reaction of analogues (**118′**) bearing a hydrogen atom or a methyl group instead of the CF$_3$ functionality on the alkene fragment was examined. When **118′** was subjected to the reaction conditions similar to those for **118**, no cyclized product was obtained. The results indicated that the CF$_3$ group played a crucial role in the 5-endo Heck-type cyclization, wherein the trifluoromethyl substituent seemed to activate the vinylic terminal carbon in **118** and stabilize the cyclized palladium intermediate **120**.

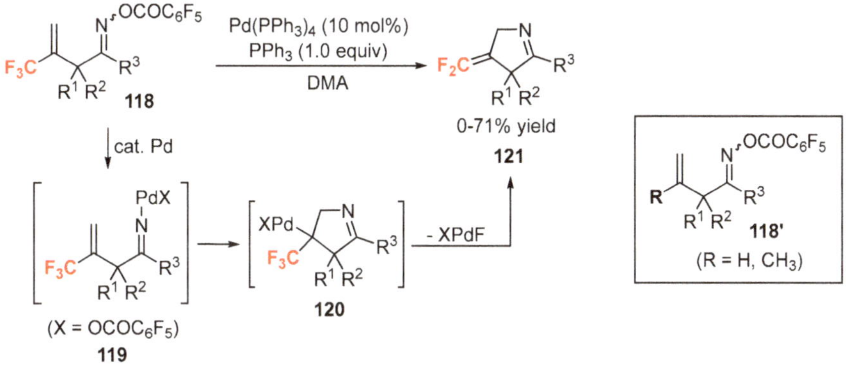

Scheme 23. Pd-catalyzed Heck-type cyclization of 2-(trifluoromethyl)allyl ketone oximes.

2.4. Other Fluorine-Containing Alkenes as the Cross-Coupling Reagents

A series of fluorinated distyrylbenzene (DSB) derivatives (**124** and **126**) were synthesized by Pd-catalyzed Mizoroki-Heck cross-couplings (Scheme 24) [81]. The reaction of 1,4-diiodobenzene (**122a**) with pentafluorostyrene (**123**) or 4-fluorostyrene (**125b**) in the presence of Pd(OAc)$_2$ gave trans-trans-1,4-bis(pentafluorostyryl)benzene (**124a**) or trans-trans-bis(4-fluorostyryl)benzene (**126a**). The coupling of 1,4-dibromo-2,5-difluorobenzene (**122b**) with styrene (**125a**), 4-fluorostyrene (**125b**), or pentafluorostyrene (**123**) using Pd(OAc)$_2$ provided 1,4-bis(styryl)-2,5-difluorobenzene (**126b**), 1,4-bis(4-fluorostyryl)-2,5-difluorobenzene (**126c**), and 1,4-bis-(pentafluorostyryl)-2,5-difluorobenzene (**124b**), respectively. The products were employed to probe the effect of fluorine substitution on the molecular properties and on the arrangement of molecules in the solid state [81]. The absorption spectroscopy showed that **124** and **126** had a λ_{max} at approximately 350 nm, and addition of dimethylaniline to the hexane solutions led to exciplex emission with λ_{max} ranging from 458 to 514 nm, depending on the positions of fluorine substitution. Moreover, the X-ray diffraction experiments for the lattice properties of **124a**, **124b**, **126b**, and **126c** indicated two possible structural motifs. One is to stack the DSB framework cofacially to form vertical "columns" within the crystal. The other is the alignment of these "columns" to maximize C–H···F electrostatic registry.

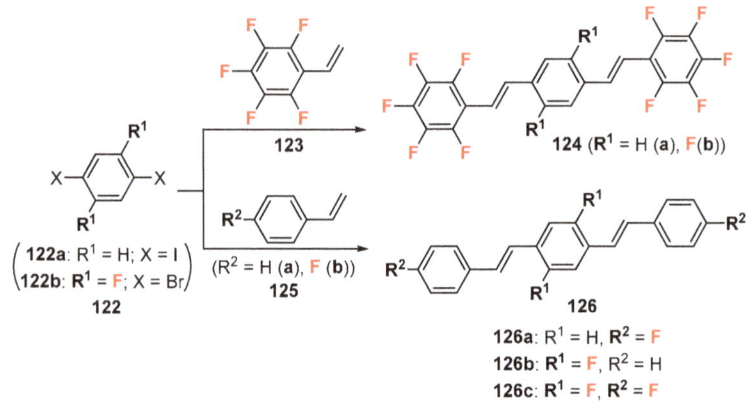
R1
X
X
R1
(122a: R1 = H; X = I
122b: R1 = F; X = Br)
122
F
F
F
F
F
123
R2
(R2 = H (a), F (b))
125
124 (R1 = H (a), F(b))
126
126a: R1 = H, R2 = F
126b: R1 = F, R2 = H
126c: R1 = F, R2 = F

Scheme 24. Synthesis of fluorinated distyrylbenzene chromophores by Heck-type reactions.

In addition, ligand-free Pd-catalyzed Mizoroki-Heck alkenylation of iodoarenes (**127**) with trifluoroethyl 2-acetoxyacrylate (**128**) was explored, which stereoselectively produced trifluoroethyl (*Z*)-2-acetoxycinnamates (**129**) in good yields (Scheme 25) [82]. Compounds **129** underwent deacetylation followed by acylation to yield isomerically pure trifluoro-ethyl (*Z*)-2-(arylacetoxy)cinnamates. *t*-BuOK-Mediated Dieckmann condensations of these products furnished a variety of pulvinone derivatives.

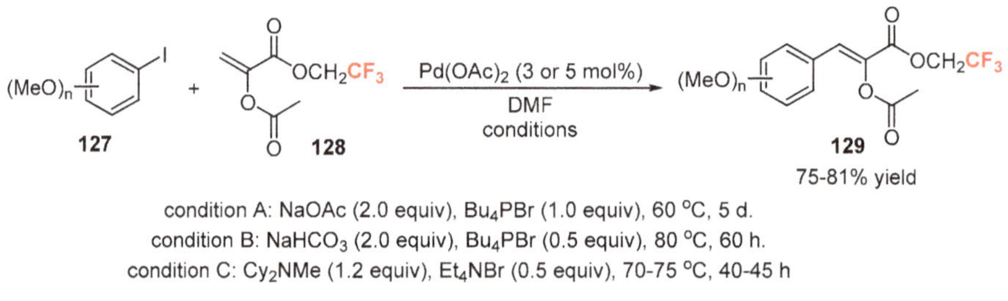
(MeO)n
127
OCH2CF3
128
Pd(OAc)2 (3 or 5 mol%)
DMF
conditions
129
75-81% yield
condition A: NaOAc (2.0 equiv), Bu4PBr (1.0 equiv), 60 °C, 5 d.
condition B: NaHCO3 (2.0 equiv), Bu4PBr (0.5 equiv), 80 °C, 60 h.
condition C: Cy2NMe (1.2 equiv), Et4NBr (0.5 equiv), 70-75 °C, 40-45 h

Scheme 25. Pd-catalyzed Mizoroki-Heck reactions of iodoarenes with trifluoroethyl 2-acetoxyacrylate.

The D-A-D-type chromophores (**132**) with a hexafluorocyclopentene thiophene "core" flanked by triphenylamine units were synthesized by facile procedures, including the Pd-catalyzed Mizoroki-Heck reactions (Scheme 26) [83]. The triphenylamine group caused the open-ring isomer of hexafluorocylopentene thiophene to close, which resulted in intramolecular *p*-conjugation extension, thus broadening the absorption spectra ranging from 200 to 850 nm with high optical densities.

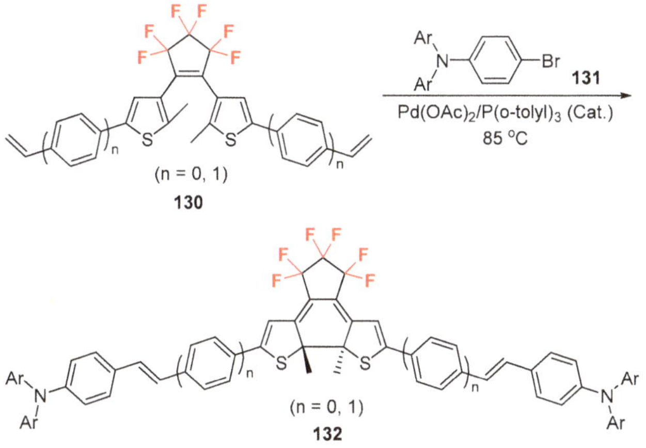

Scheme 26. Synthesis of triphenylamine derivatives with a hexafluorocyclopentene unit via the Pd-catalyzed Mizoroki-Heck cross-couplings.

3. Mizoroki-Heck Reactions of Fluorinated Halides or Pseudo-Halides with Alkenes

3.1. Fluorinated Aryl Halides or Pseudo-Halides as the Coupling Agents

In 2011, Vallribera and co-workers reported the Pd-catalyzed arylation of both electron-rich and -poor olefins (**134**) with 4-(perfluorooctyl)benzenediazonium trifluoroacetate (**133**) (Scheme 27) [84]. This Heck-type reaction supplied a variety of perfluoroalkylated aromatic compounds (**135**) by using Pd(OAc)$_2$ or 4-hydroxyacetophenone oxime-derived palladacycle (**136**) as an efficient source of Pd nanoparticles. Ligand-free Pd(OAc)$_2$ appeared to be a more general catalyst than palladacycle (**136**) for the reaction. The activity of **136** depended greatly on the nature of the alkene used. No external base was necessary for the conversion, and the reactions could be carried out at room temperature in the presence of 1 mol% of Pd(II) catalyst. Some intermediates derived from the oxidative addition of **133** to the catalyst were identified by ESI-MS experiments. The Pd(IV) species detected by this technique when **136** was used, did not seem to act as intermediates of the reaction. The TEM images of the reaction solutions showed, for the first time, the formation of nanoparticles with both catalysts under the Matsuda-Heck conditions. These Pd(0) nanoparticles might act as a reservoir of Pd(0) atoms involved in the reaction, which avoided the precipitation of black palladium and thus probably extended the lifetime of the catalyst.

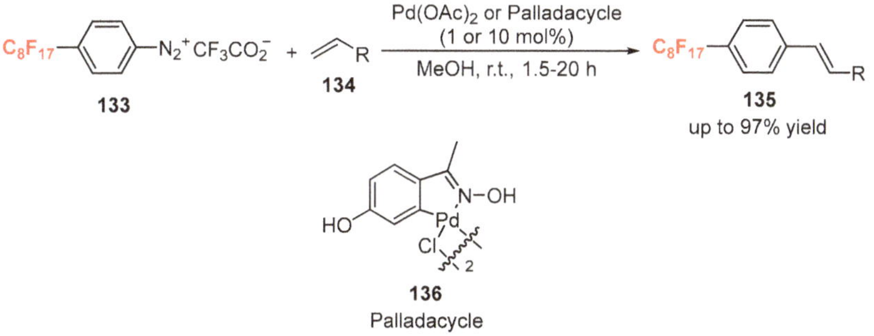

Scheme 27. The Heck-type reactions of alkenes with arenediazonium trifluoroacetate.

A series of SF$_5$-bearing alkenes (**139**, **141**, and **142**) were synthesized by employing the Matsuda-Heck reactions of 4-(pentafluorosulfanyl)benzenediazonium tetrafluoroborate (**137**) with alkenes (**138** and **140**) (Scheme 28) [85]. Compound **137** was readily synthesized and isolated as a stable salt for a wide assortment of transformations. The cross-couplings of **137** with styrene and 4-substituted styrenes in the presence of catalytic Pd(OAc)$_2$ in ethanol gave the corresponding 4′-substituted 4-(pentafluorosulfan-yl)stilbenes in good yields. The reaction run in ionic liquid [BMIM][BF$_4$] instead of EtOH resulted in lower isolated yield due to the increased formation of homo-coupling product. Reaction of **137** with methyl acrylate in the presence of Pd(OAc)2 in 95% EtOH afforded methyl *trans*-3-[4-(pentafluorosulfanyl)phenyl]prop-2-enoate in 85% yield without *cis*-isomer. However, coupling of **137** with methyl methacrylate (140) in 95% EtOH or [BMIM][BF$_4$] gave a mixture of methyl 2-methyl-3-[4-(pentafluorosulfanyl)phenyl]prop-2-enoate (**141**) and methyl 2-{[4-(pentafluorosulfanyl)phenyl]methyl}prop-2-enoate (**142**) in up to 99% yield. Furthermore, **137** enabled couplings with fluorous olefins to form the corresponding fluorinated adducts in good yields [85]. Unfortunately, the Matsuda-Heck reaction of **137** with ethyl cinnamate or norbornene in EtOH, MeOH, or [BMIM][BF$_4$] at room temperature or 70 °C failed to give the desired product [85].

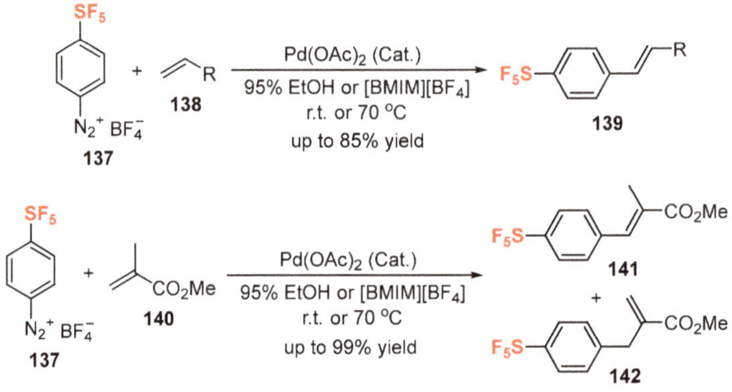

Scheme 28. The Heck-type reactions of 4-(pentafluorosulfanyl)benzenediazonium tetrafluoroborate.

An efficient catalytic system for the Mizoroki-Heck reactions of fluoroaryl halides (**143**) with alkenes (**144**) was developed (Scheme 29) [86]. Complex **145** (1 mol%) catalyzed the reaction of C$_6$F$_5$Br and styrene at 130 °C to give regioselectively *trans*-PhCH=CHAr$_F$ (**146a**, Ar$_F$ = C$_6$F$_5$) in almost quantitative yield. The studies indicated that CaCO$_3$, KF, and Na$_2$CO$_3$ were the suitable bases, and that NMP was the best solvent for the reaction. [Pd(PPh$_3$)$_4$] failed to initiate the reaction of styrene with C$_6$F$_5$Br. If C$_6$F$_5$I was employed as a substrate under the optimized conditions, much lower yield of *trans*-PhCH=CHAr$_F$ was obtained. C$_6$F$_5$Cl didn't react with styrene under the same conditions. The reaction was also efficient for other fluorinated aryl bromides such as *p*-XC$_6$F$_4$Br (X = CN, CF$_3$, OMe) and *p*-FC$_6$H$_4$Br [86]. For an activated alkene, such as methyl acrylate, the reaction proceeded fast even at 100 °C, and in this case, use of KF as a base led to higher conversion. 1-Hexene and α-methylstyrene could also be functionalized. Electron-rich alkenes such as butylvinyl ether and 2,3-dihydrofuran gave very low yields of the desired products. The reaction with styrene was not affected by oxygen or galvinoxyl, which excluded a radical addition pathway [86].

Moreover, complex **145** in CDCl$_3$ afforded immediately a 1:1.6 mixture of two isomers corresponding to the *cis* and *trans* arrangements of the two C$_6$F$_5$ groups in the dimer [86]. If **145** was dissolved in NMP at room temperature and at low concentrations, **148** was observed as the main species, determined by ^{19}F NMR spectroscopy. Upon addition of Br$^-$ anion to **148**, **149** was also detected. The equilibrium constants between **145**, **148**, and **149** in NMP at room temperature could be measured. At higher temperatures (e.g., 120 °C), the equilibria became fast, and the signals of these complexes collapsed. When the reaction of C_6F_5Br and styrene was monitored by ^{19}F NMR at 120 °C under catalytic conditions, only one set of Ar_F-Pd signals was observed. Based on these studies, a possible reaction mechanism was suggested in Scheme 29 [86]. The first step of the catalytic cycle is the coordination and insertion of the alkene by **145**, generating **151**. This step is likely to be rate-determining, which is supported by the observation of the signals of the complexes preceding this step and by the fact that the reaction rate of C_6F_5Br and styrene was simply dependent upon the concentration of styrene. Then, β-H elimination of **151** followed by reductive elimination of HX from **153** provides **146** and regenerates Pd(0) species (**147**), which is stabilized by bromide anions. The reaction outside the catalytic cycle showed that the insertion of styrene into the Pd-C_6F_5 bond of **148** was retarded by addition of NBu_4Br. If the catalytic reaction was carried out under bromide-free conditions, only 3% of **146a** was obtained and the catalyst extensively decomposed. In addition to stabilizing the Pd(0) intermediate, the bromides might increase the rate of oxidative addition. Excess styrene increased the reaction rate by facilitating the coordination-insertion process, but a large excess of styrene stopped the reaction. Addition of NBu_4Br to the latter reaction mixture could reactivate the process. It seemed that displacement of bromides on Pd(0) by alkene at high concentration severely slowed the oxidative addition, which became rate-determining instead of the coordination-insertion step [86,87]. Thus, the success of this catalysis required a compromise between the optimal conditions for the oxidative addition and the coordination-insertion.

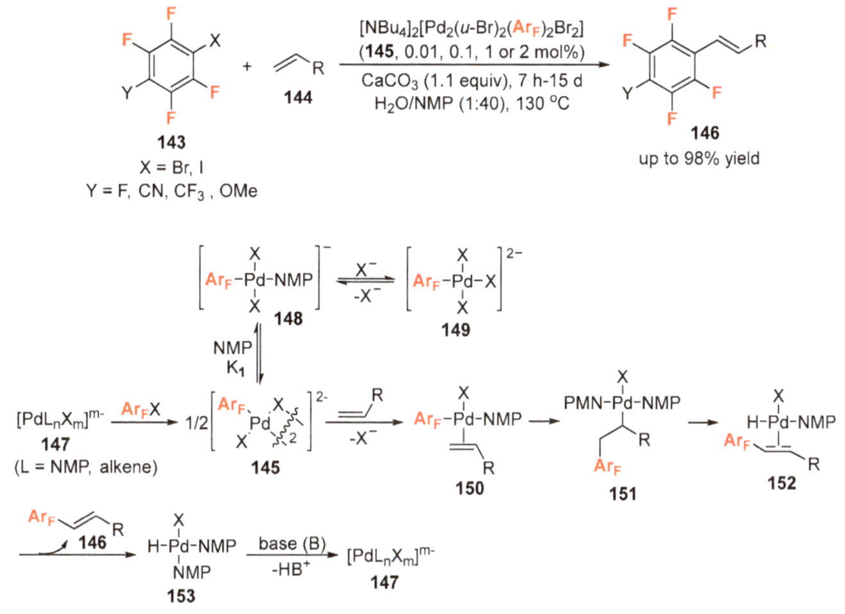

Scheme 29. The Mizoroki-Heck reactions of fluoroaryl halides with alkenes.

In 2010, Zhang and co-workers reported the first $Pd(OAc)_2$-catalyzed oxidative Heck-type reactions of the electron-deficient pentafluoroarene (**154**) with a broad range of alkenes (**155**) (Scheme 30) [88]. The reactions provided a great variety of pentafluorophenyl substituted alkenes (**156**) in moderate to high yields and with moderate to high stereo- and regioselectivities. A survey of reaction conditions indicated that the solvent, oxidants, and Pd-catalysts were critical for the reaction efficiency. A mixed solvent system of 5% DMSO in DMF was found to be the optimum reaction medium. The reaction using Ag_2CO_3 as both the base and oxidant in the presence of $Pd(OAc)_2$ (10 mol%) at 120 °C gave the best yield of the desired product. Electron-deficient olefins bearing ester, amide, phosphonate, or ketone groups worked well to give moderate to excellent yields and high stereoselectivities. Notably, nonactivated aliphatic olefins and electron-rich alkenes underwent smooth reactions in good yields, which were in sharp contrast to previous results [86,87]. In the case of polyfluoroarenes (**157**), the reactions using PivOH (1.2 equiv) instead of DMSO (5%) afforded alkenylated products (**159**) in moderate yields with moderate to good regioselectivities. In general, the reactions with electron-rich alkenes afforded higher yields than those with electron-deficient ones. The most acidic C–H bonds located between two fluorine atoms were the primary reaction sites, which provided mostly the monoalkenylated products in higher yields. It was also possible to further derivatize the polyfluoroarylated alkenes through C–H functionalization. This strategy allowed the selective installation of substituents at different positions and provided a convenient access to highly functionalized fluoroarenes by catalytic methods.

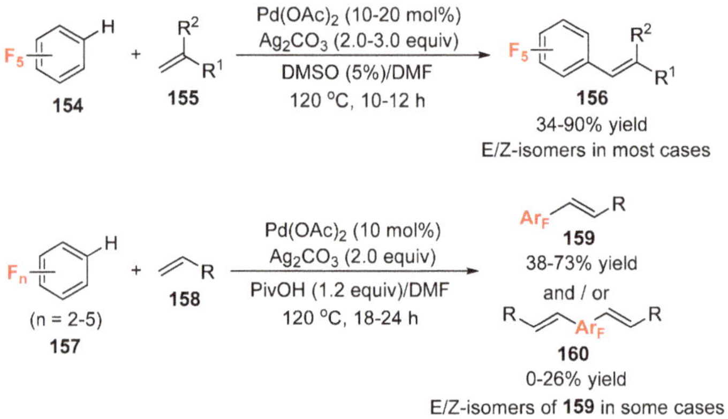

Scheme 30. Pd(OAc)$_2$-catalyzed oxidative olefination of highly electron-deficient fluoroarenes with alkenes.

Later, Zhang's research group again disclosed the thioether-promoted direct olefination of polyfluoroarenes (**161**) with alkenes (**162**) catalyzed by Pd(OAc)$_2$ (Scheme 31) [89]. Remarkably, the "inert" substrates, such as 3-substituted tetrafluorobenzenes, which previously furnished their corresponding products in low yields, afforded the alkenylated products (**163**) in high yields and with excellent stereoselectivities using only 1.0 equiv of alkene in this reaction. The results demonstrated the power of the thioether ligand PhSMe in the Pd-catalyzed reactions of polyfluoroarenes, offering a new choice to discover more efficient catalytic systems. Furthermore, a competitive reaction between electron-deficient and electron-rich alkenes with pentafluorobenzene was conducted, which provided their corresponding products in an almost 1:1 ratio, suggesting that the direct olefination of polyfluoroarenes has no bias on the nature of alkenes under these reaction conditions [89].

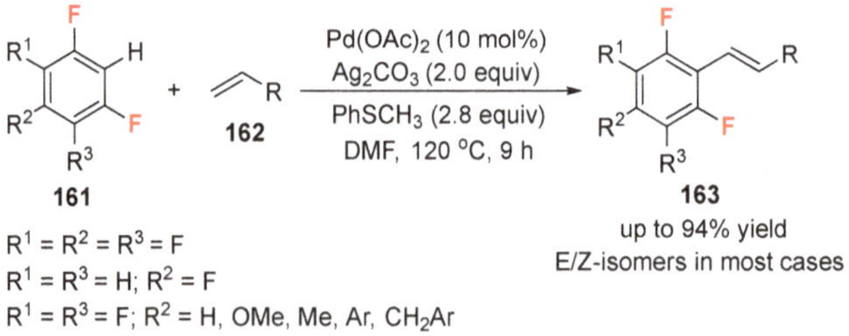

Scheme 31. Thioether-promoted direct olefination of polyfluoroarenes catalyzed by palladium.

By a similar strategy, a series of olefin-containing fluorinated benzothiadiazoles were synthesized from fluorinated benzothiadiazoles in the presence of 2.5 mol% Pd(TFA)$_2$, 3.0 equiv AgOAc, 2.5 equiv PhCO$_2$H, and 0.1 equiv benzoquinone [90]. The reactions proceeded under mild conditions and showed good functional group compatibility. The products found important applications in optoelectronic materials. Significantly, the Pd-catalyzed aerobic direct olefination of polyfluoroarenes was also explored [91]. The procedure made use of molecular O$_2$ as the terminal oxidant instead of silver(I) salt, providing a cost-efficient and environment-benign access to polyfluoroarene-alkenes. The silver species and the thioether ligand played important roles in the reactions, but the mechanisms remained unclear.

In addition, Liu and co-workers investigated the Pd(II)-catalyzed direct olefination of electron-deficient fluoroarenes (**164**) with allylic esters and ethers (**165**) (Scheme 32) [92]. In this transformation, various γ-substituted allylic esters (mixtures of E/Z- and B-isomers (**166** and **166′**)) were prepared in good to excellent yields by oxidative Heck reactions via β-H elimination, rather than β-OAc elimination. Typically, the desired products were obtained in good yields under conditions composed of Pd(OAc)$_2$ (5 mol%), AgOAc (2.0 equiv), DMSO (5%)/THF, 110 °C, and 12 h. A variety of fluorinated arenes proved to be very efficient substrates under the typical conditions, and the olefination generally took place at the more sterically accessible position. This selectivity might be attributed to the acidity of C–H bonds of the fluorinated arenes even though it was located at a more sterically hindered position. Tetrafluorobenzene and fluorinated arenes bearing an electron-donating methoxy group could also give the desired products. Yields of the products decreased with the decrease in the number of fluorine atoms substituted on the aromatic rings. Significantly, allyl acrylate and allyl methacrylate afforded the corresponding products in moderate to good yields. The oxidative olefination occurred at the allylic rather than the acrylic carbon–carbon double bond. This selectivity might be ascribed to the stability of the alkyl-Pd intermediate via chelation of Pd center by the carbonyl O-atom. This chelation effect could also result in the high regioselectivity as the (Pd)C–C(O) bond cannot rotate freely.

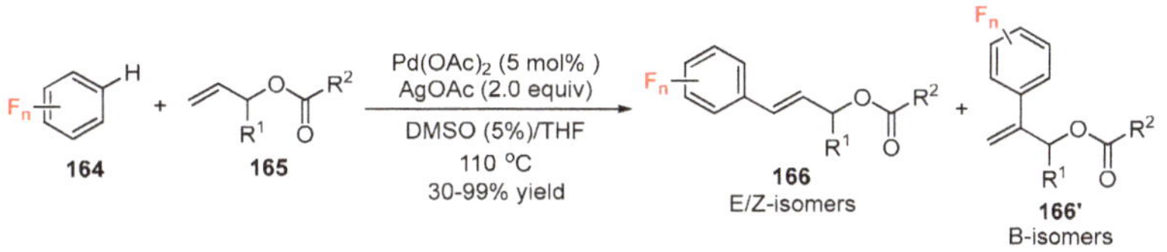

Scheme 32. The Pd-catalyzed oxidative olefination of fluoroarenes with allyl esters.

S-fluoroalkyl diphenylsulfonium triflates [Ph$_2$SR$_{fn}$][OTf] (**168**, R$_{fn}$ = CF$_3$, CH$_2$CF$_3$) were effective cross-coupling partners in the Pd-catalyzed Mizoroki-Heck-type reactions with alkenes (Scheme 33) [93]. The functionalization of various conjugated and unconjugated alkenes (**167**) by **168** in the presence of 10 mol% Pd[P(t-Bu)$_3$]$_2$ and TsOH at room temperature provided the corresponding phenylation products (**169**) in good to high yields. The bases that benefited the traditional Mizoroki-Heck reactions severely inhibited this transformation, whereas the acids significantly improved the reaction. This protocol demonstrated a new class of cross-coupling participants for Heck-type reactions [93].

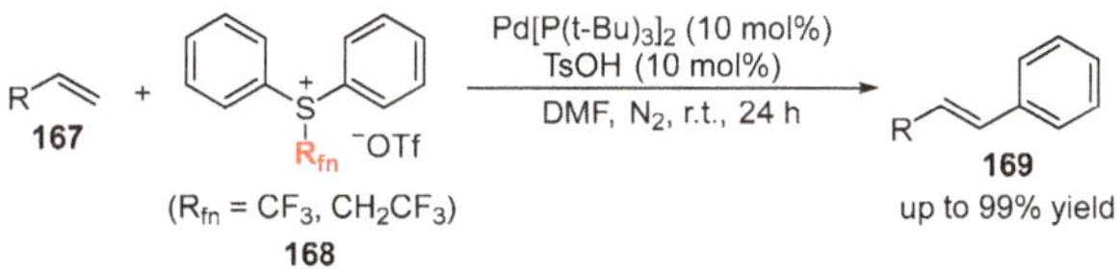

Scheme 33. Pd-catalyzed Mizoroki-Heck reactions of [Ph$_2$SR$_{fn}$][OTf] with alkenes.

Epibatidine (exo-2-(2′-chloro-5′-pyridyl)-7-azabicyclo[2.2.1]heptane), a natural compound isolated from the skin of the Ecuadorian poison frog Epipedobates tricolor, is a potent nicotinic acetylcholine receptor (nAChR) agonist [94]. In order to visualize and quantify these receptors in the human brain using positron emission tomography (PET) technique, exo-2-(2′-[^{18}F]fluoro-5′-pyridyl)-7-azabicyclo[2.2.1]heptane, a fluorine-18 ($t_{1/2}$ = 110 min) radiolabeled derivative of epibatidine, was synthesized from the nucleophilic aromatic substitution of the corresponding 2′-bromo-, 2′-iodo- or 2′-nitro exo-2-(5′-pyridyl)-7-azabicy-clo[2.2.1]heptane analogue with [^{18}F]FK-K$_{222}$ [94]. In this work, norchlorofluoroepibatidine (**173**) was employed as a reference compound, which was synthesized by reductive and stereoselective Pd-catalyzed Heck-type cross-coupling of **170** and **171** (Scheme 34) [94].

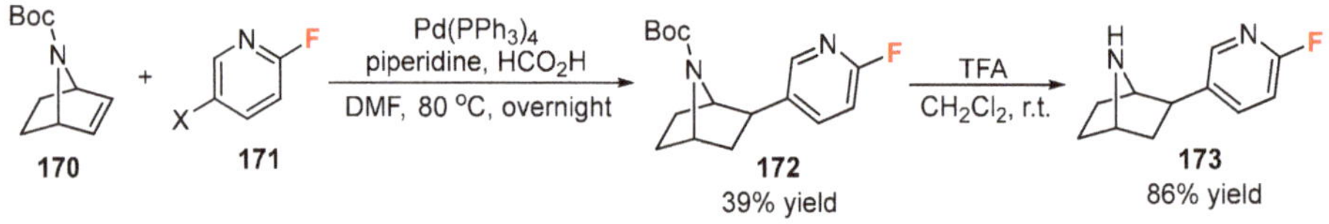

Scheme 34. Synthesis of a fluorinated derivative of epibatidine via the Mizoroki-Heck reaction.

The β-selective, chelation-controlled Heck-type reaction was found to be a convenient and versatile method to synthesize a series of new lipophilic N-alkylated nipecotic acid derivatives (**176**) bearing a vinyl ether unit embedded in the spacer and an unsymmetrical bis-aromatic residue at the terminal position of the spacer (Scheme 35) [95]. Most of the compounds displayed reasonable to good potencies and selectivities for mGAT1 (subtype of GABA transporter originating from murine cells). The influence of the presence and the position of fluorine substituents in the bis-aromatic residue concerning potency and selectivity of the compounds was defined. In general, all Z-isomers of the synthesized compounds were more potent than their corresponding E-isomers. The influence of fluorine substituents on the GAT1 uptake inhibition was generally more severe for the compounds with two aromatic rings being linked by a methanone bridge (Y = O). Substitution with fluorine in the 4-position of one aryl ring (R^4 = F) and an additional fluorine substituent at the *para*-position of the other aryl ring (R^3 = F) was the most beneficial combination, leading to one of the most potent compounds of the whole series.

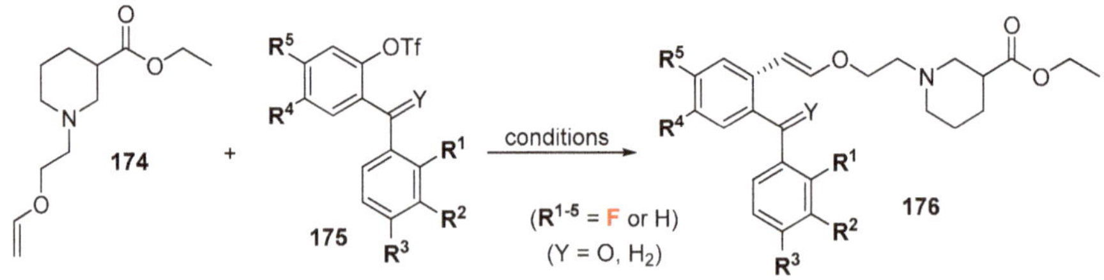

condition A: Pd(OAc)$_2$/NEt$_3$, DMF, 80 °C, 20 h, 39–78%
condition B: Pd(OAc)$_2$/PPh$_3$/NEt$_3$, DMF, 80 °C, 20 h, 19–55%
condition C: Pd(OAc)$_2$/LiCl/NaOAc/K$_2$CO$_3$, DMF/H$_2$O (v/v = 10:1), 80 °C, 20 h, 65–83%

Scheme 35. Synthesis of fluorine-containing N-substituted nipecotic acid derivatives by the Mizoroki-Heck reactions.

Additionally, four pyrimidine C-nucleosides (**180**) were built as mimics of 2′-deoxycytidine (dC) and 2′-deoxyuridine (dU). The key carbon-carbon bond formation between the readily available 2,6-substituted pyridines (**178**) and the glycal (**177**) employed palladium-catalyzed Heck-type reactions

(Scheme 36) [96]. The minor groove functional group in each derivative was replaced by a fluorine or a methyl group. Each coupling reaction yielded only the β-anomer, in part because the bulky silyl protecting group at the 3′-hydroxyl moiety precluded addition of the heterocycle from the "lower" face of the sugar. Without separating the initial coupling products (**179**) from the starting glycal (**177**), the silyl group was removed from **179** and the corresponding ketones were easily purified. Stereospecific reduction of the ketones, followed by removal of the *p*-NPE protecting group if necessary, resulted in the target compounds (**180**) (Scheme 35).

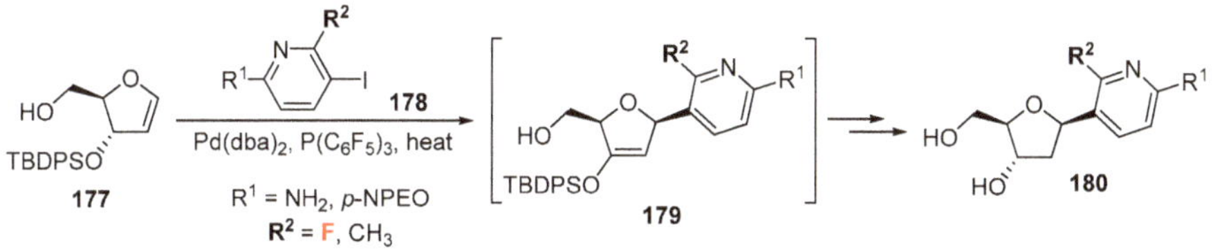

Scheme 36. Synthesis of fluorine-containing pyridine C-nucleosides via the Mizoroki-Heck reactions.

Two sets of organic dyes containing a stilbene backbone with fluorine substituents were constructed for a study on the quantum efficiency of dye-sensitized solar cells (DSSCs) (Scheme 37) [97]. Compound **181** was coupled with the commercially available 4-bromobenzaldehyde (**182a**), 4-bromo-2-fluorobenzaldehyde (**182b**), and 4-bromo-2,6-difluorobenzaldehyde (**182c**), respectively, by the Heck-type reactions to yield **183**, which were condensed with cyanoacetic acid via Knöevenagel reactions to afford the final products (H-P, H-N, F-P, FF-P, F-N, and FF-N). The results revealed that a fluorine substituent on the phenyl group *ortho* to the cyanoacrylate could enhance the light-harvesting performance in comparison with the unsubstituted one. The monofluorinated dyes showed larger short-circuit photocurrent density (Jsc) values than the non-substituted ones, due to the extension of the conjugation. However, the difluorinated dyes exhibited a lower performance because of a twisted geometry between the difluorophenyl group and the cyanoacrylate moieties, which reduced the efficiency of π-conjugation and thus decreased the Jsc value.

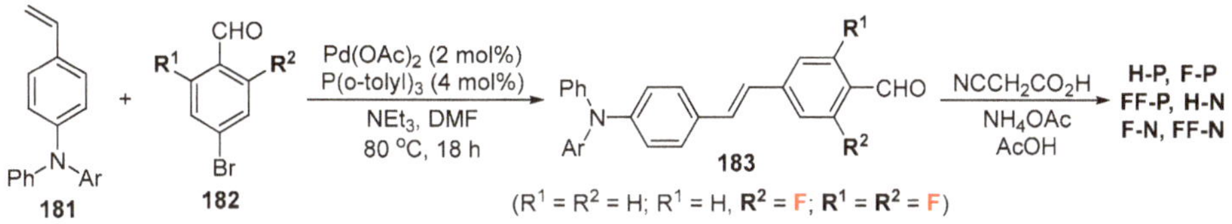

Scheme 37. Synthesis of fluorine-containing organic dyes by the Mizoroki-Heck cross-couplings.

The well-defined polyimides bearing a charge transporting (CT) and nonlinear optical (NLO) functionality in each repeat unit were prepared via two methods [98,99]. One of these methods was the Heck-type cross-coupling of a CF_3-containing brominated polyimide derivative (**184**) with styrene phosphonate **185** (Scheme 38) [98,99]. Polyimide **186** exhibited high thermal stabilities, and the electron-withdrawing phosphonate group was readily incorporated into the NLO moieties.

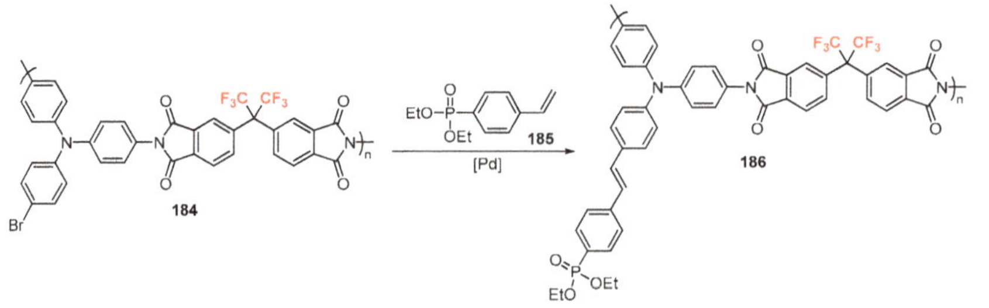

[Scheme 38 structures showing polymer 184 reacting with 185 under [Pd] catalysis to give polymer 186]

Scheme 38. Synthesis of multifunctional polymers for photorefractive applications by the Mizoroki-Heck reactions.

3.2. RCF$_2$X and R$_{fn}$X as the Cross-Coupling Reagents

Since the fluoroalkyl groups (CF$_2$R and R$_{fn}$) have found wide applications in the areas of medicinal chemistry and/or materials science, significant efforts have been devoted to the development of efficient catalytic systems and to the design of versatile reagents for mild, convenient, and direct fluoroalkylation [39–45]. In 2012, Reutrakul and co-workers described the palladium-catalyzed Heck-type reactions of [(bromodifluoromethyl)sulfonyl]benzene (**187**) with styrenes and vinyl ethers (**188**), furnishing α-alkenyl-substituted α,α-difluoromethyl phenyl sulfones (**189**) in moderate yields (Scheme 39) [100]. Although the efficiency of the reactions was ordinary, they provided an easy procedure for installation of (phenylsulfonyl)difluoromethylene group to olefins. The PhSO$_2$CF$_2$-containing compounds could be readily transformed. Reductive desulfonylation of the PhSO$_2$CF$_2$-bearing adduct (**189**) using Mg/HOAc/NaOAc afforded the corresponding CF$_2$H-substituted product [100]. Furthermore, addition of 3,5-di-*tert*-butyl-4-hydroxytoluene (BHT, 1.0 equiv) to a standard reaction mixture led to a slightly lower yield of the product, implying that the mechanistic pathway of the reaction might not proceed through a radical mechanism. This conversion represented the first report of Pd-catalyzed addition of a (phenylsulfonyl)difluoromethylene group to alkenes [100].

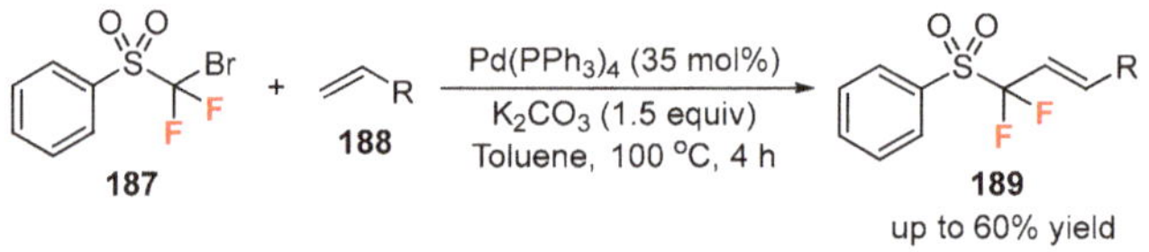

[Scheme 39 reaction: compound 187 (PhSO$_2$CF$_2$Br) + 188 (CH$_2$=CHR) → Pd(PPh$_3$)$_4$ (35 mol%), K$_2$CO$_3$ (1.5 equiv), Toluene, 100 °C, 4 h → 189, up to 60% yield]

Scheme 39. Pd-catalyzed Heck-type reactions of [(bromodifluoromethyl)-sulfonyl]benzene with alkenes.

Later, Zhang and co-workers reported a general method for palladium-catalyzed Mizoroki-Heck-type cross-couplings of alkenes (**190**) with perfluoroalkyl bromides (**191**) (Scheme 40) [101]. The reactions proceeded under mild conditions with high efficiency and broad substrate scope, and provided a variety of fluoroalkylated alkenes (**192**) in good to high yields. The optimized reaction conditions included use of 1,2-dichloroethane (DCE) as a solvent, [PdCl$_2$(PPh$_3$)$_2$] as a catalyst, and Cs$_2$CO$_3$ as a base. The bidentate ligand Xantphos was essential for the transformation as there was no reaction occurring when other phosphine ligands were used. The reaction could be extended to trifluoromethyl iodide, and other functional difluoromethyl bromides (**193**). The late-stage fluoroalkylation of bioactive compounds was also achieved in good yields, which supplied an efficient and straightforward route for application in drug discovery and development [101]. Mechanistic studies including a radical-clock experiment suggested that the free fluoroalkyl radicals (R$_{fn}$•), initiated by Pd(0)/Xantphos complex through a single electron transfer

(SET) process, might be involved in the Heck-type catalytic cycle (path A, Scheme 40). The formation of fluoroalkylated alkenes (**192** and **194**) by a bromine atom transfer from R_{fn}Br or **196** to alkene (path B, Scheme 40), followed by base-assisted elimination of the resulting benzyl bromides (**200**), could be ruled out, as the reaction of styrene with C_6F_{13}Br in the presence of [Pd(PPh$_3$)$_4$] and Xantphos without base failed to form benzyl bromide (**200**). Nevertheless, Chen and co-workers had disclosed that fluoroalkyl iodides reacted with alkyl alkenes in the presence of catalytic amounts of Pd(PPh$_3$)$_4$ to give the corresponding adducts (**200**) in high yields [102]. The reaction involved a radical-chain mechanism initiated by a single electron transfer (SET) from the Pd(0) species [102]. Furthermore, other competitive side reactions, such as oligomerization of **197** with alkene (path C, Scheme 40) and dimerization of **197** (path D, Scheme 40), were also reasonable pathways, for which the corresponding byproducts (**201** and **202**) were observed in the conversions [101].

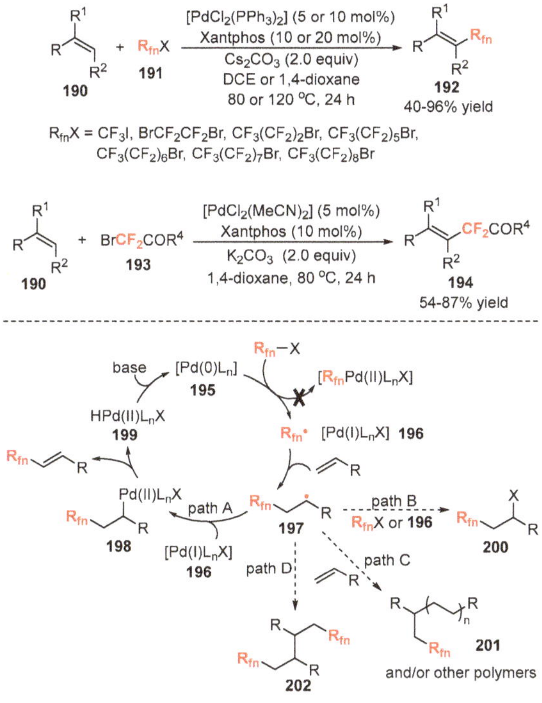

Scheme 40. Palladium-catalyzed Heck-type fluoroalkylation of alkenes.

Again, other functionalized difluoromethyl bromides (**204** and **206**) were employed as cross-coupling reagents in the Pd-catalyzed Mizoroki-Heck reactions by Zhang's research group (Scheme 41) [103,104]. With an analogous catalytic system mentioned above, a variety of difluoroalkylated alkenes (**205**) were synthesized from **203** and **204** under mild reaction conditions, showing excellent functional group compatibility (Scheme 41a) [103]. The mechanistic studies revealed again that free fluoroalkyl radicals initiated by [Pd(0)L$_n$] complexes via a SET pathway were involved in the catalytic cycle, and that the bidentate ligand Xantphos was essential for the transformation. This strategy was also successfully applied to the preparation of phosphonyldifluoromethylated alkenes (**207**) through Pd-catalyzed Heck-type reactions with bromodifluoromethylphosphonate (**206**) (Scheme 41b) [104]. The steric effect of diisopropyl (bromodifluoromethyl)phosphonate was critical for the efficiency of the reaction. Advantage of the protocol was the synthetic simplicity, providing a facile access to biologically interesting molecules.

Pd(PPh$_3$)$_4$-catalyzed Mizoroki-Heck reactions of in situ-generated benzylic iodides with styrenes (**208**) were reported by Wu and co-workers (Scheme 42) [105]. Under standard reaction conditions, a mass of perfluoroalkylated alkenes (**210**) were synthesized in moderate to excellent yields with good regio- and stereoselectivity. The transformation was totally inhibited when two equivalents of TEMPO were added, suggesting a radical mechanism (Scheme 42). Presumably, perfluoroalkyl radical (R$_{fn}\bullet$) and L$_n$Pd(I)I species are first generated from R$_{fn}$I and Pd(0)L$_n$ in the reaction. Then, the R$_{fn}\bullet$ radical adds to alkene (**208**) to afford an alkyl radical (**211**), which reacts with L$_n$Pd(I)I to give a benzylic iodide (**212**) (path A). Reaction of **212** with Pd(0)L$_n$ forms the key intermediate **213**, which might also be obtained from the direct combination of **211** with L$_n$Pd(I)I (path B). Subsequently, a Heck-type reaction of styrene with **213** generates a Pd(II) intermediate (**214**). Finally, reductive elimination of **214** produces **210** and regenerates the Pd(0)L$_n$ species under the assistant of DBU. Nonetheless, a pathway via radical addition of **211** to another molecule of styrene in which the resulting benzyl radical couples with L$_n$Pd(I)I to form **214** cannot be excluded.

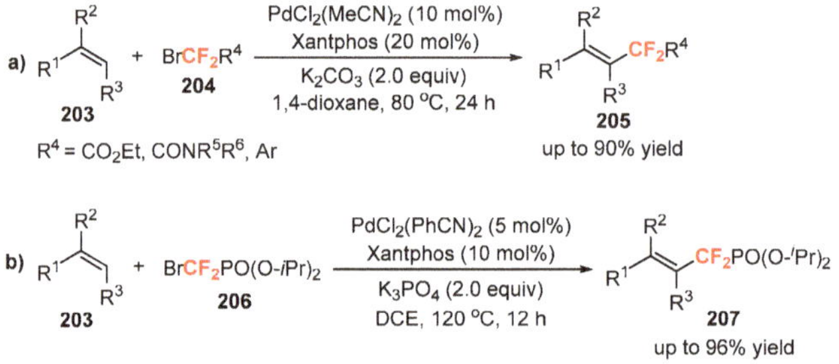

Scheme 41. Pd-catalyzed Heck-type difluoroalkylation of alkenes with functionalized difluoromethyl bromides.

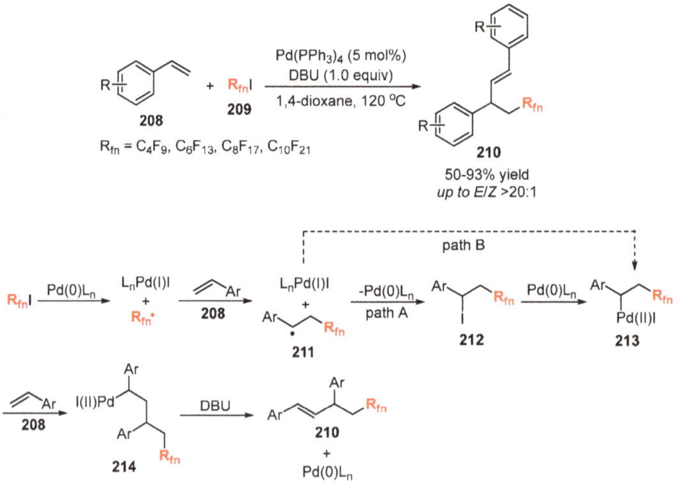

Scheme 42. Palladium-catalyzed Mizoroki-Heck reactions of perfluoroalkyl iodides and styrenes.

The Pd-catalyzed Heck-type reaction of secondary fluoroalkylated alkyl halide with alkene remains a challenge due to the sluggish oxidative addition of alkyl halide to palladium and the facile β-hydride elimination of alkylpalladium species [106]. Encouragingly, Zhang and co-workers described a convenient method for palladium-catalyzed Mizoroki-Heck-type couplings of secondary trifluoromethylated alkyl bromides (**216**) with alkenes (**215**), providing a variety of aliphatic alkenes bearing branched trifluoromethyl groups (**217**) (Scheme 43) [106]. The reactions proceeded under mild conditions and showed good functional group tolerance and high efficiency, even towards substrates bearing the sensitive hydroxy, formyl, and phthalimide groups. The reaction could also be extended to secondary difluoroalkylated alkyl iodide. The optimal reaction conditions were identified to use an assembly of PdCl$_2$(PPh$_3$)$_2$ (5 mol%), Xantphos (7.5 mol%), and KOAc (2.0 equiv) in DCE at 80 °C for 16 h. The control experiments using 1,4-dinitrobenzene as an electron transfer scavenger and the catalytic hydroquinone as a radical inhibitor suggested that a secondary trifluoromethylated alkyl radical (**218A**) via a SET pathway was likely involved in the reaction. This hypothesis was also supported by the radical clock experiment. Based on these, a plausible reaction mechanism was proposed in Scheme 43 [106]. Initially, reaction of [Pd(0)L$_n$] with **216** via a SET process generates an alkyl radical (**218A**), which adds to the carbon-carbon double bond of alkene to produce intermediate **218B**. Then, **218B** combines with [L$_n$Pd(I)Br] to give an alkylpalladium(II) complex (**219**), which undergoes β-hydride elimination to form **217** and regenerate [Pd(0)L$_n$] in the presence of a base.

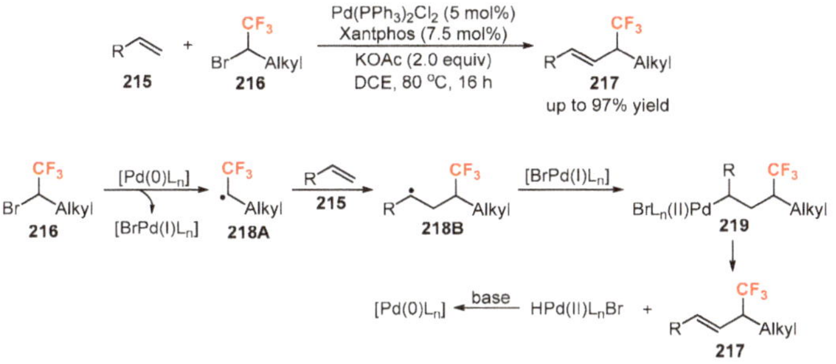

Scheme 43. Pd-catalyzed Heck-type reactions of secondary trifluoromethylated alkyl bromides with alkenes.

3.3. Fluoroalkylation Reagents as the Precursors of the Cross-Coupling Participants

In 2005, Vogel and co-worker reported the palladium-catalyzed Heck-type reactions between terminal alkenes and sulfonyl chlorides [107]. Herrmann's palladacycle (**222**) was found to be an excellent catalyst for the Mizoroki-Heck-type reactions of mono- and disubustituted olefins with arenesulfonyl chlorides. The reactions were not inhibited by radical scavenging agents. Trifluoromethanesulfonyl chloride (**220**) reacted with an excess of styrene (**221a**) in a sealed tube or microwave oven at 150 °C under similar conditions gave the desired desulfitative Mizoroki-Heck coupling product ((*E*)-3,3,3-trifluoro-1-phenylprop-1ene) (**223a**) in 50% yield (Scheme 44) [107]. The same reaction with butyl acrylate (**221b**) produced butyl (*E*)-4,4,4-trifluorobut-2-enoate (**223b**) in 32% yield.

The 3,3,3-trifluoropropenyl (CF$_3$CH=CH-) group has found important implications in medicine and material fields [108]. Conjugated aromatic systems with trifluoromethyl groups (e.g., β-trifluoromethylstyrenes) have been widely utilized in organic light emitting diodes (OLEDs) and other materials. In 2012, Prakash and co-workers described a simple and efficient elimination/Heck domino reaction sequence for the synthesis of β-trifluoromethylstyrene derivatives (**226**) (Scheme 45) [108]. The procedure involved a ligand-free palladium-catalyzed Heck reaction between aryl halides (**224**) and 3,3,3-trifluoropropene, which was generated in situ from the commercially accessible 1-iodo-3,3,3-trifluoropropane (**225**) under basic conditions, providing interesting and previously unknown β-trifluoromethylstyrenes in moderate to good yields. The reactions avoided the use of a gaseous 1,1,1-trifluoropropene reagent as one of the reactants and showed a broader substrate scope than the previously reported methods. Mechanistically, 1-iodo-3,3,3-trifluoropropane (**225**) may first undergo dehydrohalogenation to in situ form 3,3,3-trifluoropropene, which then reacts with **224** in the presence of a catalytic amount of Pd(OAc)$_2$ to yield the final products (**226**).

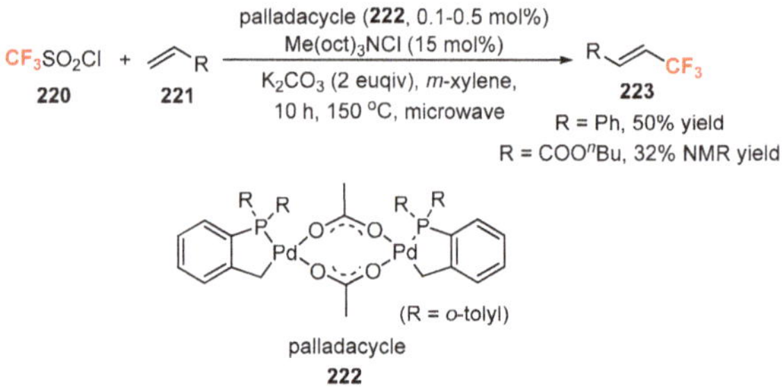

Scheme 44. Pd-catalyzed desulfitative Mizoroki-Heck type reactions of trifluoromethanesulfonyl chlorides with terminal alkenes.

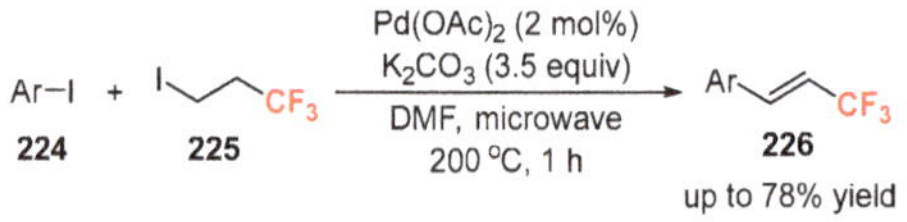

Scheme 45. The Heck-type cross-couplings of iodoarenes with 1-iodo-3,3,3- trifluoropropane.

4. Variants of the Mizoroki-Heck Reactions with the Fluorine-Containing Reagents

Bifunctionalization of alkenes triggered by the Mizoroki-Heck reactions has attracted great attention in organic synthesis. Jiang and co-workers reported the palladium-catalyzed fluoroalkylative cyclization of olefins (**227**) with R$_{fn}$I (**228**), which offered an efficient method for the construction of C$_{sp}^3$-R$_{fn}$ and C–O/N bonds in one step, affording the corresponding fluoroalkylated 2,3-dihydrobenzofuran and indolin derivatives (**229**) in moderate to excellent yields (Scheme 46) [109]. When the reaction of 2-allylphenol (**227aa**) with ICF$_2$CO$_2$Et was performed in the presence of 2,2,6,6-tetramethyl-1-piperidinoxyl (TEMPO), the formation of **229aa** (R^1 = H, R$_{fn}$ = CF$_2$CO$_2$Et) was totally inhibited and the adduct TEMPO-CF$_2$CO$_2$Et was produced in 30% yield. If 1,1-diphenylethylene was used as a radical scavenger in the same reaction, 10% of **229aa** and 51% of ethyl 2,2-difluoro-4,4-diphenylbut-3-enoate were obtained. These preliminary experiments indicated that a free •CF$_2$CO$_2$Et radical might be involved in the reaction. Thus, a plausible mechanism was suggested for the cyclization (Scheme 46) [109]. At the beginning, fluoroalkyl radical (R$_{fn}$•) and Pd(I)L$_n$I complex are formed from the interaction of Pd(0) and R$_{fn}$I via a single electron transfer process. Subsequent electrophilic radical addition of R$_{fn}$• to the C=C bond of olefin (**227a**) provides an alkyl radical (**230**). Reaction of **230** with Pd(I)L$_n$I yields an O-coordinated Pd(II) complex (**231**), which undergoes reductive elimination to afford **229a** (X = O) and regenerate the Pd(0) species. However, an alternative pathway using Pd(0) as a radical initiator followed by an intramolecular S$_N$2 substitution couldn't be excluded [109]. 2-Allylanilines could also undergo a similar mechanism.

Liang and co-workers described an efficient procedure for Pd-catalyzed domino Heck reactions/alkylation of electron-deficient polyfluoroarenes (**232**), which gave the corresponding alkylated polyfluoroarene products (**234**) in moderate to excellent yields under mild conditions (Scheme 47) [110]. The method represented a convenient, operationally simple, and useful protocol for the preparation of alkyl substituted polyfluoroarene derivatives. It was also the first example of installation of a polyfluoroarene structure involving an alkylpalladium(II) intermediate. A catalytic cycle for the conversion was proposed in Scheme 47 [110]. Initially, the in situ generated Pd(0)L$_n$ species undergoes oxidative addition of the C–I bond of 233 to form intermediate 235, which is converted through an intramolecular Heck-type reaction to yield alkylpalladium intermediate 236. Then, the silver salt abstracts iodide from 236, producing a palladium intermediate (237). Subsequently, 237 goes through a concerted metalation/deprotonation process to provide 238, which undergoes reductive elimination to produce 234 and regenerate the Pd(0)L$_n$ species for the next catalytic cycle.

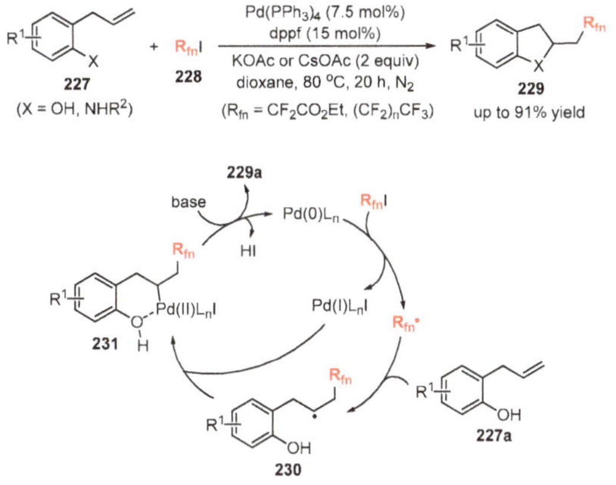

Scheme 46. Pd-catalyzed fluoroalkylative cyclization of olefins.

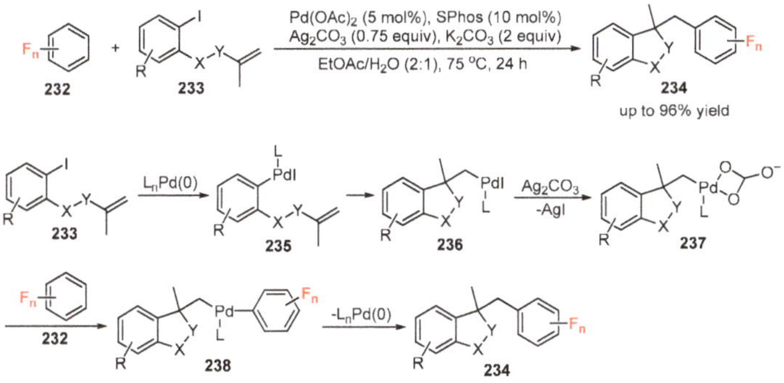

Scheme 47. Pd-catalyzed Heck/intermolecular C–H bond functionalization for the synthesis of alkylated polyfluoroarene derivatives.

Toste and co-workers disclosed the Pd-catalyzed 1,2-fluoroarylation of styrenes (239) with arylboronic acids (240) and Selectfluor® using amides (e.g., 8-aminoquinoline (AQ)) as the directing groups (Scheme 48) [111]. This strategy was also successfully used for the asymmetric 1,2-fluoroarylation of styrenes, furnishing chiral monofluorinated compounds (242) in good yields and with high enantioselectivies. Later, palladium-catalyzed 1,1-fluoroarylation of amino-alkenes (243) using arylboronic acids (244) as an arene source and Selectfluor® as a fluorine source was developed by the same research group (Scheme 48) [112]. The transformation likely proceeded through an oxidative Heck mechanism to afford 1,1-difunctionalized alkenes (245) in one pot [112], which was different from the pathway proposed for the 1,2-fluoroarylation [111]. The 1,1-fluoroarylation could also be extended to an asymmetric transformation, generating chiral benzylic fluorides (246) in good to excellent enantioselectivies [112]. These reactions promised powerful strategies for the difunctionalization of alkenes to chiral fluorinated molecules.

Palladium-catalyzed tandem C–C and C–F bond formation for bifunctionalization of allenes (**247**) was explored by Doyle and co-workers (Scheme 49) [113]. The intramolecular Heck/fluorination cascade provided monofluoromethylated heteroarenes (**248**), while the intermolecular variant for the three-component couplings of allenes (**249**), aryl iodides (**250**), and AgF gave the corresponding linear or branched monofluoromethyl isomers (**251** or **252**). The regioselectivity of the latter case was dramatically dependent upon the structure of allene. The mechanistic studies indicated that a palladium fluoride, generated from the halide exchange with AgF, was the key intermediate in the reaction (Scheme 49). Nevertheless, the exact mechanism of the reaction remained unclear [113].

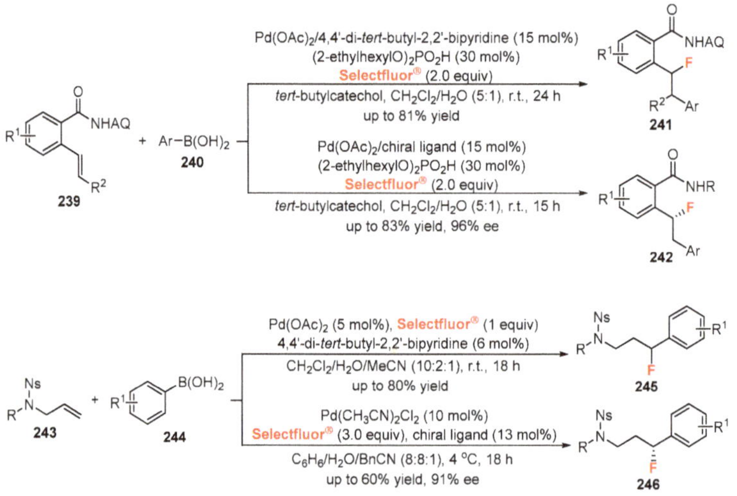

Scheme 48. Pd-catalyzed enantioselective fluoroarylation of alkenes.

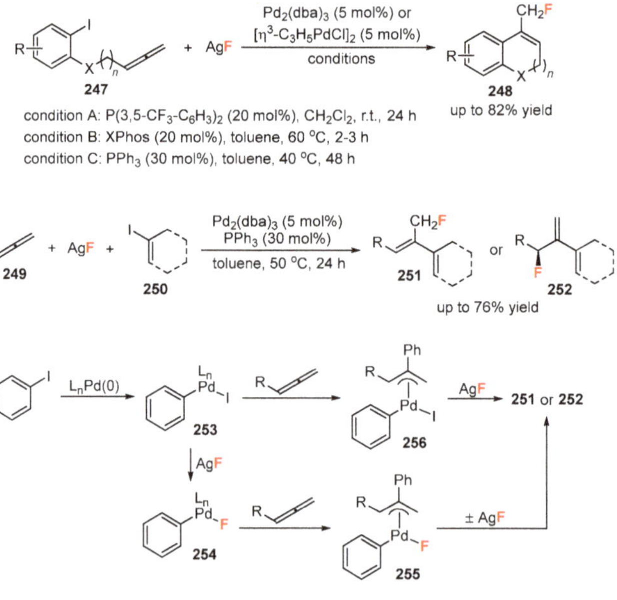

Scheme 49. Pd-catalyzed cascade carbofluorination of allenes.

5. Conclusions

In summary, the Mizoroki-Heck reaction has become a very powerful tool for building complex molecules [1–6]. The straightforward incorporation of fluorine-containing moieties by Mizoroki-Heck reactions has proved to be attractive and advantageous, as these reactions efficiently construct the carbon-carbon double bonds bearing fluorinated functionalities by simple procedures. The key to the high efficiency of the reactions is the choice of suitable fluorine-containing reagents and catalytic systems. Furthermore, the fluorinated compounds have been extensively used in the areas of materials and life sciences due to the significant impacts of fluorine substitution on the physiochemical and biological properties of the molecules. Since most fluorinated organic compounds have to be manually synthesized, the development of general and selective fluorination/fluoroalkylation methods with broad functional group tolerance that enable mild and late-stage functionalization of complex molecules is a perpetual topic in the realm of synthetic organic chemistry. Mizoroki-Heck cross-coupling has been evidenced as one of the most convenient methods for the preparation of fluorine-substituted alkenes or variants under mild conditions, which are very useful building blocks in the synthesis of functional molecules. This review highlights the palladium-catalyzed Mizoroki-Heck reactions using fluorine-containing reagents as the coupling participants. The variants of the reactions with fluorine-containing reagents catalyzed by palladium catalysts are also briefly introduced. It is undoubted that new Mizoroki-Heck reactions using fluorine-containing coupling partners will be continuously and prosperously developed. Challenging unrealized programs, such as the incorporation of elusive fluorinated functionalities (e.g., SeR_{fn}, SR_{fn}, and OR_{fn} groups) into the organic frameworks, the facile bifunctionalization of alkenes with rigid fluorine-containing reagents, and the asymmetric synthesis of complex fluorinated molecules via the Mizoroki-Heck cross-coupling reactions, will draw more attention in the future.

Acknowledgments: We thank the Army Medical University, the National Natural Science Foundation of China (21602165), the "Hundred Talent" Program of Hubei Province, and the Wuhan Youth Chen-Guang Project (2016070204010113) for financial support.

Author Contributions: Jing Yang, Hua-Wen Zhao, Jian He and Cheng-Pan Zhang analyzed the data and wrote the paper.

Conflicts of Interest: The authors declare no conflict of interest.

References

1. Beletskaya, I.P.; Cheprakov, A.V. The Heck reaction as a sharpening stone of palladium catalysis. *Chem. Rev.* **2000**, *100*, 3009–3066. [CrossRef] [PubMed]
2. Whitcombe, N.J.; Hii, K.K.; Gibson, S.E. Advances in the Heck chemistry of aryl bromides and chlorides. *Tetrahedron* **2001**, *57*, 7449–7476. [CrossRef]
3. Farina, V. High-turnover palladium catalysts in cross-coupling and Heck chemistry: A critical overview. *Adv. Synth. Catal.* **2004**, *346*, 1553–1582. [CrossRef]
4. Nicolaou, K.C.; Bulger, P.G.; Sarlah, D. Palladium-catalyzed cross-coupling reactions in total synthesis. *Angew. Chem. Int. Ed.* **2005**, *44*, 4442–4489. [CrossRef] [PubMed]
5. Phan, N.T.S.; Van Der Sluys, M.; Jones, C.W. On the nature of the active species in palladium catalyzed Mizoroki-Heck and Suzuki-Miyaura couplings-homogeneous or heterogeneous catalysis, a critical review. *Adv. Synth. Catal.* **2006**, *348*, 609–679. [CrossRef]
6. Li, H.; Johansson, S.; Carin, C.C.; Colacot, T.J. Development of preformed Pd Catalysts for cross-coupling reactions, beyond the 2010 Nobel Prize. *ACS Catal.* **2012**, *2*, 1147–1164. [CrossRef]
7. Heck, R.F. Acylation, methylation, and carboxyalkylation of olefins by group VIII metal derivatives. *J. Am. Chem. Soc.* **1968**, *90*, 5518–5526. [CrossRef]
8. Mizoroki, T.; Mori, K.; Ozaki, A. Arylation of olefin with aryl iodide catalyzed by palladium. *Bull. Chem. Soc. Jpn.* **1971**, *44*, 581. [CrossRef]
9. Heck, R.F.; Nolley, J.P. Palladium-catalyzed vinylic hydrogen substitution reactions with aryl, benzyl, and styryl halides. *J. Org. Chem.* **1972**, *37*, 2320–2322. [CrossRef]

10. Cacchi, S.; Morera, E.; Ortar, G. Palladium-catalysed vinylation of enol triflates. *Tetrahedron Lett.* **1984**, *25*, 2271–2274. [CrossRef]
11. Nilsson, P. Reaction with nonaromatic alkenyl halides or alkenyl sulfonates. *Sci. Synth.* **2013**, *3*, 285–302.
12. Weimar, M.; Fuchter, M.J. Reaction with nonaromatic halides, sulfonates, or related compounds. *Sci. Synth.* **2013**, *3*, 137–168.
13. Miura, M.; Hashimoto, H.; Itoh, K.; Nomura, M. Palladium-catalyzed desulfonylative coupling of arylsulfonyl chlorides with acrylate esters under solid-liquid phase transfer conditions. *Tetrahedron Lett.* **1989**, *30*, 975–976. [CrossRef]
14. Dubbaka, S.R.; Vogel, P. Organosulfur compounds: Electrophilic reagents in transition-metal-catalyzed carbon-carbon bond-forming reactions. *Angew. Chem. Int. Ed.* **2005**, *44*, 7674–7684. [CrossRef] [PubMed]
15. Dubbaka, S.R.; Zhao, D.; Fei, Z.; Vollaa, C.M.R.; Dyson, P.J.; Vogel, P. Palladium-catalyzed desulfitative Mizoroki-Heck coupling reactions of sulfonyl chlorides with olefins in a nitrile-functionalized ionic liquid. *Synlett* **2006**, 3155–3157. [CrossRef]
16. Blaser, H.-U.; Spencer, A. The palladium-catalysed arylation of activated alkenes with aroyl chlorides. *J. Organomet. Chem.* **1982**, *233*, 267–274. [CrossRef]
17. Liu, C.; Meng, G.; Szostak, M. N-Acylsaccharins as amide-based arylating reagents via chemoselective N–C cleavage: Pd-catalyzed decarbonylative Heck reaction. *J. Org. Chem.* **2016**, *81*, 12023–12030. [CrossRef] [PubMed]
18. Schmidt, A.F.; Kurokhtina, A.A.; Larina, E.V.; Yarosh, E.V.; Lagoda, N.A. Direct kinetic evidence for the active anionic palladium(0) and palladium(II) intermediates in the ligand-free Heck reaction with aromatic carboxylic anhydrides. *Organometallics* **2017**, *36*, 3382–3386. [CrossRef]
19. Masllorens, J.; Moreno-Manas, M.; Pla-Quintana, A.; Roglans, A. First Heck reaction with arenediazonium cations with recovery of Pd-triolefinic macrocyclic catalyst. *Org. Lett.* **2003**, *5*, 1559–1561. [CrossRef] [PubMed]
20. Felpin, F.-X.; Nassar-Hardy, L.; Le Callonnec, F.; Fouquet, E. Recent advances in the Heck-Matsuda reaction in heterocyclic chemistry. *Tetrahedron* **2011**, *67*, 2815–2831. [CrossRef]
21. Oger, N.; d'Halluin, M.; Le Grognec, E.; Felpin, F.-X. Using aryl diazonium salts in palladium-catalyzed reactions under safer conditions. *Org. Process Res. Dev.* **2014**, *18*, 1786–1801. [CrossRef]
22. Moriarty, R.M.; Epa, W.R.; Awasthi, A.K. Palladium-catalyzed coupling of alkenyl iodonium salts with olefins: A mild and stereoselective Heck-type reaction using hypervalent iodine. *J. Am. Chem. Soc.* **1991**, *113*, 6315–6317. [CrossRef]
23. Szabó, K.J. Mechanism of the oxidative addition of hypervalent iodonium salts to palladium(II) pincer-complexes. *J. Mol. Catal. A Chem.* **2010**, *324*, 56–63. [CrossRef]
24. Hwang, L.K.; Na, Y.; Lee, J.; Do, Y.; Chang, S. Tetraarylphosphonium halides as arylating reagents in Pd-catalyzed Heck and cross-coupling reactions. *Angew. Chem. Int. Ed.* **2005**, *44*, 6166–6169. [CrossRef] [PubMed]
25. Tian, Z.-Y.; Hu, Y.-T.; Teng, H.-B.; Zhang, C.-P. Application of arylsulfonium salts as arylation reagents. *Tetrahedron Lett.* **2018**, *59*, 299–309. [CrossRef]
26. Tasker, S.Z.; Gutierrez, A.C.; Jamison, T.F. Nickel-catalyzed Mizoroki-Heck reaction of aryl sulfonates and chlorides with electronically unbiased terminal olefins: High selectivity for branched products. *Angew. Chem. Int. Ed.* **2014**, *53*, 1858–1861. [CrossRef] [PubMed]
27. Maity, S.; Dolui, P.; Kancherla, R.; Maiti, D. Introducing unactivated acyclic internal aliphatic olefins into a cobalt catalyzed allylic selective dehydrogenative Heck reaction. *Chem. Sci.* **2017**, *8*, 5181–5185. [CrossRef] [PubMed]
28. Phipps, R.J.; McMurray, L.; Ritter, S.; Duong, H.A.; Gaunt, M.J. Copper-catalyzed alkene arylation with diaryliodonium salts. *J. Am. Chem. Soc.* **2012**, *134*, 10773–10776. [CrossRef] [PubMed]
29. Dughera, S.; Barbero, M. Gold catalyzed Heck-coupling of arenediazonium o-Benzenedisulfonimides. *Org. Biomol. Chem.* **2018**, *16*, 295–301.
30. Zhu, K.; Dunne, J.; Shaver, M.P.; Thomas, S.P. Iron-catalyzed Heck-type alkenylation of functionalized alkyl bromides. *ACS Catal.* **2017**, *7*, 2353–2356. [CrossRef]
31. Wang, S.-S.; Yang, G.-Y. Recent developments in low-cost TM-catalyzed Heck-type reactions (TM = transition metal, Ni, Co, Cu, Fe). *Catal. Sci. Technol.* **2016**, *6*, 2862–2876. [CrossRef]
32. Beletskaya, I.P.; Cheprakov, A.V. Modern Heck reactions. *RSC Catal. Ser.* **2015**, *21*, 355–478.

33. Hajipour, A.R.; Rezaei, F.; Khorsandi, Z. Pd/Cu-free Heck and Sonogashira cross-coupling reaction by Co nanoparticles immobilized on magnetic chitosan as reusable catalyst. *Green Chem.* **2017**, *19*, 1353–1361. [CrossRef]
34. Kurandina, D.; Parasram, M.; Gevorgyan, V. Visible light-induced room-temperature Heck reaction of functionalized alkyl halides with vinyl arenes/heteroarenes. *Angew. Chem. Int. Ed.* **2017**, *56*, 14212–14216. [CrossRef] [PubMed]
35. Wang, G.-Z.; Shang, R.; Cheng, W.-M.; Fu, Y. Irradiation-induced Heck reaction of unactivated alkyl halides at room temperature. *J. Am. Chem. Soc.* **2017**, *139*, 18307–18312. [CrossRef] [PubMed]
36. O'Hagan, D. Understanding organofluorine chemistry. An introduction to the C–F bond. *Chem. Soc. Rev.* **2008**, *39*, 308–319. [CrossRef] [PubMed]
37. Eisenstein, O.; Milani, J.; Perutz, R.N. Selectivity of C–H activation and competition between C–H and C–F bond activation at fluorocarbons. *Chem. Rev.* **2017**, *117*, 8710–8753. [CrossRef] [PubMed]
38. Sowaileh, M.F.; Hazlitt, R.A.; Colby, D.A. Application of the pentafluorosulfanyl group as a bioisosteric replacement. *ChemMedChem* **2017**, *12*, 1481–1490. [CrossRef] [PubMed]
39. Hiyama, T.; Kanie, K.; Kusumoto, T.; Morizawa, Y.; Shimizu, M. *Organofluorine Compounds: Chemistry and Applications*; Springer: Berlin, Germany, 2000.
40. Kirsch, P. *Modern Fluoroorganic Chemistry: Synthesis, Reactivity, Applications*, 2nd ed.; Wiley-VCH: Weinheim, Germany, 2013.
41. Zhou, Y.; Wang, J.; Gu, Z.; Wang, S.; Zhu, W.; Aceñ, J.; Soloshonok, V.A.; Izawa, K.; Liu, H. Next generation of fluorine-containing pharmaceuticals, compounds currently in phase II–III clinical trials of major pharmaceutical companies: New structural trends and therapeutic areas. *Chem. Rev.* **2016**, *116*, 422–518. [CrossRef] [PubMed]
42. Berger, A.A.; Völler, J.-S.; Budisa, N.; Koksch, B. Deciphering the fluorine code-the many hats fluorine wears in a protein environment. *Acc. Chem. Res.* **2017**, *50*, 2093–2103. [CrossRef] [PubMed]
43. Zhang, Q.; Kelly, M.A.; Bauer, N.; You, W. The curious case of fluorination of conjugated polymers for solar cells. *Acc. Chem. Res.* **2017**, *50*, 2401–2409. [CrossRef] [PubMed]
44. Kusoglu, A.; Weber, A.Z. New insights into perfluorinated sulfonic-acid ionomers. *Chem. Rev.* **2017**, *117*, 987–1104. [CrossRef] [PubMed]
45. Amii, H.; Uneyama, K. C–F Bond activation in organic synthesis. *Chem. Rev.* **2009**, *109*, 2119–2183. [CrossRef] [PubMed]
46. Yerien, D.E.; Bonesi, S.; Postigo, A. Fluorination methods in drug discovery. *Org. Biomol. Chem.* **2016**, *14*, 8398–8427. [CrossRef] [PubMed]
47. Yang, L.; Dong, T.; Revankar, H.M.; Zhang, C.-P. Recent progress on fluorination in aqueous media. *Green Chem.* **2017**, *19*, 3951–3992. [CrossRef]
48. Preshlock, S.; Tredwell, M.; Gouverneur, V. ^{18}F-Labeling of arenes and heteroarenes for applications in positron emission tomography. *Chem. Rev.* **2016**, *116*, 719–766. [CrossRef] [PubMed]
49. Marciniak, B.; Walkowiak-Kulikowsk, J.; Koroniak, H. On the halofluorination reactions of olefins as selective, and an efficient methodology for the introduction of fluorine into organic molecules. *J. Fluor. Chem.* **2017**, *203*, 47–61. [CrossRef]
50. Song, X.; Xu, C.; Wang, M. Transformations based on ring-opening of gem-difluorocyclopropanes. *Tetrahedron Lett.* **2017**, *58*, 1806–1816. [CrossRef]
51. Zhang, C.-P.; Chen, Q.-Y.; Guo, Y.; Xiao, J.-C.; Gu, Y.-C. Progress in fluoroalkylation of organic compounds via sulfinatodehalogenation initiation system. *Chem. Soc. Rev.* **2012**, *41*, 4536–4559. [CrossRef] [PubMed]
52. Heitz, W.; Knebelkamp, A. Synthesis of fluorostyrenes via palladium-catalyzed reactions of aromatic halides with fluoroolefins. *Makromol. Rapid Commun.* **1991**, *12*, 69–75. [CrossRef]
53. Patrick, T.B.; Agboka, T.Y.; Gorrell, K. Heck reaction with 3-fluoro-3-buten-2-one. *J. Fluor. Chem.* **2008**, *129*, 983–985. [CrossRef]
54. Rousee, K.; Bouillon, J.P.; Couve-Bonnaire, S.; Pannecoucke, X. Stereospecific synthesis of tri- and tetrasubstituted α-fluoroacrylates by Mizoroki-Heck reaction. *Org. Lett.* **2016**, *18*, 540–543. [CrossRef] [PubMed]
55. Hirotaki, K.; Hanamoto, T. Mizoroki-Heck reaction of (1-fluorovinyl)methyldiphenylsilane with aryl iodides. *J. Org. Chem.* **2011**, *76*, 8564–8568. [CrossRef] [PubMed]

56. Patrick, T.B.; Blay, A.A. Heck reaction with ethyl (*E*)- and (*Z*)-3-fluoropropenoate. *J. Fluor. Chem.* **2016**, *189*, 68–69. [CrossRef]
57. Ichikawa, J.; Sakoda, K.; Mihara, J.; Ito, N. Heck-type 5-*endo-trig* cyclizations promoted by vinylic fluorines: Ring-fluorinated indene and 3*H*-pyrrole syntheses from 1,1-difluoro-1-alkenes. *J. Fluor. Chem.* **2006**, *127*, 489–504. [CrossRef]
58. Sakoda, K.; Mihara, J.; Ichikawa, J. Heck-type 5-*endo-trig* cyclization promoted by vinylic fluorines: Synthesis of 5-fluoro-3*H*-pyrroles. *Chem. Commun.* **2005**, *0*, 4684–4686. [CrossRef] [PubMed]
59. Krutak, J.J.; Burpitt, R.D.; Moore, W.H.; Hyatt, J.A. Chemistry of ethenesulfonyl fluoride. Fluorosulfonylethylation of organic compounds. *J. Org. Chem.* **1979**, *44*, 3847–3858. [CrossRef]
60. Qin, H.-L.; Zheng, Q.; Bare, G.A.L.; Wu, P.; Sharpless, K.B. A Heck–Matsuda process for the synthesis of β-arylethenesulfonyl fluorides: Selectively addressable bis-electrophiles for sufex click chemistry. *Angew. Chem. Int. Ed.* **2016**, *55*, 14155–14158. [CrossRef] [PubMed]
61. Zha, G.F.; Zheng, Q.; Leng, J.; Wu, P.; Qin, H.L.; Sharpless, K.B. Palladium-catalyzed fluorosulfonylvinylation of organic iodides. *Angew. Chem. Int. Ed.* **2017**, *56*, 4849–4852. [CrossRef] [PubMed]
62. Karimi, B.; Behzadnia, H.; Elhamifar, D.; Akhavan, P.F.; Esfahani, F.K.; Zamani, A. Transition-metal-catalyzed oxidative Heck reactions. *Synthesis* **2010**, *2010*, 1399–1427. [CrossRef]
63. Lee, A.L. Enantioselective oxidative boron Heck reactions. *Org. Biomol. Chem.* **2016**, *14*, 5357–5366. [CrossRef] [PubMed]
64. Ma, W.; Gandeepan, P.; Li, J.; Ackermann, L. Recent advances in positional-selective alkenylations: Removable guidance for twofold C–H activation. *Org. Chem. Front.* **2017**, *4*, 1435–1467. [CrossRef]
65. Chinthakindi, P.K.; Govender, K.B.; Kumar, A.S.; Kruger, H.G.; Govender, T.; Naicker, T.; Arvidsson, P.I. A synthesis of "dual warhead" β-aryl ethenesulfonyl fluorides and one-pot reaction to beta-sultams. *Org. Lett.* **2017**, *19*, 480–483. [CrossRef] [PubMed]
66. Zha, G.-F.; Bare, G.A.L.; Leng, J.; Shang, Z.-P.; Luo, Z.; Qin, H.-L. Gram-scale synthesis of β-(hetero)arylethenesulfonyl fluorides via a Pd(OAc)$_2$ catalyzed oxidative Heck process with DDQ or AgNO$_3$ as an oxidant. *Adv. Synth. Catal.* **2017**, *359*, 3237–3242. [CrossRef]
67. Sokolenko, L.V.; Yagupolskii, Y.L.; Vlasenko, Y.G.; Babichenko, L.N.; Lipetskij, V.O.; Anselmi, E.; Magnier, E. Arylation of perfluoroalkyl vinyl sulfoxides via the Heck reaction. *Tetrahedron Lett.* **2015**, *56*, 1259–1262. [CrossRef]
68. Fuchikami, T.; Yatabe, M.; Ojima, I. A facile synthesis of *trans*-β-trifluoromethylstyrene, *trans*-2,3,4,5,6-pentafluorostilbene and their derivatives. *Synthesis* **1981**, *1981*, 365–366. [CrossRef]
69. Chen, W.; Xu, L.; Xiao, J. A general method to fluorous ponytail-substituted aromatics. *Tetrahedron Lett.* **2001**, *42*, 4275–4278. [CrossRef]
70. Birdsall, D.J.; Hope, E.G.; Stuart, A.M.; Chen, W.; Hu, Y.; Xiao, J. Synthesis of fluoroalkyl-derivatised BINAP ligands. *Tetrahedron Lett.* **2001**, *42*, 8551–8553. [CrossRef]
71. Feng, J.; Cai, C. An efficient synthesis of perfluoroalkenylated aryl compounds via pincer-Pd catalyzed Heck couplings. *J. Fluor. Chem.* **2013**, *146*, 6–10. [CrossRef]
72. Su, H.L.; Balogh, J.; Al-Hashimi, M.; Seapy, D.G.; Bazzi, H.S.; Gladysz, J.A. Convenient protocols for Mizoroki-Heck reactions of aromatic bromides and polybromides with fluorous alkenes of the formula H$_2$C=CH(CF$_2$)$_{n-1}$CF$_3$ (n = 8, 10). *Org. Biomol. Chem.* **2016**, *14*, 10058–10069. [CrossRef] [PubMed]
73. Jian, H.; Tour, J.M. Preparative fluorous mixture synthesis of diazonium-functionalized oligo(phenylenevinylene)s. *J. Org. Chem.* **2005**, *70*, 3396–3424. [CrossRef] [PubMed]
74. Darses, S.; Pucheault, M.; Genêt, J.P. Efficient access to perfluoroalkylated aryl compounds by Heck reaction. *Eur. J. Org. Chem.* **2001**, *2001*, 1121–1128. [CrossRef]
75. Sugiyama, Y.; Endo, N.; Ishihara, K.; Kobayashi, Y.; Hamamoto, H.; Shioiri, T.; Matsugi, M. Development of efficient processes for multi-gram scale and divergent preparation of fluorous-Fmoc reagents. *Tetrahedron* **2015**, *71*, 4958–4966. [CrossRef]
76. Sakaguchi, Y.; Yamada, S.; Konno, T.; Agou, T.; Kubota, T. Stereochemically defined various multisubstituted alkenes bearing a tetrafluoroethylene (-CF$_2$CF$_2$-) fragment. *J. Org. Chem.* **2017**, *82*, 1618–1631. [CrossRef] [PubMed]
77. Konno, T.; Yamada, S.; Tani, A.; Miyabe, T.; Ishihara, T. A novel synthesis of trifluoromethylated multi-substituted alkenes via regio- and stereoselective heck reaction of (*E*)-4,4,4-trifluoro-1-phenyl-2-buten-1-one. *Synlett* **2006**, *2006*, 3025–3028. [CrossRef]

78. Konno, T.; Yamada, S.; Tani, A.; Nishida, M.; Miyabe, T.; Ishihara, T. Unexpected high regiocontrol in Heck reaction of fluorine-containing electron-deficient olefins—Highly regio- and stereoselective synthesis of β-fluoroalkyl-α-aryl-α,β-unsaturated ketones. *J. Fluor. Chem.* **2009**, *130*, 913–921. [CrossRef]
79. Li, Y.; Tu, D.-H.; Gu, Y.-J.; Wang, B.; Wang, Y.-Y.; Liu, Z.-T.; Liu, Z.-W.; Lu, J. Oxidative Heck reaction of fluorinated olefins with arylboronic acids by palladium catalysis. *Eur. J. Org. Chem.* **2015**, *2015*, 4340–4343. [CrossRef]
80. Ichikawa, J.; Nadano, R.; Ito, N. 5-Endo Heck-type cyclization of 2-(trifluoromethyl)allyl ketone oximes: Synthesis of 4-difluoromethylene-substituted 1-pyrrolines. *Chem. Commun.* **2006**, 4425–4427. [CrossRef] [PubMed]
81. Renak, M.L.; Bartholomew, G.P.; Wang, S.; Ricatto, P.J.; Lachicotte, R.J.; Bazan, G.C. Fluorinated distyrylbenzene chromophores: Effect of fluorine-regiochemistry on molecular properties and solid-state organization. *J. Am. Chem. Soc.* **1999**, *121*, 7787–7799. [CrossRef]
82. Bernier, D.; Brückner, R. Novel synthesis of naturally occurring pulvinones: A Heck coupling, transesterification, and dieckmann condensation strategy. *Synthesis* **2007**, 2249–2272. [CrossRef]
83. Fan, C.; Yang, P.; Wang, X.; Liu, G.; Jiang, X.; Chen, H.; Tao, X.; Wang, M.; Jiang, M. Synthesis and organic photovoltaic (OPV) properties of triphenylamine derivatives based on a hexafluorocyclopentene "core". *Sol. Energy Mater. Sol. Cells* **2011**, *95*, 992–1000. [CrossRef]
84. Salabert, J.; Sebastián, R.M.; Vallribera, A.; Roglans, A.; Nájera, C. Fluorous aryl compounds by Matsuda-Heck reaction. *Tetrahedron* **2011**, *67*, 8659–8664. [CrossRef]
85. Okazaki, T.; Laali, K.K.; Bunge, S.D.; Adas, S.K. 4-(Pentafluorosulfanyl)benzenediazonium tetrafluoroborate: A versatile launch pad for the synthesis of aromatic SF_5 compounds via cross coupling, azo coupling, homocoupling, dediazoniation, and click chemistry. *Eur. J. Org. Chem.* **2014**, 1630–1644. [CrossRef]
86. Albéniz, A.C.; Espinet, P.; Martínruiz, B.; Milstein, D. Catalytic system for Heck reactions involving insertion into Pd-(perfluoro-organyl) bonds. *J. Am. Chem. Soc.* **2001**, *123*, 11504–11505. [CrossRef] [PubMed]
87. Albeniz, A.C.; Espinet, P.; Martin-Ruiz, B.; Milstein, D. Catalytic system for the Heck reaction of fluorinated haloaryls. *Organometallics* **2005**, *24*, 3679–3684. [CrossRef]
88. Zhang, X.; Fan, S.; He, C.Y.; Wan, X.; Min, Q.Q.; Yang, J.; Jiang, Z.X. $Pd(OAc)_2$ catalyzed olefination of highly electron-deficient perfluoroarenes. *J. Am. Chem. Soc.* **2010**, *132*, 4506–4507. [CrossRef] [PubMed]
89. Wu, C.Z.; He, C.Y.; Huang, Y.; Zhang, X. Thioether-promoted direct olefination of polyfluoroarenes catalyzed by palladium. *Org. Lett.* **2013**, *15*, 5266–5269. [CrossRef] [PubMed]
90. Xiao, Y.-L.; Zhang, B.; He, C.-Y.; Zhang, X. Direct olefination of fluorinated benzothiadiazoles: A new entry to optoelectronic materials. *Chem. Eur. J.* **2014**, *20*, 4532–4536. [CrossRef] [PubMed]
91. He, C.-Y.; Qing, F.-L.; Zhang, X. Pd-catalyzed aerobic direct olefination of polyfluoroarenes. *Tetrahedron Lett.* **2014**, *55*, 2962–2964. [CrossRef]
92. Li, Z.; Zhang, Y.; Liu, Z.-Q. Pd-catalyzed olefination of perfluoroarenes with allyl esters. *Org. Lett.* **2012**, *14*, 74–77. [CrossRef] [PubMed]
93. Wang, S.-M.; Song, H.-X.; Wang, X.-Y.; Liu, N.; Qin, H.-L.; Zhang, C.-P. Palladium-catalyzed Mizoroki-Heck-type reactions of $[Ph_2SR_{fn}][OTf]$ with alkenes at room temperature. *Chem. Commun.* **2016**, *52*, 11893–11896. [CrossRef] [PubMed]
94. Dolci, L.; Dolle, F.; Valette, H.; Vaufrey, F.; Fuseau, C.; Bottlaender, M.; Crouzel, C. Synthesis of a fluorine-18 labeled derivative of epibatidine for in vivo nicotinic acetylcholine receptor PET imaging. *Bioorg. Med. Chem.* **1999**, *7*, 467–479. [CrossRef]
95. Quandt, G.; Höfner, G.; Wanner, K.T. Synthesis and evaluation of N-substituted nipecotic acid derivatives with an unsymmetrical bis-aromatic residue attached to a vinyl ether spacer as potential GABA uptake inhibitors. *Bioorg. Med. Chem.* **2013**, *21*, 3363–3378. [CrossRef] [PubMed]
96. Sun, Z.; Ahmed, S.; McLaughlin, L.W. Syntheses of pyridine C-nucleosides as analogues of the natural nucleosides dC and dU. *J. Org. Chem.* **2006**, *71*, 2922–2925. [CrossRef] [PubMed]
97. Lin, Y.-D.; Chow, T.J. Fluorine substituent effect on organic dyes for sensitized solar cells. *J. Photochem. Photobiol. A Chem.* **2012**, *230*, 47–54. [CrossRef]
98. Belfield, K.D.; Najjar, O.; Sriram, S.R. Synthesis of polyurethanes and polyimides for photorefractive applications. *Polymer* **2000**, *41*, 5011–5020. [CrossRef]
99. Belfield, K.D.; Chinna, C.; Najjar, O.; Sriram, S.; Schafer, K.J. Methodology for the synthesis of new multifunctional polymers for photorefractive applications. *ACS Symp. Ser.* **1999**, *726*, 250–263.

100. Surapanich, N.; Kuhakarn, C.; Pohmakotr, M.; Reutrakul, V. Palladium-mediated Heck-type reactions of [(bromodifluoromethyl)sulfonyl]benzene: Synthesis of α-alkenyl- and α-heteroaryl-substituted α,α-difluoromethyl phenyl sulfones. *Eur. J. Org. Chem.* **2012**, *2012*, 5943–5952. [CrossRef]
101. Feng, Z.; Min, Q.Q.; Zhao, H.Y.; Gu, J.W.; Zhang, X. A general synthesis of fluoroalkylated alkenes by palladium-catalyzed Heck-type reaction of fluoroalkyl bromides. *Angew. Chem. Int. Ed.* **2015**, *54*, 1270–1274. [CrossRef] [PubMed]
102. Chen, Q.-Y.; Yang, Z.-Y.; Zhao, C.-X.; Qiu, Z.-M. Studies on fluoroalkylation and fluoroalkoxylation. Part 28. Palladium(0)-induced addition of fluoroalkyl iodides to alkenes: An electron transfer process. *J. Chem. Soc. Perkin Trans. 1* **1988**, *0*, 563–567. [CrossRef]
103. Zhang, X.; Zhang, F.; Min, Q.-Q. Palladium-catalyzed Heck-type difluoroalkylation of alkenes with functionalized difluoromethyl bromides. *Synthesis* **2015**, *47*, 2912–2923. [CrossRef]
104. Feng, Z.; Xiao, Y.-L.; Zhang, X. Palladium-catalyzed phosphonyldifluoromethylation of alkenes with bromodifluoromethylphosphonate. *Org. Chem. Front.* **2016**, *3*, 466–469. [CrossRef]
105. Ai, H.-J.; Cai, C.-X.; Qi, X.; Peng, J.-B.; Zheng, F.; Wu, X.-F. Palladium-catalyzed Heck reaction of in-situ generated benzylic iodides and styrenes. *Tetrahedron Lett.* **2017**, *58*, 3846–3850. [CrossRef]
106. Fan, T.; Meng, W.-D.; Zhang, X. Palladium-catalyzed Heck-type reaction of secondary trifluoromethylated alkyl bromides. *Beilstein J. Org. Chem.* **2017**, *13*, 2610–2616. [CrossRef] [PubMed]
107. Dubbaka, S.R.; Vogel, P. Palladium-catalyzed desulfitative mizoroki-heck couplings of sulfonyl chlorides with mono- and disubstituted olefins: Rhodium-catalyzed desulfitative Heck-type reactions under phosphine- and base-free conditions. *Chem. Eur. J.* **2005**, *11*, 2633–2641. [CrossRef] [PubMed]
108. Prakash, G.K.S.; Krishnan, H.S.; Jog, P.V.; Iyer, A.P.; Olah, G.A. A domino approach of Heck coupling for the synthesis of β-trifluoromethylstyrenes. *Org. Lett.* **2012**, *14*, 1146–1149. [CrossRef] [PubMed]
109. Liao, J.; Fan, L.; Guo, W.; Zhang, Z.; Li, J.; Zhu, C.; Ren, Y.; Wu, W.; Jiang, H. Palladium-catalyzed fluoroalkylative cyclization of olefins. *Org. Lett.* **2017**, *19*, 1008–1011. [CrossRef] [PubMed]
110. Wu, X.-X.; Chen, W.-L.; Shen, Y.; Chen, S.; Xu, P.-F.; Liang, Y.-M. Palladium-catalyzed domino Heck/intermolecular C–H bond functionalization: Efficient synthesis of alkylated polyfluoroarene derivatives. *Org. Lett.* **2016**, *18*, 1784–1787. [CrossRef] [PubMed]
111. Talbot, E.P.A.; Fernandes, T.d.A.; McKenna, J.M.; Toste, F.D. Asymmetric palladium-catalyzed directed intermolecular fluoroarylation of styrenes. *J. Am. Chem. Soc.* **2014**, *136*, 4101–4104. [CrossRef] [PubMed]
112. He, Y.; Yang, Z.; Thornbury, R.T.; Toste, F.D. Palladium-catalyzed enantioselective 1,1-fluoroarylation of aminoalkenes. *J. Am. Chem. Soc.* **2015**, *137*, 12207–12210. [CrossRef] [PubMed]
113. Braun, M.-G.; Katcher, M.H.; Doyle, A.G. Carbofluorination via a palladium-catalyzed cascade reaction. *Chem. Sci.* **2013**, *4*, 1216–1220. [CrossRef]

© 2018 by the authors. Licensee MDPI, Basel, Switzerland. This article is an open access article distributed under the terms and conditions of the Creative Commons Attribution (CC BY) license (http://creativecommons.org/licenses/by/4.0/).

Review

Heck Reaction—State of the Art

Sangeeta Jagtap

Department of Chemistry, Baburaoji Gholap College, Sangvi, Pune 411027, India; sangeetajagtap@rediffmail.com; Tel.: +91-20-2728-0204

Received: 20 August 2017; Accepted: 6 September 2017; Published: 11 September 2017

Abstract: The Heck reaction is one of the most studied coupling reactions and is recognized with the Nobel Prize in Chemistry. Thousands of articles, hundreds of reviews and a number of books have been published on this topic. All reviews are written exhaustively describing the various aspects of Heck reaction and refer to the work done hitherto. Looking at the quantum of the monographs published, and the reviews based on them, we found a necessity to summarize all reviews on Heck reaction about catalysts, ligands, suggested mechanisms, conditions, methodologies and the compounds formed via Heck reaction in one review and generate a resource of information. One can find almost all the catalysts used so far for Heck reaction in this review.

Keywords: Heck reaction; reviews; C-C coupling; catalysis; mechanism; application

1. Introduction

The new era of research started after the introduction of coupling reactions, i.e., carbon–carbon bond forming reactions like Heck [1], Suzuki [2], Sonogoshira [3], Negishi [4], Kumada [5], Stille [6], Tsuji-Trost [7], etc. These reactions have played an enormously decisive and important role in shaping chemical synthesis and have revolutionized the way one thinks about synthetic organic chemistry. The Heck reaction (Equation (1)) is used extensively in many syntheses, including agrochemical, fine chemicals, pharmaceutical, etc. The reaction was introduced by Mizoroki [8] and Heck [9] independently more than four decades ago. It has drawn much attention due to high efficiency and simplicity. Heck methodology is attractive from a synthetic point of view because of its high chemoselectivity and mild reaction conditions along with low toxicity and cost of the reagent if, specifically, the catalyst is recycled.

$$R\text{-olefin} + Ar\text{-X (aryl halide)} \xrightarrow[\text{Base, Solvent}]{\text{Catalyst}} R\text{-}Ar + BHX^{\oplus\ominus} \qquad (1)$$

The Heck reaction is described as a vinylation or arylation of olefins where a large variety of olefins can be used, like derivatives of acrylates, styrenes or intramolecular double bonds. The aryl halide variants developed in addition to typical aryl bromides and iodides are aromatic triflates, aroyl chlorides, aryl sulfonyl chlorides, aromatic diazonium salts, aroyl anhydrides, aryl chlorides and arylsilanols. The catalyst is the essential part of a reaction where a variety of metals along with a huge range of ligands is studied. Significant progress for the preparation and characterization of variety of ligands and catalysts has been made for avoiding protection and deprotection procedures, therefore allowing for syntheses to be carried out in fewer steps. Development, novel catalytic properties and extensive mechanistic studies are summarized in several reviews based on seminal work of many researchers and reviewers as described in following sections. Palladium is usually the preferred metal as it tolerates a wide variety of functional groups and it has a remarkable ability to assemble C-C bonds between appropriately functionalized substrates. Most palladium based methodologies proceed with stereo- and regioselectivity and with excellent yields. Generally, the less crowded structure is preferred during Heck reaction, and often favours a trans product. Few mechanisms are also supported by the discussion on the regioselectivity and stereoselectivity of Heck reaction.

Sometimes compounds such as TBAB (Tetra butyl ammonium bromide) are added in the reaction mixture along with organic or inorganic bases needed for the sequestration of acid generated. Typical solvents for the Heck reaction are dipolar aprotic solvents like DMF (dimethyl formamide) and NMP (N-methyl-2-pyrrolidone), however, the reaction is also performed in many other different solvents and in fact there are large numbers of reviews dedicated to the use of various solvents. Besides these, many manuscripts describe the reaction in the absence of one of the component (other than substrates) such as ligand free, organic solvent free and so on. Recent publications also describe how to recover used catalyst specially using aqueous medium which implies the potential for greener approach for organic reactions.

There are a large number of excellent reviews available on Heck reaction and related topics based on applications, mechanism, quest for high turn over numbers (TONs), asymmetric synthesis, separation techniques, etc. In addition to this, there are a few manuscripts that give a general overview about Heck reaction, some of them are detailed and some are not so descriptive. Many reviews discuss various catalysis techniques with certain aspects where along with the Heck reaction, other coupling reactions are also discussed, however for the present review, only the Heck related chemistry is considered. It should be noted that for this review only the reviews and few related articles covering Heck chemistry are considered. Consideration of book articles, although informative, is beyond the scope of this review.

2. Reviews on Catalyst Development

The majority of the work on Heck reaction has been focused on the catalyst development. There are numerous reviews based on this and hence are further categorized for better understanding.

2.1. Overviews

When the mechanism was not fully explored, the reaction and its possible mechanistic pathways are reviewed by one of its pioneers, R.F. Heck [10–12] by focusing on possibilities of the intermediate formed during Heck reaction by considering various examples for its support. These reviews mention the working reaction conditions explored for the reaction and focuses on the requirements of the substrates, bases, solvents and the conditions to be used. It was observed that the reaction is usually regioselective and stereospecific and is tolerant of almost every functional group. Data have proved that the direction of addition of the organopalladium complex to the olefin depends upon the steric and electronic influence of the substituent present. The direction of addition is most often dominated by steric effects where the organic group is seen to be added to the least substituted carbon of the double band.

Initially, the intermediate mono-organopalladium(II) species, $RPdL_2X$, were prepared by exchange reactions of mercurials in general, with palladium(II) salts (Scheme 1a), however it suffered from the disadvantage that, many such main group organometallics were not easily accessible and moreover they were needed to be used in stoichiometric amounts in the synthesis of organic compounds. Hence, the finely divided palladium metal or palladium(0) phosphine complexes were used for such reactions (Scheme 1b).

a. $RHgCl + PdCl_2 + 2L \longrightarrow RPdL_2Cl + HgCl_2$
L=Ligand

b. $RX + Pd(PR'_3)_n \longrightarrow RPd(PR'_3)_2X + (n-2)PR'_3$

Scheme 1. Initial work by Heck [10] to get an active species (**a**) $RPdL_2X$ and (**b**) $RPd(PR'_3)_2X$.

Effects of different triaryl phosphines have also been studied. It is mentioned that, generally the reaction does not require anhydrous or anaerobic conditions although it is advisable to limit access of oxygen when arylphosphines are used as a component of the catalyst. Aryl, heterocyclic, benzyl, or vinyl halides are commonly employed often with bromides and iodides in comparison to halides with an easily eliminated beta-hydrogen atom (i.e., alkyl derivatives) since they form only olefins by elimination under the normal reaction conditions. Generally used bases are secondary or tertiary amine, sodium or potassium acetate, carbonate, or bicarbonates. The catalyst is commonly palladium acetate, although palladium chloride or preformed triarylphosphine palladium complexes, as well as palladium on charcoal, have been used.

Triarylphosphine or a secondary amine is required when organic bromides are used although a reactant, product, or solvent may serve as the ligand for reactions involving organic iodides. Solvents such as acetonitrile, dimethylformamide, hexamethylphosphoramide, N-methylpyrrolidinone, and methanol have been used, but are often not necessary. The procedure is applicable to a very wide range of reactants and yields are generally good to excellent. Temperature used is in the range of 50 to 160 °C, where the reaction proceeds homogeneously. When nucleophilic secondary amines are used as co-reactants with most vinylic halides, a variation occurs that often produces tertiary allylic amines as major products. A comparison of Heck reaction with similar types of reaction where the organic halide is replaced by other reagents such as organometallics, diazonium salts, or aromatic hydrocarbons has also been discussed.

De Meijere and Meyer [13] have reviewed mainly the Heck couplings with various oligohaloarenes along with an overall discussion on development in mechanism and catalysts until 1994. The review stresses upon the careful choice of substrates and skilful tailoring of reaction conditions leading to impressive sequences.

As exemplified in the review, the Heck reaction, together with other mechanistically related palladium catalysed transformations with arene, alkene and alkyne derivatives, opens the door to a tremendous variety of elegant and highly convergent routes to structurally complex molecules. The reaction is not hindered by heteroatoms such as oxygen, nitrogen, sulphur and phosphorus in most of the cases. With Heck reaction, a range of chemoselective and regioselective monocouplings of highly functionalized substrates with unsymmetrical and multisubstituted reaction partners could be achieved. A number of examples are given that demonstrate the cascade reactions in which three, four, five, and even eight new C-C bonds are formed to yield oligofunctional and oligocyclic products like 1–6 (Figure 1) with impressive molecular complexity. Cases are presented to establish the reactivity for enantioselective construction of complex natural products with quaternary stereocenters as exemplified by the synthesis of (R,R)-crinan 7, picrotoxinin 8, and morphine 9 (Figure 2).

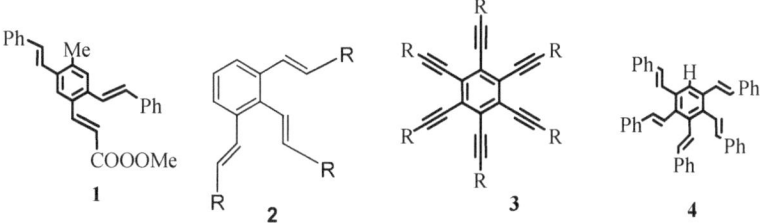

Figure 1. *Cont.*

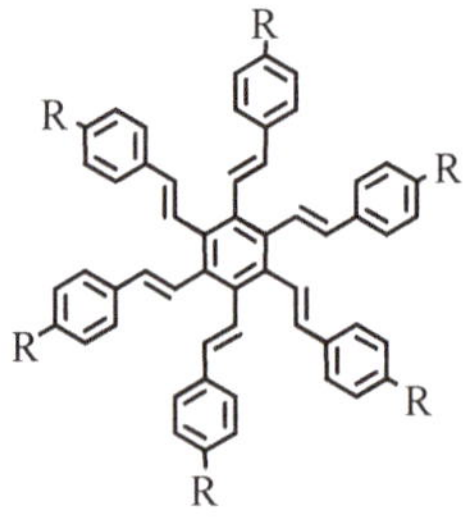

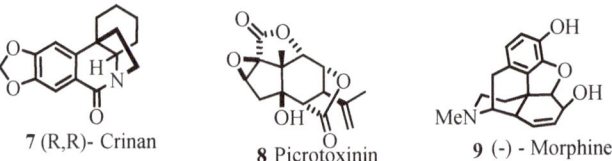

Figure 1. Oligofunctional and oligocyclic products formed via Heck reaction.

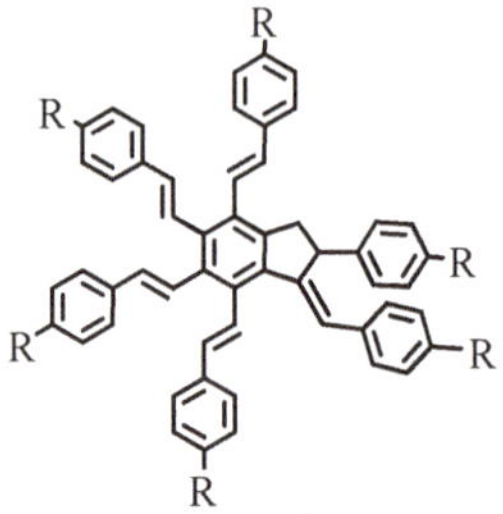

7 (R,R)- Crinan **8 Picrotoxinin** **9 (-) - Morphine**

Figure 2. Enantioselective construction of complex natural products reviewed by de Meijere and Meyer [13].

Several examples for syntheses of natural products **10–27** (Figure 3) have been provided that involve different types of Heck transformations as one of the important key steps like,

- Heck reactions with intermolecular asymmetric induction where compounds with ee (enantiomeric excess) up to 99%.
- Multiple component reactions and domino coupling reactions performed with several alkenes and various haloarenes.
- Pd-catalysed heteroannelations.
- Intramolecular Heck reactions.
- Palladium catalysed additions to triple bonds and the coupling products derived from norbornene and dicyclopentadiene that serve as precursors for diverse polycyclic aromatic compound.

Thus, the review gives the expanse of Heck reaction which is one of the important synthetic methods available to organic chemists.

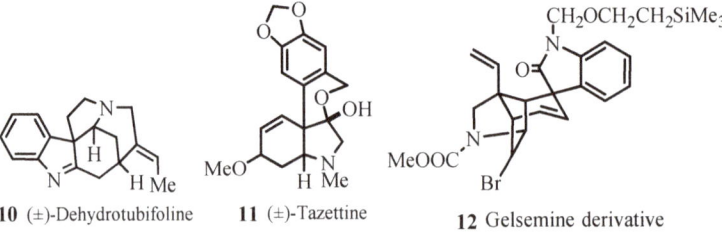

10 (±)-Dehydrotubifoline **11** (±)-Tazettine **12** Gelsemine derivative

Figure 3. Cont.

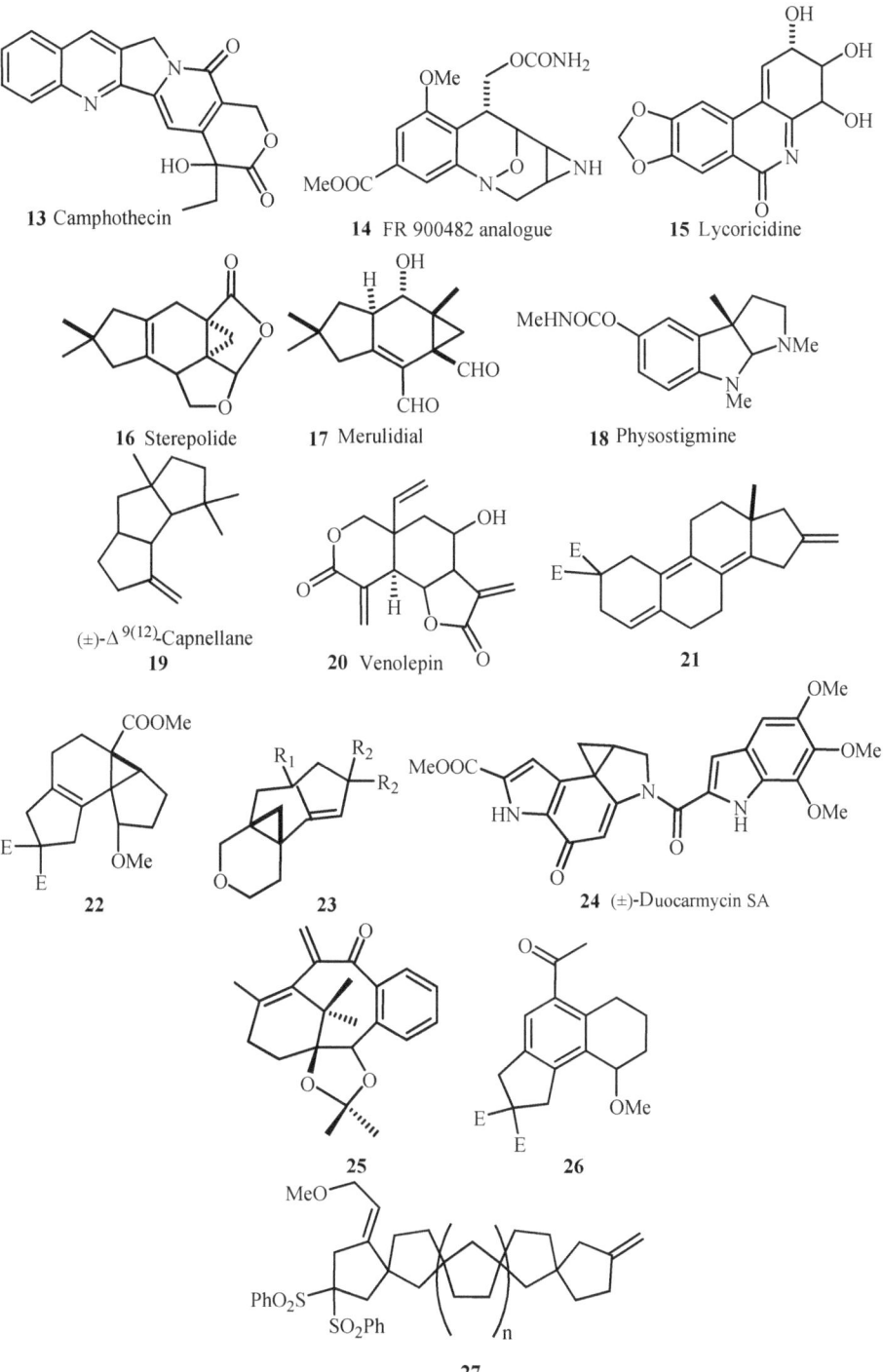

Figure 3. Natural products involving synthesis via different types of Heck transformations.

Kobetić and Biliškov [14] have summarized the basics of Heck reaction mechanism, various compositions of applicable catalyst, their developments and applications of Heck reaction until then. Sequential development of both homogeneous and heterogeneous catalysis has been discussed giving examples. Use of palladium with carbon, metal oxides and their salts, silica, porous glass tube, glass beads and guanidinium phosphane clay, zeolites and molecular sieves, MCM 41, polymer, dendrimer, etc., are elaborated with their reaction conditions and yield. Many of these catalysts shows good activity and few of them have proved to be good in recycle study as well. The review also takes an account of substrates, solvents and reaction conditions used. There are few examples of cascade and multiple coupling, synthesis of natural and biologically active compounds and enantioselective Heck-type reactions. Catalysts **28–36** (Figure 4) are seen to give yields in the range of 48 to 93%. Out of all these palladacycles, **36** is the most studied and most reviewed catalyst. In the review, the use of a phase transition catalyst (quaternary ammonium salts), and the solid base that accelerates the Heck reaction are considered to be big breakthrough discoveries.

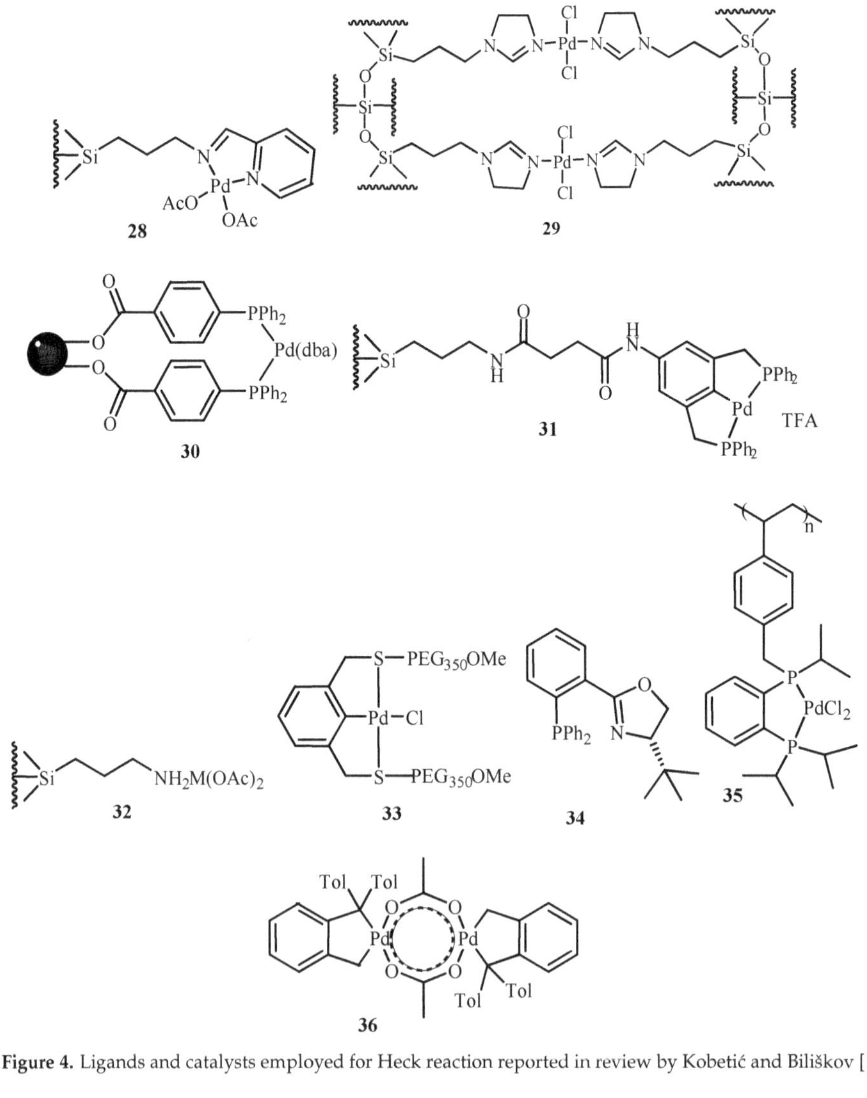

Figure 4. Ligands and catalysts employed for Heck reaction reported in review by Kobetić and Biliškov [14].

A review by Sahu and Sapkale [15] discusses the mechanism of palladium catalysed coupling reactions, details about palladium metal, its reactivity and use for reaction like Heck, Suzuki, Negishi, Hiyama, Stille, Fukuyama, Sonogoshira, Buchwald Hartwig and their variants.

The reactivity of palladium is explained on the basis of few points such as

- Electronegativity of palladium develops a polarised Pd-X bond leading to relatively strong Pd-H and Pd-C bonds.
- Its variable oxidation states such as Pd(0), Pd(II) and sometimes Pd(IV) are essential for processes such as oxidative addition, transmetalation and reductive elimination.
- Oxidation states such as Pd(I), Pd(III) and Pd(IV) are also known, where Pd(IV) species are essential in C-H activation mechanisms.

Although not in detail, references for Heck reaction in ionic liquid and for Amino-Heck are given along with the mention of application of Heck reaction in the synthesis of an alkaloid Rhazinal **37** (Figure 5), an antimitotic agent obtained from the stem extract of Kopsia teoi and a promising starting point for anticancer agent. MOM-Rhazinal (MOM = Methoxymethyl ether) derivative is made from the derivative of pyrrol where intramolecular Heck reaction takes place in presence on Pd(OAc)$_2$.

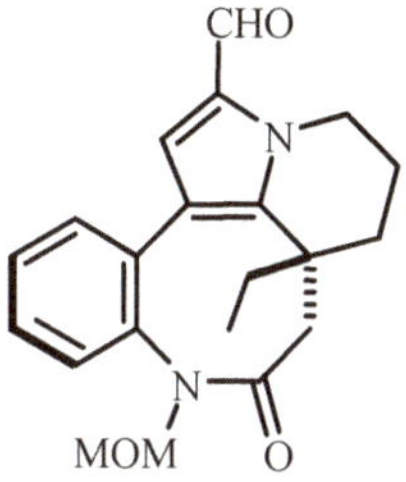

Figure 5. MOM-Rhazinal (MOM: Methoxymethyl ether).

2.2. Catalysts Variants

Every research in catalysis mostly proceeds with the aim of improving the overall catalytic activity for a particular reaction. A progressive development targeting the increase in activity has been discussed thoroughly for Heck reaction as well. Researchers work hard to find cheaper ligands, catalyst system with higher activity so as to minimize the load of palladium and ligand and to find suitable systems for the processing of aryl chlorides too.

2.2.1. Overall Progress

Beletskaya and Cheprakov [16] have published a critical and profound review on Heck reaction that systematically gives details of the work done up to 2000. It is still one of the most comprehensive reviews on Heck reactions.

A number of examples are presented to elaborate the use of various catalysts and ligands for Heck reactions like palladacycles, pincers, carbene complexes, etc. The review supports the mechanism involving Pd(0)/Pd(II) cycle by giving many examples, however the possibility of a mechanism involving Pd(IV)/Pd(II) cycle for few ligands has also been considered. For the reaction of styrene and aryl halide with a mechanism involving a Pd(0)/Pd(II) cycle, regiochemistry of the addition to styrene depends on the solvent and the nature of the anionic ligand. At higher solvent polarity and the weaker coordination ability of ligand there is a greater contribution of a truly cationic form of palladium complex, and the higher relative yield of 1,1-diphenylethylene instead of stilbene.

The review elaborates on the necessity of the stability of a catalyst, particularly for recyclable catalytic systems. Use of bidentate phosphines ligands are the better option for this, where use of excess ligand is not necessary to make stable complexes, and a ratio of 1:1 leading to (L-L)Pd complex is effective for more stable catalyst and thus leading to higher TONs.

The problem of catalyst deactivation, various solvent systems such as aqueous, molten salts, fluorous, supercritical, subcritical fluids, Heck reaction under pressure and microwave conditions are also discussed. It has been seen that high pressure has a beneficial effect on Heck reactions where oxidative addition and migratory insertions steps of the Heck cycles have a negative activation volume and thus are likely to be accelerated by pressure. However, the PdH elimination can be retarded by high pressure, leading to a change of the product distribution while potentially extending the lifetime of palladium catalyst. For microwave assisted Heck reaction, very fast heating by means of microwaves lead to shortening of the reaction time, while the yields and selectivity do not greatly differ when compared with the same reactions carried out using conventional heating.

Many examples of recyclable (phase-separation) catalysis of liquid-liquid and solid-liquid systems are discussed in the review. Heck chemistry with less usual leaving groups like diazonium salts, thallium(III) and lead(IV) derivatives and acid chlorides and anhydrides are given that provides an alternative leaving groups so as to have more reactive substrates and milder procedures. A section on reactions using metals other than palladium like Cu, Ni, Co, Rh, Ir is present in review that says 'though none of them can rival palladium in synthetic versatility, some features may complement Heck chemistry to provide either cheaper catalysts or catalysts capable of effective processing of some specific substrates'.

A number of simple N-heterocyclic carbene (NHC) palladium-based complexes have emerged as effective catalysts for a variety of cross-coupling reactions and have been proved to be excellent substitutes for phosphines ligands in homogeneous catalysis in a wide range of catalytic processes. Their efficiency is not limited to their binding ability to any transition metal as they also bind to main group elements such as beryllium, sulphur, and iodine. Because of their such specific coordination chemistry, they both stabilize and activate metal centres in quite different key catalytic steps of organic syntheses.

Hillier et al. [17] have reviewed the catalytic cross-coupling reactions mediated by palladium with nucleophilic NHC as ancillary ligands of diazabutadienes **38–40** (Figure 6) mostly based on their own work.

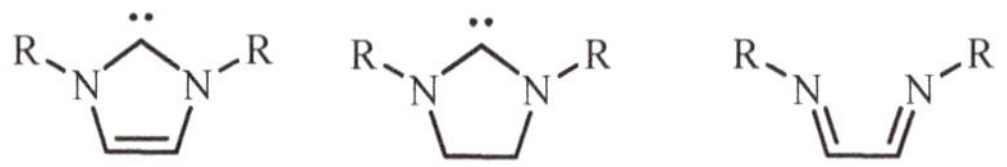

R = for 38, 39, 40 = 2,4,6-trimethylphenyl, 2,6-di-iso-propylphenyl, cyclohexyl
R = for 38 = 4-methylphenyl, adamantly, 2,6-dimethylphenyl

Figure 6. N-heterocyclic carbene (NHC) ligands of diazabutadienes reviewed by Hillier et al. [17].

On application of these catalysts to Heck reactions, excellent yields except few were seen. In all cases, the *trans* products were selectively obtained. However, no activity was observed for aryl chloride substrates. It was also observed that the reactions involving the less reactive aryl halides require bulky electron-donating phosphines. Even excess phosphine and higher Pd loading is required as they are prone to decomposition under Heck conditions increasing the cost of large-scale processes.

Herrmann [18] has reviewed N-heterocyclic carbenes as ligands where ligands and metal complexes **41–51** (Figure 7) are found to be active towards Heck reaction giving TONs up to 1.7×10^6. These ligands are preferred mainly for the reason that they not only bind to transition metal but also to main group elements. They are robust and found to have high thermal and hydrolytic durability, easy accessibility and require lower amount of loading. They could be derivatized in future to have water-soluble catalysts (two-phase catalysis), immobilization, and in chiral modifications.

However, it was observed that it is necessary to have bulky NHC ligands for successful reactions with catalysts made of these ligands.

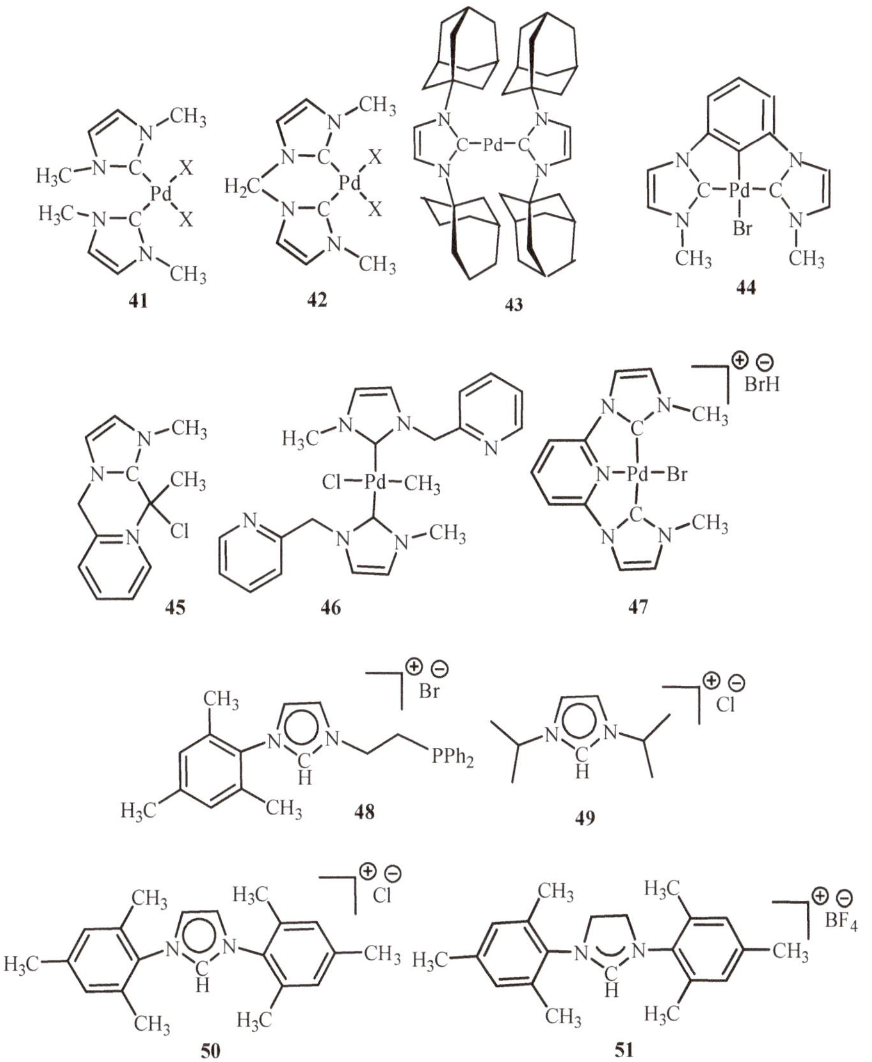

Figure 7. N-heterocyclic carbene ligands and metal complexes reviewed by Herrmann [18].

Herrmann et al. [19] presents an eloquent summary about catalytic applications of palladium complexes with phosphorus ligands containing a metallated sp^3-carbon centre (palladacycles) **36**, **52** and with N-heterocyclic carbene ligands **41–46** and **53–62** (Figure 8) for C-C and C-N coupling reactions of aryl halides. The activity of **36** was found to get increased to TONs of up to 4×10^4 after addition of tetrabutyl ammonium bromide in the reaction of aryl chlorides like 4-chloroacetophenone with n-butyl acrylate under standard conditions. Recycling of this catalyst could be achieved by using non-aqueous ionic liquids (NAILs) as media.

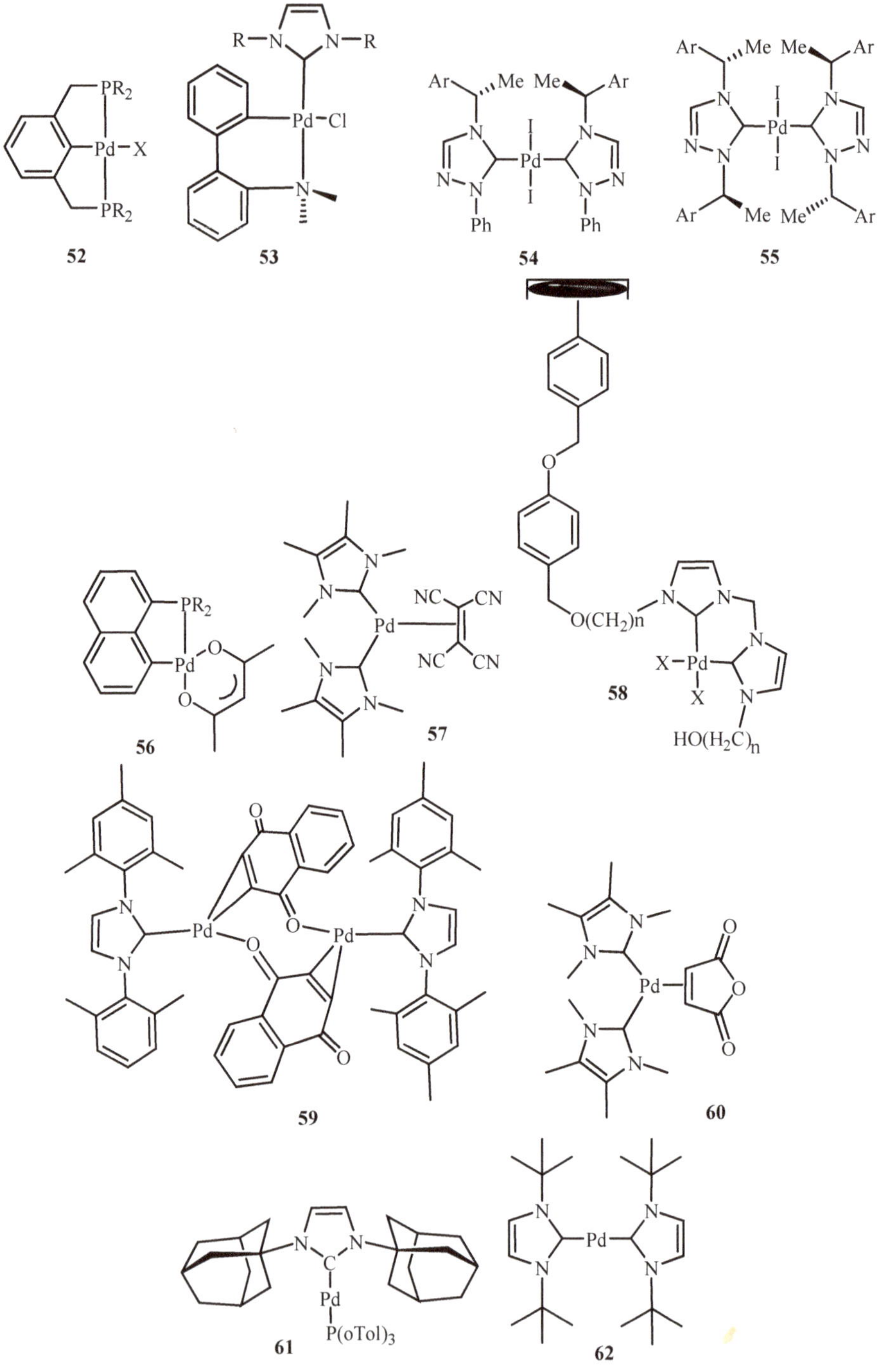

Figure 8. Palladium complexes with phosphorus and N-heterocyclic carbene ligands reviewed by Herrmann et al. [19].

A number of evidences are given for mechanistic discussions and about their role in the catalytic cycle. The catalytic activity of these complexes strongly depends on the steric bulk of the NHC ligand. Structural versatility is a great advantage of N-heterocyclic carbenes where chirality, functionalization, immobilisation and chelate effects can be achieved by easy means. Few complexes contain both NHC and phosphine ligands leading to combine the advantages of stability of bis(carbene) complexes with the good activity of phosphine complexes in C-C coupling reactions.

Beletskaya and Cheprakov [20] have given a critical survey on the application of palladacycles as catalysts for cross-coupling and similar reactions where the advantages and limitations of palladacycle catalysts are discussed. The advantages being the slow release of Pd(0) that helps to suppress unwanted processes of nucleation and growth of large inactive Pd metal particles. Bulky ligands with electron-rich phosphines or heterocyclic carbenes are essential and should be used for desired selectivity. However, the phosphine-free chemistry has limited scope. These ligands can be combined with palladacycles into hybrid catalysts, which retain the advantages of both. Such complexes are usually not more active than non-palladacyclic complexes with the same ligands. A review of various palladacycles with subclass of phosphine-free catalysts such as phosphine-derived palladacycles **36, 56, 63–68**, phosphite palladacycles **69–71**, CN-palladacycles including imine palladacycles **72–78**, oxime palladacycles **79–82** and some miscellaneous CN- **83–91**, CS- and CO-palladacycles **92–94**, pincer palladacycles **95–97**, hybrid palladacyclic catalysts **98–102** and couple of palladacycles as structurally defined catalysts **103, 104** (Figure 9) is taken. It was seen that in most of the cases palladacycles serve as a source of highly active zero-valent palladium species.

Pincer catalysts are bis-chelated palladacycles of XCX (X = P, N, S) type. They are extremely stable. It was observed that using pincer catalysts, electron rich phosphine or NHC ligands, the Heck reaction can be carried out with although cheap but otherwise unreactive, chloroarenes. However, based on a survey over the application of palladacycles in catalysis, it was stated the initial promises have not been fulfilled. The catalysts, often announced as outstanding because of very high catalytic activity in several test reactions, very rarely find applications in preparative chemistry. Neither enantioselectivity nor recyclability has been realized. Dozens of palladacycles of all imaginable classes have been studied in various cross-coupling reactions, but none appeared to be the well-defined catalyst, as was thought earlier.

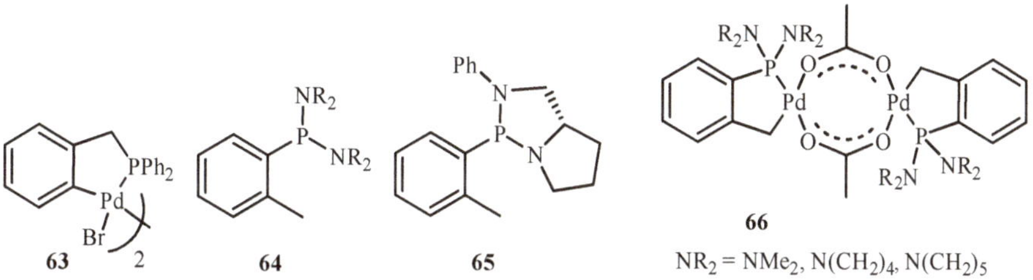

Figure 9. *Cont.*

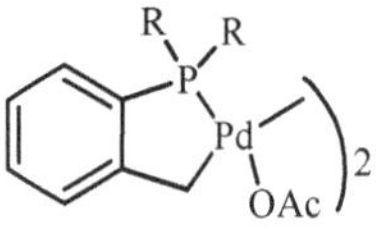
67

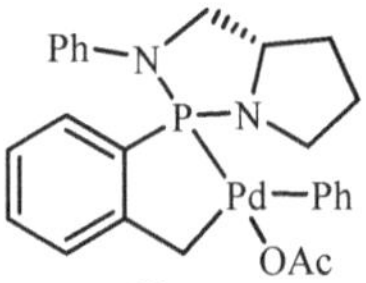
68

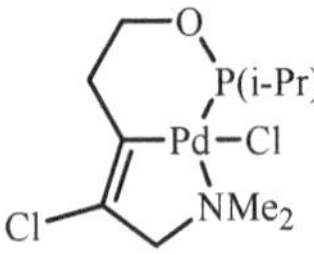
69

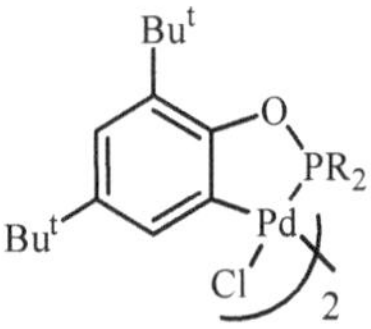
R = Ph, iPr
70

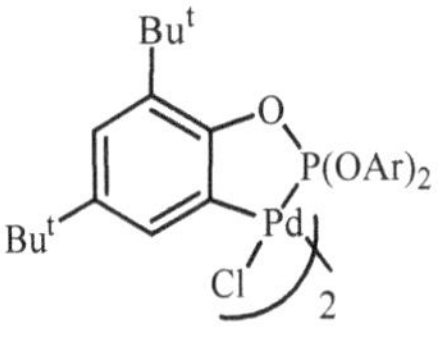
Ar = 2,4-(t-Bu)$_2$C$_6$H$_3$
71

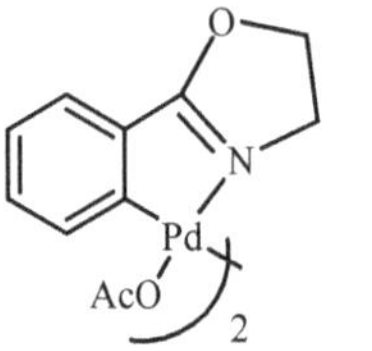
72

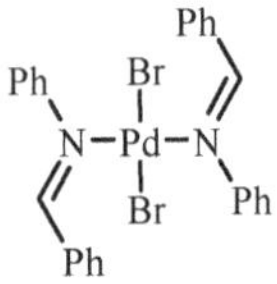
73

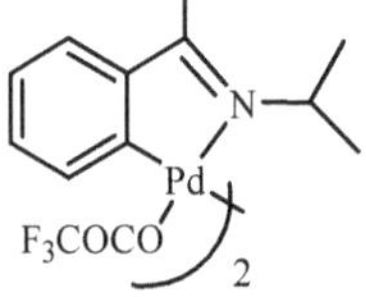
74

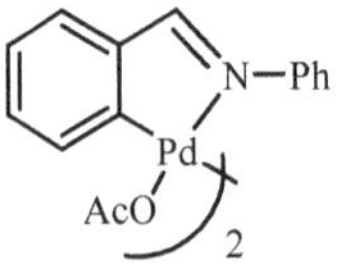
75

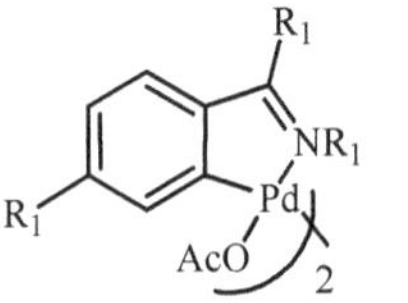
76

R$_1$ = (CH$_2$)$_3$C$_6$F$_{17}$

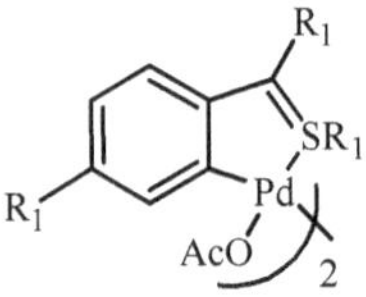
77

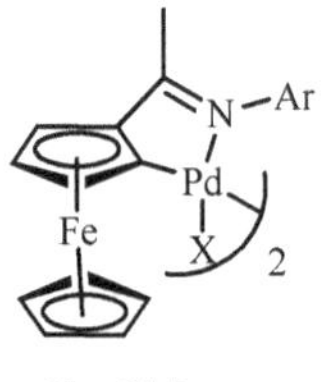
X = Cl, I
78

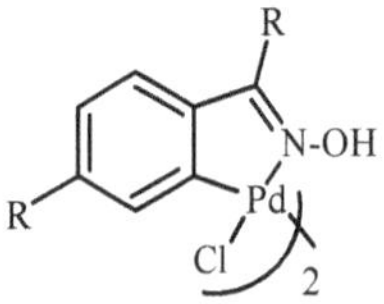
79

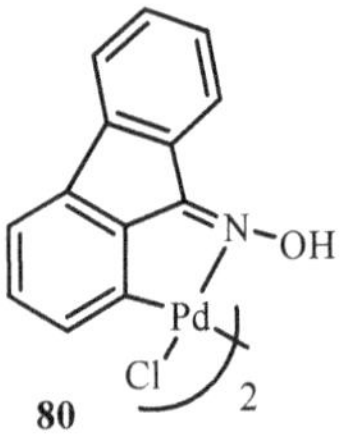
80

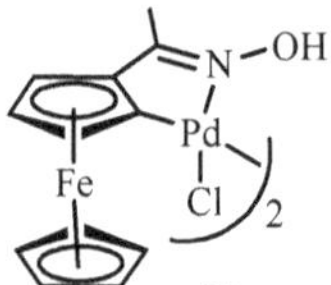
81

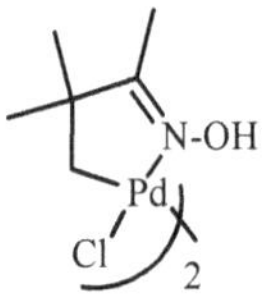
82

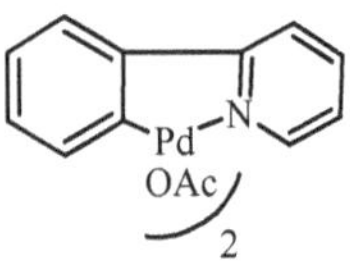
83

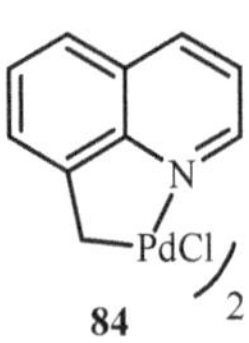
84

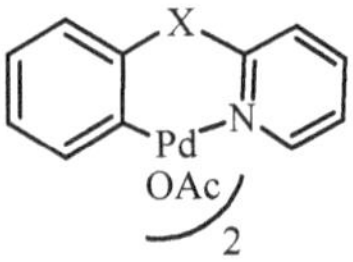
X = CH$_2$, O
85

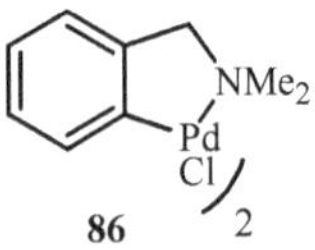
86

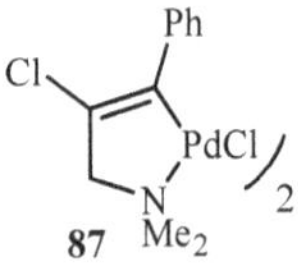
87

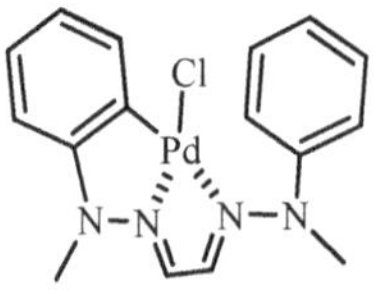
88

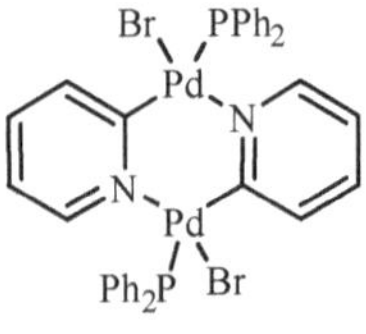
89

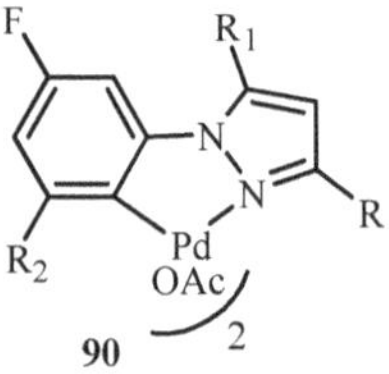
90
R, R$_1$ = Me, CF$_3$
R$_2$ = H, F

Figure 9. *Cont.*

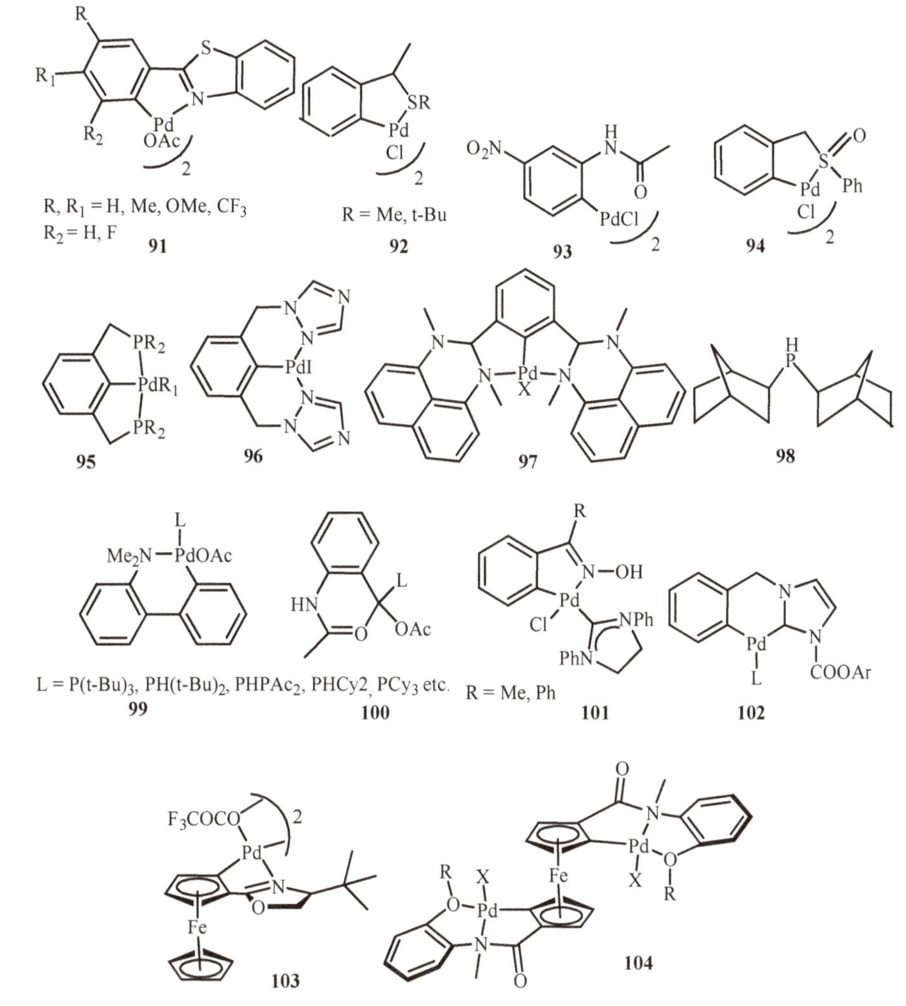

Figure 9. Ligands and palladacycles reviewed by Beletskaya and Cheprakov [20].

A review by Zafar et al. [21] is about the progress of palladium compounds as a catalyst for Heck-Mizoroki and Suzuki-Miyaura coupling reactions. Some synthesized palladium compounds and their progress in terms of ligand modification and other associated parameters up to early 2014 for Heck-Mizoroki and Suzuki-Miyaura coupling reactions are summarized in it. It was observed that the electron donating ligands such as carbenes, phosphines and nitrogen donor ligands can increase the activity, selectivity and stability of its catalysts. Palladium nanoparticles and palladacycle can also act as precatalysts for Heck and Suzuki coupling reactions. The technological hurdles in using homogeneous catalysts are minimized by putting the catalyst on a polymer support.

2.2.2. Homogeneous Catalysis

Quest for High TON

Herrmann et al. [22] have reviewed applications of metal complexes **36**, **45**, **52**, **56**, **105** and **106** (Figure 10) in Heck type reactions where detailed information about developments in palladium catalytic systems and their successful approach towards activation of less reactive substrates like aryl chlorides is mentioned. Palladacycles are found to be active against a broad spectrum of reactions and have advantages like active towards more economic aryl halides, high activity at low palladium:ligand ratio (1:1) and improved thermal stability and life-time in solution. The possible mechanisms involving Pd(0)/Pd(II) or Pd(II)/Pd(IV) catalytic cycles has also been reviewed for this class of catalyst. The mechanism of Heck type reactions catalysed by palladacycles is thought to involve active Pd(0) species however the possibility of a Pd(II)/Pd(IV) mechanism, working in competition is also not ruled out. In fact, it is stated that, Pd(II):Pd(IV) could only be a side mechanism which cannot solely be responsible for the high turnovers.

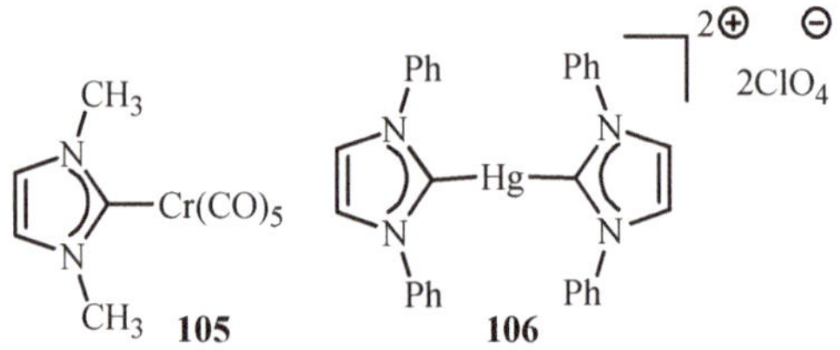

Figure 10. Metal complexes reviewed by Herrmann et al. [22].

Farina [23] has discussed the problems associated like scope, purity, impurity profile, throughput and time for developing high-turnover catalysts for the cross-coupling and Heck reactions. Numerous examples are given to illustrate the developments in the area of palladacycles and coordinatively unsaturated Pd catalysts featuring bulky phosphanes of high denticity like **107–109** (Figure 11) including **36, 52, 70, 71, 76, 78, 79, 90, 91, 95** and few more similar compounds.

These palladacycles are reviewed from a mechanistic and synthetic standpoint, and compared with more traditional catalysts obtained from conventional mono- and polydentate N and P-based ligands, as well as Pd catalysts without strong ligands, such as Pd colloids or heterogeneous catalysts and polymer supported catalysts. Carbene ligands, though less documented until then were believed to have potential for high-TON research because they are more robust than most phosphines ligands.

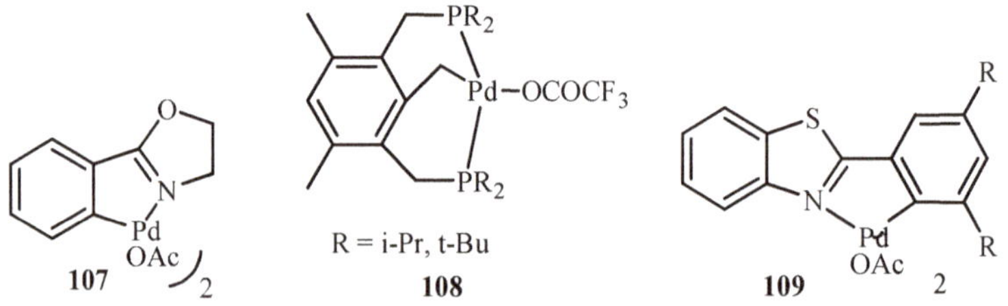

Figure 11. Palladacycles and coordinatively unsaturated Pd catalysts featuring bulky phosphanes of high denticity in review by Farina [23].

Reactions Involving Aryl Chlorides and Bromides

For Heck reactions, aryl bromides and chlorides are the most desirable substrates being cheaper and more readily available; however, not many are used as substrates as they are lower in reactivity due to their stronger C-X bonds and by far having lower TONs by several orders of magnitude. This section describes few reviews that draw attention to future challenges in this area by highlighting advances concerning Heck reactions using aryl bromides and chlorides. A review by Whitecombe and others [24] discuss the development of catalytic systems that can activate unreactive aryl halide towards Heck catalysis. The lesser yields are attributed to the poor reactivity of aryl halides due to their C-X bonds strength as well as the higher temperature used to activate aryl bromides and chlorides resulting in the catalyst decomposition where P-C bond cleavage takes place eventually leading to metal precipitation. This is demonstrated by giving a number of examples using monodentate phosphine ligands, or chelating diphosphine, phosphite, phosphonium salt. However, there are examples of successful reactions taking place under similar conditions with palladacycles, N-heterocyclic carbene complexes and pincers (all previously discussed structures) as they are stable at higher temperature.

The reactivity of ArX is found to be somewhat better if electron withdrawing groups are present in arylhalides; however, reactivity decreases if an electron rich substituent is present typically on aryl halide. The mechanistic study discussed indicates that, the prediction of single mechanism is no longer adequate and there are still unknown factors at play in Heck mechanism. The review exhaustively elaborates the mechanism involving classical Pd(0)/Pd(II) cycles and also Pd(II)/Pd(IV) cycle (this is discussed in detail in present review under mechanism section). In addition to these catalytic systems, heterogeneous catalyst and the catalysts using metals other than palladium, like Ni, Cr, Fe and Co, are described where the yield obtained using these metals are reasonable; however, they are used less than Pd. Use of conditions such as molten salts or phosphine free conditions facilitates the product formation at a higher rate.

A review by Littke and Fu [25] describes the progressive developments in the area of palladium-catalysed couplings of aryl chlorides. It was always observed that the palladium-catalysed coupling processes show poor reactivity towards aryl chlorides, although they are more attractive substrates than the corresponding bromides, iodides, and triflates in terms of cost and availability. Traditional palladium triarylphosphane catalysts are effective only for the coupling of certain activated aryl chlorides (for example, heteroaryl chlorides and substrates bearing electron-withdrawing groups), but not for aryl chlorides in general. However, catalysts using bulky, electron-rich phosphine and carbene ligands have proved to be mild and versatile for aryl chloride coupling.

In a similar review by Zapf and Beller [26], ligands and palladacycles **51**, **99** and **110** to **115** (Figure 12) are reviewed for their activity towards the C-C and C-N coupling reactions of aryl halides, especially aryl chlorides. An important advantage of this class of ligands is their significantly increased stability towards air and moisture. Due to their basicity and steric bulkiness, they constitute excellent ligands for palladium-catalysed coupling reactions. It was found that the palladacycles with high ligand–palladium ratios are suitable for aryl-X activation reactions at elevated temperatures. However, specially designed basic and sterically demanding phosphines show superior performance under milder reaction conditions. In addition, when monocarbene palladium(0) quinone complexes were tested for the Heck reaction using tetra-n-butylammonium bromide as an ionic liquid, a reasonable to good activity for both electron deficient and electron rich aryl chlorides was obtained.

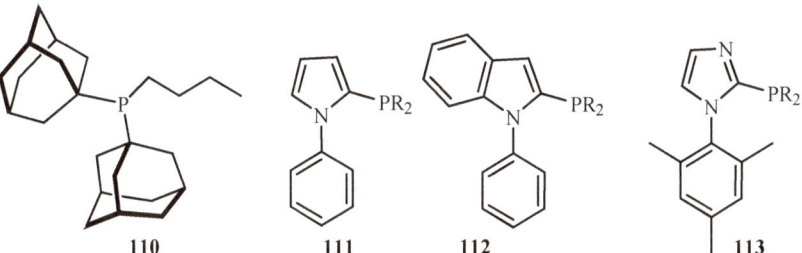

Figure 12. *Cont.*

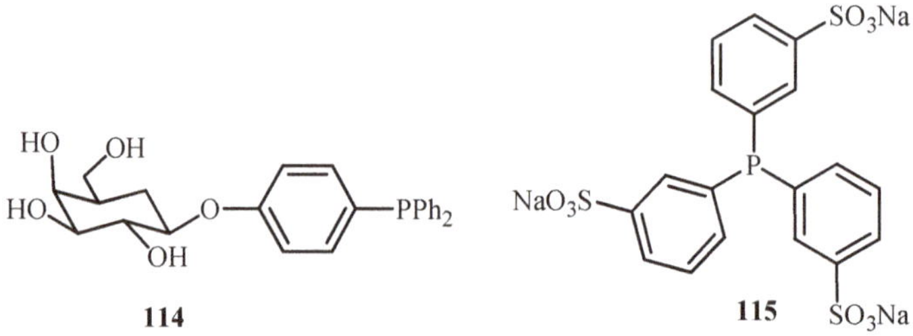

Figure 12. Ligands reviewed by Zapf and Beller [26].

2.2.3. Heterogeneous Catalysis

Carbon—carbon bond coupling reactions—Suzuki, Heck and Stille—using catalyst on solid support has been reviewed by Franzén [27]. Metal-catalysed coupling reactions are reportedly efficient and reliable methods for the introduction of new carbon-carbon bonds onto molecules attached to a solid support. A concise summary of the use of these reactions, in the field of solid phase organic synthesis resulting in small organic molecule libraries is presented. This involves the palladium-catalysed intramolecular Heck reaction for the solid-phase synthesis of indole analogues **116**, cyclic tetrapeptide derivative via macrocyclization **117**, 2-substituted benzofuran carboxylic acids **118**, phenyl acetylene oligomers **119**, α,β-unsaturated methyl ester phenyl Sulfonide **120**, fused bicyclic amino acid derivatives **121**, β-keto esters **122** and in the generation of 1,2-disubstituted olefin libraries **123** (Figure 13).

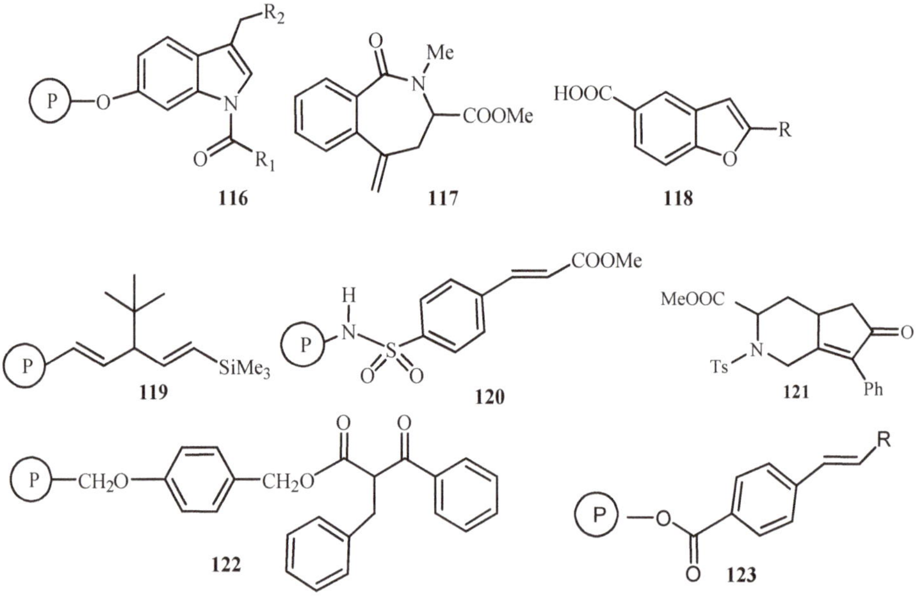

Figure 13. Molecule made via solid-phase synthesis using Heck reaction reviewed by Franzén [27].

A review by Biffis et al. [28] articulately describes the application of palladium metal catalysts to Heck reaction. The major advantages of supported palladium metal are the simplification of the work-up procedure and the possibility of facile recovery of the precious metal. Initially, the review gives a brief outline of the historical development of heterogeneous catalysis as applied to the Heck reaction followed by the discussion on both supported metal catalysts and stabilized colloidal palladium catalysts. Heterogeneous catalysts supported over different kinds of supports (carbon, inorganic oxides, molecular sieves, polymeric materials, etc.) are reviewed with particular attention to the metal leaching and the nature of catalysis. Ample of examples are given to suggest the presence of leaching, precautions to be taken to prevent leaching or use of methods like re-capture of the leached palladium species with some support. The data provide convincing evidences suggesting the catalytic cycle is sustained mainly by soluble species leached out from the starting solid material after using palladium metal catalysts, either supported or colloidal. The advantage of using heterogeneous catalysis is that it does not necessarily require ligand; the reaction temperature can be high enough for speeding up of the reaction and more of all, the recycling of catalyst is possible although not to a large extent due to extensive metal leaching, metal phase restructuring, structural damage of the support, fouling by carbonaceous deposits, etc., nevertheless systematic tailoring of support can overcome these problems.

Recently, the state of the art, benefits, and challenges of coupling microwave heating with heterogeneous Pd/C catalysis are discussed in the review by Cini et al. [29]. Microwave dielectric heating allowed a significant acceleration of the C-C coupling reaction rate, shortening the reaction time from hours to minutes. The deactivation of catalyst was not observed when Pd/C-catalysed Mizoroki-Heck reaction was carried out under microwave heating and the recycling of catalyst was also possible. Palladium supported on macroscopic pattern-vertically-aligned carbon nanotubes (Pd/VA-CNTs (carbon nanotubes)) catalyst exhibits higher activity in comparison to Pd supported on simple activated charcoal under the same reaction conditions.

2.2.4. Industrial Catalysis

A review on 'Palladium-Catalysed C-C Coupling: Then and Now' by Barnard [30] focuses on some of the early work in palladium-catalysed C-C bond formation and change in the methodologies during its developments. It talks about how catalyst has been developed from simple Pd compounds (chloride, acetate) with ligands like triphenylphosphine for variety of aryl halides with even sterically restricted partners, up to the development and applicability of stable, bulky and efficient ligands like PCy_3, $P(tBu)_3$, biphenyldialkylphosphine, palladacycle, pincer and carbene complexes that can generate highly active species.

The efforts made to develop supported Pd catalysts suitable for recycle, involving both ligandless and anchored ligand systems, the quest for achieving improved conditions allowing high turnover numbers for aryl bromides and reaction of less reactive aryl chlorides are also discussed. It also mentions about two important reviews, i.e., by Corbet and Mignani [31] on the range of patented cross-coupling technologies and by Yin and Liebscher [32] on C-C coupling by heterogeneous Pd catalysts.

2.2.5. Nano Catalysis

Narayanan [33] has highlighted some of the advances in the application of noble metal nanoparticles as catalysts for Suzuki and Heck reactions. Metal nanoparticles suspended in colloidal solutions and those adsorbed onto bulk supports are attractive catalysts for a wide variety of organic and inorganic reactions, compared to bulk catalysts as they have a high surface-to-volume ratio and very active surface atoms. Important aspects such as shape dependence on the catalytic activity, novel types of supported metal nanoparticles as nanocatalysts and the use of bi-metallic, tri-metallic and multi-metallic nanoparticles as catalysts for the Suzuki and Heck cross-coupling reactions are considered.

A review by Cai et al. [34] provides a summary of bimetallic nanomaterial-catalysed organic transformations and expresses the potential for such bimetallic nanoparticle catalysis to have significant reaction scope, especially with palladium such as magnetically separable "quasi-homogeneous" Pd-Ni nanoalloys, Au-Pd particles confined in silica nanorattles, carbon-supported bimetallic Pd-M (M = Ag,

Ni, and Cu) nanoparticles, etc. It was observed that the carbon-supported bimetallic nanoparticles Pd-Cu/C prepared by γ-irradiation at room temperature exhibits high catalytic efficiency in the Suzuki- and Heck-type coupling reactions.

Baboo [35] has reviewed the multimetallic nanomaterial based catalysis for the reactions like oxidation, hydrogenation, coupling reactions, *viz.* Heck, Suzuki and Sonogoshira, Hydrodechlorination, amidation, reductive amination and hydrogenolysis. Although it is known that nanomaterial based catalysts can easily be separated and reused with same catalytic activities, researchers have paid more attention to the use of multimetallic nano catalysts that show excellent performance than their monometallic nano catalysts. It is mentioned that, despite the great success of bimetallic nanomaterials in terms of their application to oxidation, hydrogenation, and coupling reactions, they have not yet found a wider application in the reactions for the synthesis of complex molecule. The review talks about the use of Pd-based bimetallic nanomaterials and magnetically separable "quasi-homogeneous" Pd-Ni nanoalloys for coupling reactions.

Recently, Labulo et al. [36] have reviewed the CNTs as efficacious supports for palladium-catalysed carbon–carbon cross-coupling reactions. Such catalysts have shown superior catalytic performance and better recyclability for these reactions as they impart stability to the palladium catalyst. The wide variety of surface functionalization techniques for CNTs that improve their properties as catalyst supports, as well as the methods available for loading the catalyst nanoparticles onto the CNTs with a particular focus on the effect of the solvent, base and catalyst loading has been discussed in detail in this review. It was observed that the yield is largely affected by the choice of solvent and base employed for the catalytic reaction. An improved yield could be achieved with para- and meta-substituted aryl halides and not much improvement is seen with those substituted at the ortho-position. Although this catalyst possesses excellent catalytic activities compared with commercial Pd/C catalysts, it suffers from the problem of leaching.

2.3. Array of Heck Type Reactions

This section deals with the Heck reaction performed with some modifications. Reviews are arranged in the way they fit best into one of the category however there are few with interlinks.

2.3.1. Dehydrogenative/Oxidative Heck Reaction

When Ar-H is used at the place of Ar-X in Heck reaction it is called as dehydrogenative Heck reaction also known as Oxidative Heck reaction or Fujiwara-Moritani [37] reaction or simply Fujiwara reaction sometimes. The fact is this version of the Mizoroki–Heck reaction was the first catalytic Heck reaction to be discovered in 1968, where the palladium(II)-catalysed arylation of olefins from phenylmercuric chloride, using catalytic amounts of copper(II) chloride assisted by oxygen for the regeneration of palladium(II) was carried out. This reaction is closely related to the classical Mizoroki–Heck reaction, where instead of initiation by oxidative addition process, it follows transmetallation step or a C-H activation step (Scheme 2).

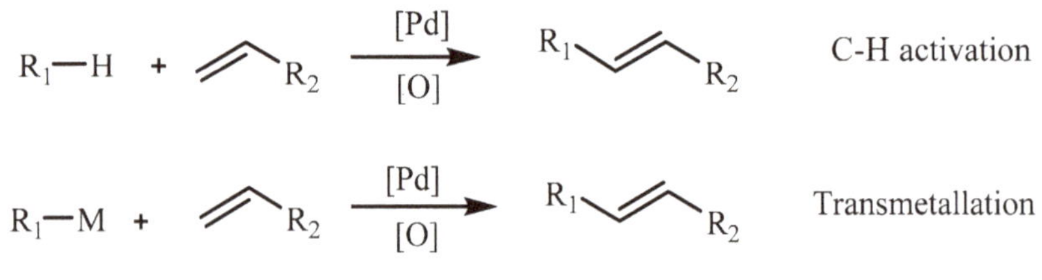

Scheme 2. Reaction scheme for C-H activation and transmetallation.

This is one of the important reactions with respect to atom economy principle since it directly uses the Ar-H compounds thus eliminating additional synthetic step in many total syntheses for halogenations, i.e., in other words, C-H activation of arenes eliminates the need for halogen. The best example of this reaction is the synthesis of benzofuran and dihydrobenzofuran. These structures are important components of numerous biologically active compounds and are preferred to be prepared by dehydrogenative Heck reaction.

Gligorich and Sigman [38] presented a review on advancements and challenges of palladium(II) catalysed oxidation reactions with molecular oxygen as the sole oxidant, where reviewers have discussed some oxidative Heck reactions. The review paper highlights some of the developments until then in direct molecular oxygen-coupled Pd(II)-catalysed oxidation reactions. Although there are reports with positive results for the development of more efficient ligand-modulated oxidative Heck reactions that can be performed under an air atmosphere at room temperature, the asymmetric oxidative Heck reaction still suffers from many limitations and requires further studies.

Karimi et al. [39] have reviewed similar types of reactions describing the oxidative Heck reactions of organometallic compounds such as organomercuric acetates, organoboronic acids, organofluorosilicates and arylstannanes. A number of successful examples are given for the intermolecular, intramolecular, nonsymmetrical, asymmetrical oxidative Heck reaction in presence or in absence of air, with ligand-free or ligand based catalysts formed from palladium, polymer supported palladium(II), transition metal and organometallic catalysts other than palladium along with the examples of asymmetric reactions, intermolecular and intramolecular oxidative Heck reactions via C-H activation. However, the substrate scope of these transformations is still extremely limited and there is much room for the development of newer methods that working well under more convenient reaction conditions.

A review by Su and Jiao [40] reveals the development in the area of palladium catalysed oxidative Heck reactions of alkenes with organometallic compounds, which are effective arylating or akenylating agents. It was observed that the organometallic compounds specially derived from Group III to Group VII like organoboronic. organothallium, organosilicon, organotin, organotin, organophosphorus, organoantimony, organobismuth, organotellurium, hypervalent iodonium salts or simply arenes are efficient substrates for the oxidative Heck reaction. They are found to be remarkably stable and easy to prepare. In addition, various metals other than Pd, including Ru, Rh, Ir and Ni, though less explored for oxidative Heck reaction, are also discussed. The yields of product obtained using them are reasonable up to 72%. The mechanism based on many experimental studies has been reported where the evidence of detection of single charged cationic palladium (II) complexes are given.

A review by Le Bras and Muzart [41] highlights the same subject particularly with respect to its progress of procedures. Numerous results of stoichiometric as well as catalytic palladium mediated arylations of various arenes and heteroarenes are presented. Most of these reactions use an excess of either the arene or the alkene, often with a relatively high Pd catalyst loading and require a terminal oxidant other than molecular oxygen. This becomes an expensive issue and can be a challenge for applications on large scales and even their compatibility with the atom economy principle. Thus, there is wide scope for researchers to work in this area.

Yet another review by Le Bras and Muzart [42] covers the palladium-catalysed annelations of internal alkynes through reactions leading to the loss of only two hydrogens from the substrates. This process is explained with the mechanism that involves (i) dual C-H bond activation; (ii) both C-H and N-H bond activation; (iii) successive amino (or oxy) palladation and C-H bond activation; or (iv) C-H bond activation followed by a Heck-type process. Though not much work has been done in this area, such sustainable processes will become valuable tools for the synthesis of diverse carbocycles and heterocycles.

A review by Lee [43] highlights the use of the oxidative Heck reaction (also referred to as oxidative boron Heck when boron is used in transmetallation reaction (Scheme 2)) in enantioselective Heck-type couplings. This technique overcomes several limitations of the traditional Pd(0)-catalysed Heck coupling and has subsequently allowed for intermolecular couplings of challenging systems such as cyclic enones, acyclic alkenes, and even site selectively on remote alkenes. This has also enabled enantioselective intermolecular couplings of more challenging systems such as desymmetrisation of quaternary centres, cyclic enones, acyclic alkenes and even site selectively on remote alkenes via a redox–relay coupling. A number of examples have been cited along with probable mechanisms for its justification.

2.3.2. Reductive Heck Reaction

The term reductive Heck or reductive arylation is used when the intermediate forms the conjugate addition product in palladium-catalysed Mizoroki-Heck reactions instead of giving substitution product via β-hydride elimination (Scheme 3). Reductive Heck product is a regularly observed side product and the extent of its formation in the reaction varies greatly with base, temperature, substrate and solvent.

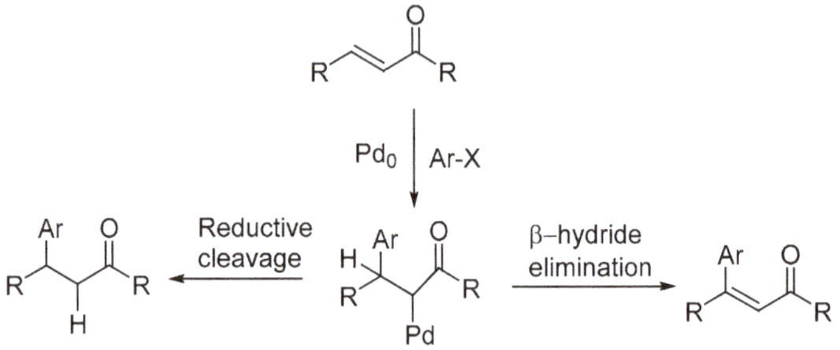

Scheme 3. Collapsing of intermediate to form conjugate addition product or substitution product via β-hydride elimination.

Xiaomei et al. [44] have reviewed the progress in reductive Heck reaction, where unsaturated halides react with olefinic compounds at the similar Heck conditions to form the addition products. Discussion on catalytic systems, mechanism, applications and limitation in organic chemistry is given.

2.3.3. Intramolecular Heck Reaction

Intramolecular Heck reaction is generally more efficient than the intermolecular Heck reactions for many reasons like, in the intermolecular Heck reactions, only mono- and disubstituted olefins can participate, while in the intramolecular case tri- and tetrasubstiuted olefins can readily get inserted; regiocontrol of olefin addition is difficult in the intermolecular for electronically neutral olefins whereas the regioselectivity in intramolecular process is governed by ring-size of the newly formed cycle and is generally directed by steric considerations giving highly regioselective couplings. In addition, construction of cyclic compounds containing an endo- or exo-cyclic double bond can efficiently be brought about by using the intramolecular Heck reaction.

A review by Guiry and Kiely [45] summarizes the development of the intramolecular Heck chemistry and the methodology used for the construction of carbocycles and heterocycles along with the mechanism. The review is mainly focused on the optimization of palladium catalysts derived from various diphosphine and phoshinamine ligands for the preparation of a variety of cyclic products like cis-decalins, hydrindans, indolizidines, diquinanes and the synthesis of quaternary carbon centres. A number of examples of the application of intramolecular Heck cyclization as the key step in the preparation of many complex natural product syntheses are given. Some of their own work on ligand synthesis used for such reactions is also discussed.

A review by Oestreich [46] summarizes the exciting development of the enantioselective intermolecular Heck reaction (particularly the Heck-Matsuda [47] reaction that uses arene diazonium salts as an alternative to aryl halides and triflates) of acyclic alkenes and the synthetic utility of enantioselective intermolecular couplings of cyclic alkenes. Many such examples that give very high enantioselectivity for intermolecular Heck reactions are cited.

2.3.4. Asymmetric Heck Reaction

A breakthrough in Heck reaction took place when in 1989 Shibasaki [48] and Overman [49] independently reported the first examples of asymmetric Heck reactions. They used chiral ligands BINAP and (R,R)-DIOP for such reactions, respectively. A classic example of the intramolecular Heck reaction in action is Overman's synthesis of (−)-scopadulcic acid A **124** (Figure 14), where a tandem double Heck cyclisation (6-exo-trig followed by a 5-exo-trig) rapidly accesses the carbon skeleton found in the natural product.

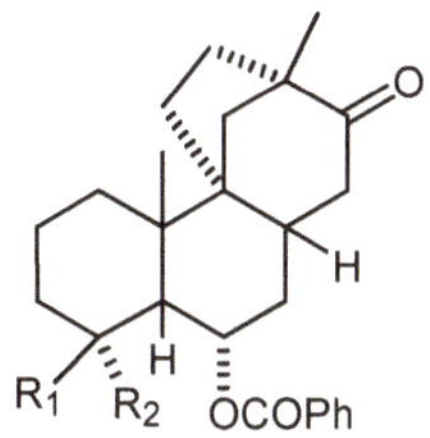

Scopadulcic acid A : R_1 = COOH, R_2 = CH_2OH
Scopadulcic acid B : R_1 = Me, R_2 = COOH

Figure 14. Scopadulcic acid.

From 2003, the catalytic asymmetric variant of Heck reaction emerged as a reliable method for enantioselective carbon-carbon bond formation. Dounay and Overman [50] reviewed the application of catalytic asymmetric Heck cyclization in natural product total synthesis for the formation of tertiary and quaternary stereocenters by considering synthesis of some terpenoids (like ernolepin, Oppositol and Prepinnaterpene, Desmethyl-2-methoxycalamenene, Capnellenols, $\Delta^{9(12)}$-Capnellene, Kaurene and Abietic Acid, Retinoids); Alkaloids (like Lentiginosine and Gephyrotoxin 209D, 5-Epiindolizidine 167B and 5E,9Z-indolizidine, 223AB. Physostigmine and Physovenine, Quadrigemine C and Psycholeine, Idiospermuline, Spirotryprostatin, Eptazocine); Polyketides (like Halenaquinone and Halenaquinol, Xestoquinone, Wortmannin). A brief discussion on understanding of the mechanisms of catalytic reactions is provided necessary for the rational development of efficient asymmetric processes.

A review by Diéguez [51] covers reports on the ligands derived from carbohydrates for asymmetric catalysis until 2003. High enantioselectivities have been observed by using bidentate ligands, usually diphosphines and phosphine-oxazoline ligands. They also have excellent control on selectivity, based on the properties of the ligand. Carbohydrate ligands have proved to be some of the most versatile ligands for enantioselective catalysis. Examples of numerous carbohydrate based ligands are discussed in the review, however, for Pd-catalysed Heck reaction, only the following kinds of carbohydrate derivative ligands **125–127** (Figure 15) have been seen to be efficiently applied.

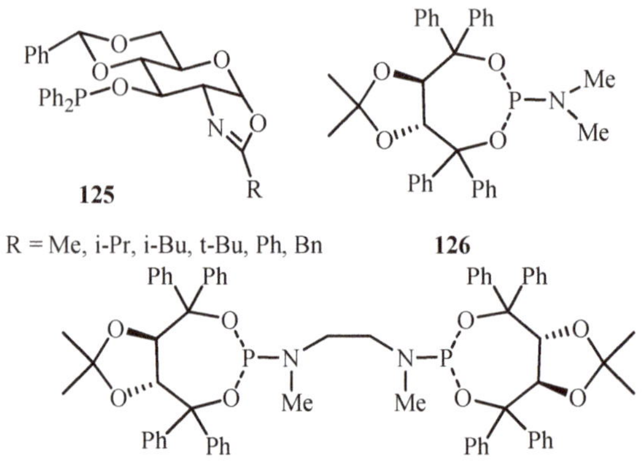

125

R = Me, i-Pr, i-Bu, t-Bu, Ph, Bn

126

127

Figure 15. Carbohydrate based ligands reviewed by Diéguez [51].

McCartney and Guiry [52] comprehensively reviewed the asymmetric inter- and intramolecular Heck (Scheme 4) and related reactions since their original development until 2011 with respect to substrate scope, reactivity, regio- and enantioselectivity. The formation of products is supported by predicting mechanism. The classification is based on the nature of ligands in terms of their denticity, chirality and nature of donor atoms involved so as to understand the continued development of ligand architectural design and their application. The review addresses the significant improvements in reaction times, a disadvantage of Heck reactions performed in classical way, use of microwave-assisted protocols and ligand design. The asymmetric Fujiwara-Moritani and oxidative boron Heck-type reactions and the recent additions to Heck type processes, are also discussed.

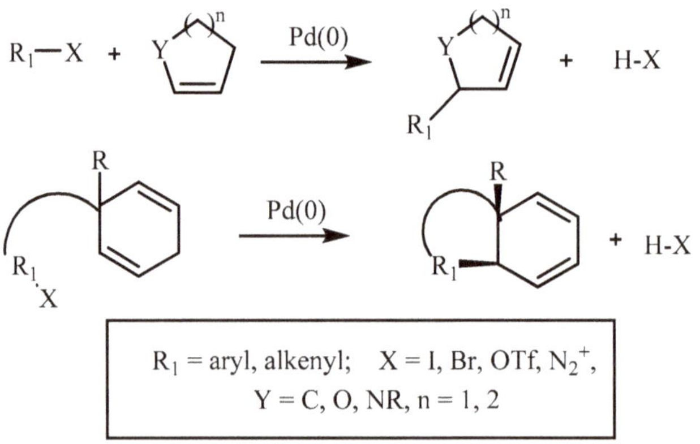

$R_1 = $ aryl, alkenyl; $X = $ I, Br, OTf, N_2^+, $Y = $ C, O, NR, n = 1, 2

Scheme 4. Asymmetric inter- and intramolecular Heck reaction.

2.3.5. Heck Reactions Involving Heteroatom

A review by Daves and Hallberg [53] highlights typical heteroatom-substituted olefins to organopalladium reagents, where a new carbon-carbon bond is formed via 1,2-addition of an organopalladium species to the often strongly polarized carbon-carbon double bond of a heteroatom substituted olefin. These types of reactions afford versatile and efficient synthetic routes to a wide variety of compounds. Heck and other similar types of reactions involving intermediate palladium complex undergoing 1,2-additions have been discussed. Compounds with heteroatoms O, N, S, P are considered to explore their reactivity towards such addition reaction. A number of examples with cyclic enol ethers, glycals, and cyclic enol ethers derived from carbohydrates, acyclic enol ethers like thioenol ethers, enol carboxylates, enamides and enamines, vinylsilanes and vinylphosphonates are given. Product formation is explained based on well accepted mechanism for Heck reaction involving Pd(0) complex. Increased understanding of the pathways by which these reactions occur, that include delineation of the factors determining reaction regio- and stereochemistry, and control of competing modes of σ-organopalladium adduct decomposition to products, make possible the utilization of these reactions in the synthesis of complex structures.

Bedford [54] has reviewed the use of palladacyclic catalysts in C-C and C-heteroatom bond-forming reactions. Palladacycles are air and moisture stable inexpensive catalysts, can easily be handled and stored, and act as clean sources of low-coordinate palladium(0). They as pre-catalysts play a significant role in a range of C-C and C-heteroatom bond forming reactions. Activity of the catalysts comprised of P-C palladacyclic, Palladium P,C,P-pincer complexes, L-C palladacyclic, L,L,C- and L,C,L-pincer based (where L = N, S) have been compared for the C-C and C-heteroatom bond-forming reactions including discussion on the likely nature of the true active catalysts produced in situ. These catalysts show the TON ranging from a few thousand up to a few million. Consideration to the Aryl chlorides as substrates and the mechanism involving palladacycle for such reaction is thoroughly discussed. It was found that the complexes **36**, **63** and **70** show reasonable activity with some, usually less electronically challenging substrates. The possibility of recyclability of a few catalysts for Heck and other reactions has been discussed. It was predicted previously that when the catalysts are immobilized on solid supports they could work as recyclable catalyst. Polystyrene-immobilized catalyst **74** shows comparable activity to homogeneous analogues in the Heck coupling of iodobenzene with styrene. However, in recycle study the filtrate after reaction and not the recovered catalyst that showed the activity comparable to that obtained in the first run. This activity was concluded to be due to formation of nanoparticulate palladium which is stabilized by an ammonium salt.

2.4. Methodology

Various attempts have been made by researchers to improve the catalyst activity, lowering the cost of catalyst; hence different methodologies are adopted using non-conventional techniques for Heck reactions. Reviews based on these methodologies are described here.

2.4.1. For Improvement in Activity

Use of tetraalkylammonium salts to improve the yields in Heck reactions particularly is considered to be a remarkable contribution in catalysis. The combination of such salts (phase-transfer catalysts) and insoluble bases accelerates the rate of reaction to great extent even at lower reaction temperatures and is commonly known as Jeffery conditions. A review of these salts in Heck type reactions is taken by Jeffery [55]. It was seen that, under appropriate conditions, tetraalkylammonium hydrogensulfate can just be as efficient as tetraalkylammonium chloride or bromide for facilitating Heck-type reactions. Thus, an appropriate selection of the catalyst system (Pd/Base/QX) can allow this type of reactions to be efficiently realized at will, in a strictly anhydrous medium or in a water-organic solvent mixture or in water alone.

2.4.2. For Lowering the Cost

Tucker and de Vries [56] describe their own efforts in the area of palladium- and nickel-catalysed aromatic substitution carbon-carbon bond formation reactions in the review titled 'homogeneous catalysis for the production of fine chemicals'. The main focus of this review is on low cost and low waste production methods. A number of examples are discussed to prove methodology for lowering production cost such as reducing the amount of catalyst, eliminating use of ligands or use of cheaper phosphite or phosphoramidite ligands, carrying out reactions at lower temperatures, replacing palladium by nickel, replacing aryl bromides or iodides with the cheaper chlorides and simplified work-up. For waste free production methods, use of aromatic anhydrides as aryl donor is suggested. Methods for recycling of palladium in ligand-free Heck and Suzuki reactions is described that involves treatment of palladium black, precipitating at the end of the reaction, with a small excess of I_2 prior to its re-use in the next run.

Typically, the Heck reactions are catalysed by palladium, a precious metal. However, in many cases low cost transition metals are found to play a similar role which is reviewed by Wang and Yang [57]. This review focuses on low-cost transition metal catalysed Heck-type reactions. The cited examples indicate that some low-cost transition metals like Ni, Co, Cu, Fe are active for Heck-type reactions. Ni is found to give best performance among them; in fact, sometimes the results with Ni are comparable to that with Pd. Co and Cu exhibit outstanding activity to alkyl electrophile involved Heck- type reactions. This is attributed to their abilities to generate alkyl radical. It was also observed in a few cases that under certain conditions these low-cost transition metals may show unique catalytic properties, which are absent in Pd-mediated systems. Two mechanisms have been predicted for Heck-type reactions using these transition metals, cationic mechanism and radical mechanism based on the reacting species. Phenyl or benzyl halides predominantly take the cationic mechanism, whereas alkyl halides usually follow the radical mechanism, especially in Co- and Cu- catalysed systems because Co and Cu have good capacity of producing radical from alkyl halides. However, the investigations on low-cost transition metals catalysed Heck-type reactions highlight the extension of the substrate scope and draw a little attention to the development of catalysts, ligands and solvents. Hence, more study is needed in this area.

2.4.3. Non-Conventional Methodologies

Beletskaya et al. [58] have published a critical overview on unconventional methodologies for transition-metal catalysed Heck reaction highlighting the efforts and interest in developing more efficient processes according to the new requirements of chemistry. A vast array of non-conventional methodologies is described considering different parameters involved in the reaction like substrates, catalytic system, solvent, reaction conditions, or work-up. However, it is also stated that, although large numbers of interesting methodologies are available from an academic point of view, they are useless from a practical point of view. Hence, there is a need for reconsideration of a few points while performing Heck reactions at the industrial level such as:

- Most of the methodologies deal with the simpler reactions between the reactive substrates like aryl iodides or activated aryl bromides and acrylates in contrast to the more desirable, but less reactive aryl chlorides or other olefinic substrates such as electron-rich olefins.
- More attention should be paid at the workup and separation of by-products formed during achieving a high regio- and stereoselective products.
- Whenever possible, commercially available starting materials, reagents, ligands, catalysts, or solvents must be used and proper time should be given to experimental work needed to prepare the different reaction components.
- The catalyst must be recyclable and/or display high TONs.
- Heterogeneous catalysis or a use of ligandless catalysts, recoverable ligands or stabilized nanoparticles are preferred for better recovery and lower cost provided metal leaching is prevented.
- An ideal recyclable catalyst is the one in which filtration of the reaction mixture produces a catalytically active solid and an inactive filtrate.
- Use of high temperature with some stable catalysts may prove to be unfavourable for the selectivity of the product.
- The expensive equipment or reaction medium utilized in some methodologies cannot compensate for the little or no improvement observed in many cases with respect to the conventional methodologies; in addition, the application of these methodologies is normally restricted to a small scale.

- Reproducibility, atom-economy, low-cost, scalable and practical procedures, are needed to extend the methodologies from the academic laboratory to the industrial plant.

Further research must be undertaken in order to clarify the reaction mechanisms involved in the different processes, which remain unclear in most cases; it is crucial to have a better knowledge of the nature and properties of the real catalytic species in order to improve any given reaction.

2.5. Selectivity

Under the section Asymmetric Heck reaction (Sections 2–4) reviews dealing with the generation of asymmetric centre by means of Heck reactions were summarized whereas under this section the progress of ligands and catalysts for the development of various regioselective and enantioselective transformations via Heck reactions are discussed.

A review by Tietze et al. [59] on 'enantioselective palladium catalysed transformation' discusses many other important organic reactions including Heck reaction where enantioselectivity in intramolecular and intermolecular Heck reaction is discussed with lot of examples collected right from its first example of such kind by Shibasaki [48] and Overman [49]. A number of examples with new improved ligands for development of various enantioselective transformations has been discussed aiming at higher (preferably over 95%) ee values. The review asserts that there is no field where enantioselective Pd catalysis cannot be employed. However, the disadvantage of such catalysis is the high price of Pd and the usually small turnover numbers, making the processes too expensive for industrial use. Nevertheless, novel chiral Pd catalysts resulting in high turnover number can be synthesized to overcome this issue in addition to a broad range of enantioselective transformations suitable for the chemical industry.

Almost at the same time, Shibasaki et al. [60] reviewed a similar topic particularly aiming at the asymmetric Heck reaction. Since a variety of carbocyclic, heterocyclic and spirocyclic systems can be constructed, the asymmetric Heck reaction becomes a powerful method for the synthesis of both tertiary and quaternary chiral carbon centres, with an enantiomeric excess often in the range of 80% to 99%. The scope of the reaction with respect to the product alkene isomerization was limited due to regioselectivity, and was predicted to be solved by development of new generation of ligands dissociating more rapidly from the products, thus improving both enantio- and regiocontrol.

Oestreich [61] describes the evolution of inter- as well as intramolecular Heck reactions from regio- to diastereo- and finally to enantioselective transformations with a special reference to heteroatom-directed Heck reactions in his review. The concept of "Chelation Control" **128**, **129** (Figure 16) controls regio- and stereoselectivity with the aid of attractive interactions between substrate and reagent/catalyst is discussed.

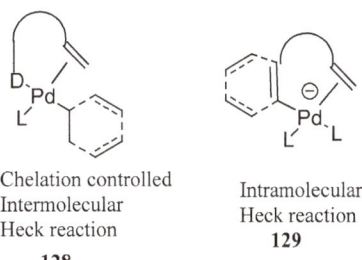

Chelation controlled
Intermolecular
Heck reaction
128

Intramolecular
Heck reaction
129

Figure 16. Chelation controlled inter- and intramolecular Heck reaction.

Several examples of removable catalyst-directing groups developed for the preferential regioselective intermolecular arylation of alkenes are given such as amino-directed intermolecular Heck reaction of vinyl ethers. Mono- versus bidentate phosphines influence a regiochemical switch based on the bite angle and the mechanistic rationale for inverted regioselectivity is also discussed. Review covers the syntheses of stereodefined, multi-arylated alkenes, the diastereoselective construction of tertiary and quaternary carbon centres, and also the combination of substrate with catalyst-control in an enantioselective transformation. Many examples have been discussed to show the neighbouring-group effects playing important role in Heck chemistry and expect few more discoveries in this field. The example of a substrate-controlled enantioselective reaction illustrates that the enantioselection is sometimes discriminated not only by the chiral reagent but also by a suitably located donor.

2.6. Aqueous Media

Nowadays, use of water has become increasingly popular for fine synthetic chemistry in industry for the reasons: water is non-toxic, nonflammable, inexpensive, and environmentally friendly solvent and there by use of organic solvent can be avoided. In addition, one of the major drawbacks of homogeneous metal catalysis lies in the separation of the reaction product from the catalyst and requires costly procedures. The concept of transition metal catalysis in water is used where the catalyst is easily recovered by separation of the aqueous and organic phase if a biphasic system is used. Its popularity has been increased since the development of the Ruhrchemie–Rhone Poulenc process using a modified water-soluble rhodium complex in the hydroformylation methodology. However, the limitations of aqueous phase catalysis are:

- Stability of substrate or product in water.
- Partial solubility of substrate in the aqueous phase to avoid mass transfer limitation.
- Necessity of preparation of water soluble ligands or dispersing agents to maintain catalyst in aqueous phase.
- Challenges for future developments in this area to develop catalysts with scope and activity comparable to the best organic-phase catalyst systems.

Nevertheless, there is still strong interest in developing efficient and recoverable catalysts for use in pharmaceutical and other fine chemical synthetic processes, as can be seen from following reviews.

Genet and Savignac [62] have reviewed the palladium cross-coupling reactions carried out in aqueous medium. Reactions like Heck, Sonogashira, Tsuji-Trost, Suzuki, Stille as well as protecting group chemistry in aqueous media are discussed in the review. Although palladium is known to be unstable in aqueous medium, there are reports of the excellent compatibility of water-soluble palladium catalysts with water- soluble phosphines such as TPPTS (3,3′,3″-Phosphanetriyltris(benzenesulfonic acid) trisodium salt) 115, TPPMS (Sodium Diphenylphosphinobenzene-3-sulfonate) 130 and salts of acid or amines 131, 132 (Figure 17), offering new opportunities for such reactions at mild conditions and with new selectivity.

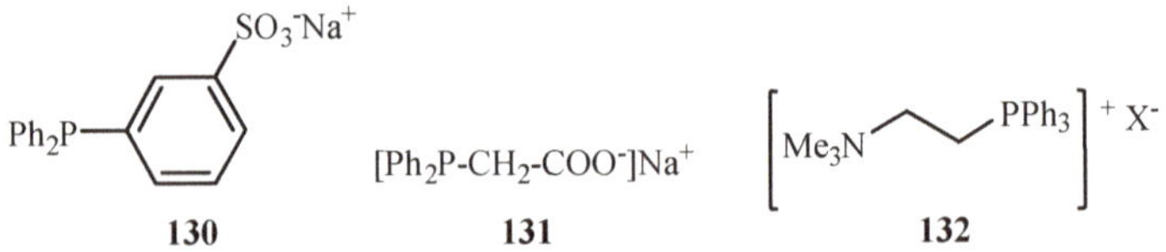

Figure 17. Water-soluble phosphines reviewed by Genet and Savignac [62].

It was seen that the careful selection of reaction conditions, co-solvents and catalysts, is very important for long life catalyst and new selectivities. The advantages of the two-phase aqueous system, i.e., easy separation of the products and recycling the expensive palladium can be obtained by using palladium catalysed reactions with water-soluble phosphines that has increased the potential of modern palladium catalysis.

Li [63] has reviewed many organic reactions in aqueous media where water serves as a medium for various palladium-catalysed reactions of aryl halides with acrylic acid or acrylonitrile to give the corresponding coupling products in high yields. In addition, reactions at superheated and microwave heating conditions with bulky phosphine ligands along with arenediazonium salts instead of aryl halides in the Heck-type reaction are mentioned. The reason for a high yield of coupling product for reaction involving the use of Pd(OAc)$_2$ with water-soluble ligand, TPPTS, and formation of complex mixture was attributed to high dielectric constant of water. Examples of many transition metals other than palladium in water have been cited.

Velazquez and Verpoort [64] have reviewed the reports on the use of N-heterocyclic carbene transition metal complexes for catalysis in water. The typical phosphine and amine-type ligands can thus be displaced by this type of catalysis because of their higher stability and reactivity. Palladium complex **133** and a complex of ligand **134** (Figure 18) are used for catalysing the Heck reaction successfully in water.

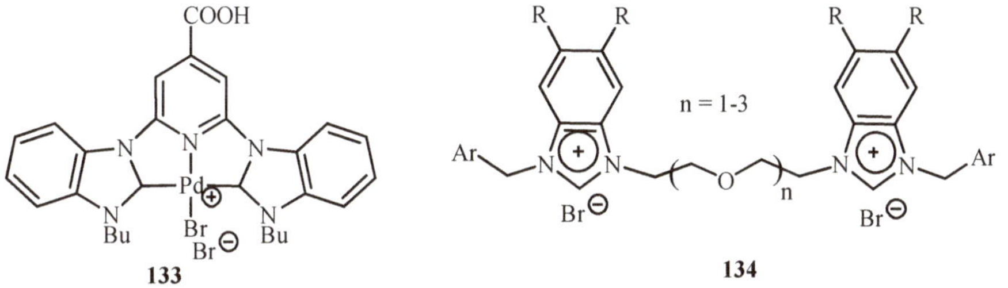

Figure 18. N-heterocyclic carbene ligand and transition metal complex for catalysis in water reviewed by Velazquez and Verpoort [64].

A review by Hervé and Len [65] is based on both Heck and Sonogashira cross-coupling of nucleosides following two important aspects of the green chemistry, i.e., use of aqueous medium and no protection/deprotection steps. It focuses on the study of C5-modified pyrimidines and C7-deaza or C8-modified purines where these chemical modifications have been developed using palladium cross-coupling reactions. The review encompasses variations of the starting materials, alkene and alkyne, nature of the solvent, palladium source and ligand at either room temperature or higher temperature. Heck cross-couplings were performed using Na$_2$PdCl$_4$ (80 mol %) and Pd(OAc)$_2$ (5–10 mol %) in the presence of TPPTS as ligand in a mixture of CH$_3$CN/H$_2$O and in sole water. Using these procedures, yields up to 98% have been reported.

3. Reviews on Mechanism

The mechanism of Heck reaction is discussed and reviewed by many researchers since 1972 when for the first time it was given in complete detail by Heck and Nolley [9]. There are articles that argue the case for having a Pd(0)/Pd(II) mechanism while others favour the Pd(II)/Pd(IV) mechanism based on evidential data. Few discuss the specific intermediate formed during a particular mechanism while others demonstrate the presence of different intermediate. This section deals with reviews on such reviews on mechanism of Heck reaction.

A review by Cabri and Candiani [66] explains the basis of a common mechanistic hypothesis for the coordination-insertion process of unsaturated systems on palladium(II) complexes based on the results obtained until then by several research groups on the Heck reaction. All of these results are explained by the commonly accepted mechanism based on Pd(0)/Pd(II) cycle (Scheme 5).

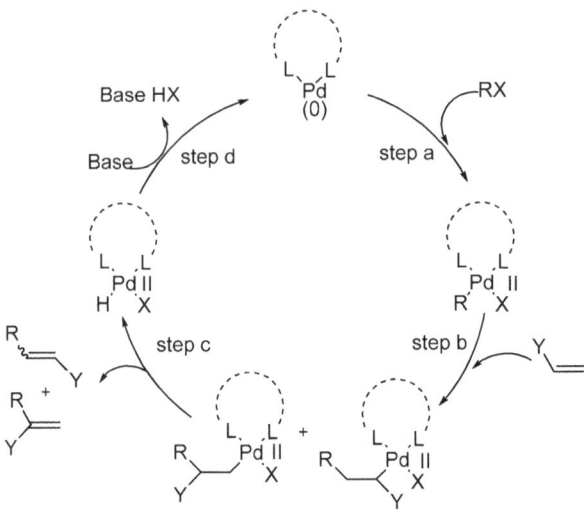

Scheme 5. Mechanism based on Pd(0)/Pd(II) cycle by Cabri and Candiani [66].

This mechanistic model gives a better understanding of the scope and limitations of the Heck reaction. The steps involved in the mechanism of Heck reaction are:

step (a) Oxidative addition
step (b) Coordination-insertion
step (c) β-Hydride elimination-dissociation
step (d) Recycling of the $L_2Pd(0)$

Several factors like leaving groups, neutral ligands, additives (like bases, salts, etc.), and olefin substituents play important roles in overall reaction course via coordination-insertion process as shown in Scheme 6.

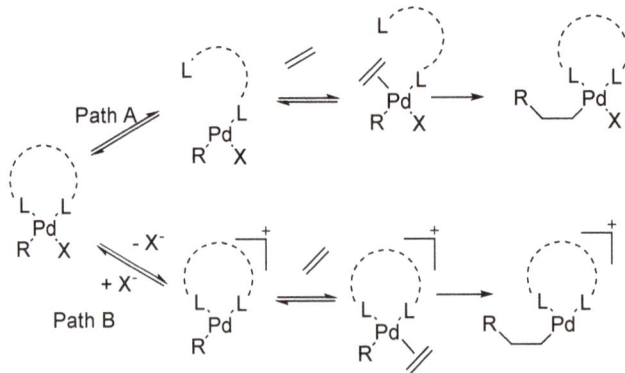

Scheme 6. Olefin substituents via coordination-insertion process.

It is essential that the insertion process requires a coplanar assembly of the metal, ethylene, and the hydride and hence the insertion process is stereoselective and occurs in a *syn* manner. In addition, the energy barrier for the generation of the reactive configuration in a tetracoordinated complex is low with respect to a pentacoordinated one. Therefore, pentacoordinated species are possibly not involved in the coordination process. The β-hydride elimination is also stereoselective and occurs in a *syn* manner. Its efficiency is related to the dissociation of the olefin from the palladium(II)-hydride complex. It was also observed that, the presence of a base is necessary in order to transform the $L_2Pd(H)X$ into the starting $L_2Pd(0)$ complex and complete the catalytic cycle.

The mechanism is also supported by the discussion on the regioselectivity and stereoselectivity of Heck reaction. It was observed that the preferential formation of the branched products in the arylation of heterosubstituted olefins like enol ethers, enol amides, vinyl acetate, allyl alcohols, and homoallyl alcohols was independent of the substituents on the aromatic ring, the reaction temperature, and the solvent and is related only to the coordination-insertion pathway, though the stereoselectivity was dependent on the added base. In the intramolecular asymmetric Heck reaction, sometimes the enantioselectivity was related to the geometry of the substrate olefin.

A review by Crisp [67] examines the implications of details on the mechanism of the Heck reaction using both traditional and non-traditional catalytic systems. It is mentioned that, while studying the intermediate formed during oxidative addition of aryl halide to Pd complex, the conventionally thought intermediate $ArPdL_2X$ is not produced rather an intermediate $ArPd(PPh_3)_2(OAc)$ is formed. The formation of this species is also supported by Amator and co-workers [68]. It was observed that the styrene on reaction with preformed $PhPd(PPh_3)_2(OAc)$ during the carbon–carbon bond forming step of the Heck reaction in DMF at room temperature forms stilbene whereas on reaction with $PhPdI(PPh_3)_2$ under similar conditions, stilbene was not formed. Further, when acetate anion was added to a mixture of $PhPdI(PPh_3)_2$ and styrene, the formation of stilbene was observed at room temperature. These observations are consistent with the dissociation of acetate ion from $PhPd(PPh_3)_2(OAc)$ to form an equilibrium mixture containing the cationic complex $[PhPd(PPh_3)_2]^+$ (Scheme 7).

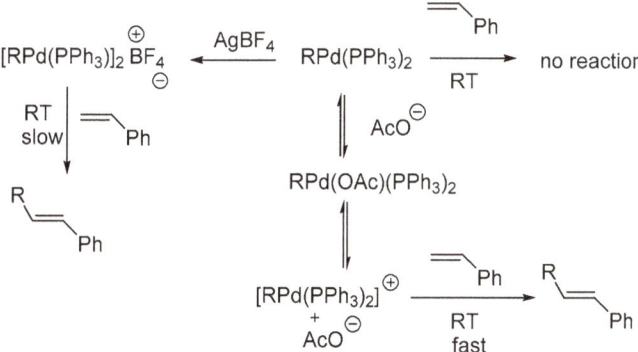

Scheme 7. Intermediates generated in Heck reaction mechanism.

The observations using various reaction conditions for inter and intramolecular Heck couplings, in presence of different ligands (PPh_3/P(o-tolyl)$_3$/Chelating phosphine ligands), bases (organic/inorganic), solvent (polar/nonpolar) and/ or TBAB are cited. In addition to this, the recent modifications to traditional reaction conditions are also reviewed and interpreted that, for intermolecular couplings involving reactive electrophiles (aryl or vinyl iodides) and alkenes containing an electron withdrawing group, a traditional catalyst system such as $Pd(OAc)_2$ with 2–4 equivalents of L [L = PPh_3 or P(o-tolyl)$_3$] or PdL_2Cl_2 or PdL_4 along with organic or inorganic base should suffice. Such systems require temperatures in the range 50–100 °C. In order to lower the temperature, the most effective protocol is to add R_4NX (X = Cl, Br) and use an aqueous solvent with K_2CO_3 as the base. For aryl or vinyl triflates with alkenes containing an electron withdrawing group, a traditional catalyst system can also be used. For alkenes not containing an electron withdrawing group, a halide free condition, achieved by either using aryl or vinyl triflates as the electrophile or adding a halide sequestering agent (Ag$^+$) for aryl or vinyl halides, will be advantageous. However, for electrophiles (like aryl bromides with electron donating groups or aryl chlorides) undergoing oxidative addition more slowly, high temperatures (above 120 °C) are usually required and a stable metal-ligand system that does not decompose at higher temperature is essential for prolonged life of the catalyst and hence PPh$_3$ will not be suitable ligand in such cases. For intramolecular Heck cyclizations, the reaction conditions appear to vary depending on the ring size, stereochemistry of the alkene and whether a tertiary or quaternary centre is being formed. The presence of halides does not appear to impede the cyclization at elevated temperatures and can be beneficial for high ee's in asymmetric Heck couplings. The use of halide free conditions can produce rapid Heck couplings but variable ee's are seen for asymmetric cyclization.

The Amatore group of researchers have invested much effort into investigations of the mechanism of Heck reaction. Kinetic evidences and electrochemical techniques like steady state voltammetry along with spectroscopic techniques like UV (Ultra violet spectroscopy), NMR (Nuclear Magnetic Resonance Spectroscopy) was used for the elucidation of mechanisms of palladium-catalysed reactions [69]. Their work emphasizes the crucial role played by the anions born by the precursors of palladium(0) complexes and rationalize empirical findings dispersed in literature concerning the specificity of palladium catalytic systems. A new adequate catalytic cycle has been proposed for Heck reactions where the fundamental role of chloride or acetate ions brought by palladium(II) complexes, precursors of palladium(0) complexes, is established (Scheme 8).

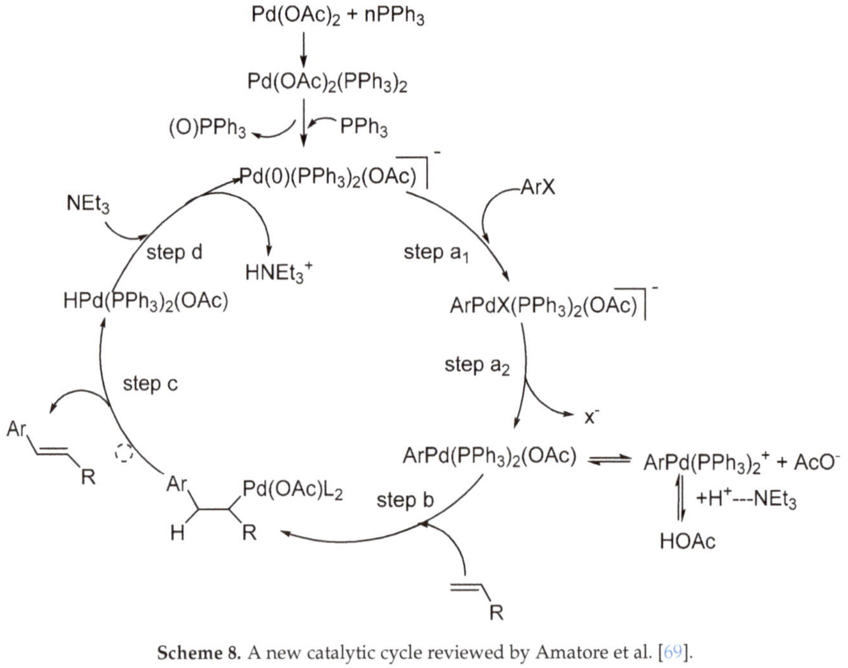

Scheme 8. A new catalytic cycle reviewed by Amatore et al. [69].

It is proposed that the new reactive anionic palladium(0) complexes species are formed where palladium(0) is ligated by either chloride ions Pd(0)(PPh$_3$)$_2$Cl$^-$ or by acetate ions Pd(0)(PPh$_3$)$_2$(OAc)$^-$. The reactivity of such anionic palladium(0) complexes in oxidative addition to aryl iodides strongly depends on the anion ligated to the palladium(0). The structure of the arylpalladium(II) complexes formed in the oxidative addition also depends on the anion.

The existence of intermediate anionic pentacoordinated arylpalladium(II) complexes ArPdI(Cl)(PPh$_3$)$_2$$^-$ **135** and ArPdI(OAc)(PPh$_3$)$_2$$^-$ **136** (Figure 19) are also predicted based on evidences. Their stability and reactivity is discussed depending on the presence of chloride or acetate anion. ArPdI(Cl)(PPh$_3$)$_2$$^-$ is found to be more stable and it affords trans ArPdI(PPh$_3$)$_2$, whereas, ArPdI(OAc)(PPh$_3$)$_2$$^-$ is quite unstable and rapidly affords the stable trans ArPd(OAc)(PPh$_3$)$_2$ complex.

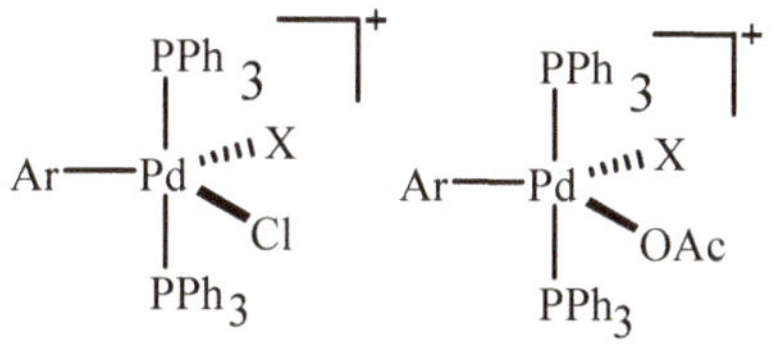

Figure 19. Intermediate anionic pentacoordinated arylpalladium(II) complexes formed in Heck reaction.

Another review by Amator and Jutand [70] is the resurgence on the mechanism of Heck reaction. It aims at giving the evidences of anionic Pd(0) and Pd(II) intermediates in palladium-catalysed Heck and cross-coupling reactions. It was proved that the anions of PdCl$_2$L$_2$ and Pd(OAc)$_2$, precursors of palladium(0) play a crucial role during the reaction. Thus, Pd(0)L$_2$, postulated as the common catalyst in catalytic systems, is not formed as a main intermediate, instead, previously unsuspected reactive tricoordinated anionic complex species Pd^{0}L$_2$Cl$^-$ and Pd^{0}L$_2$(OAc)$^-$ are found to be the effective catalysts. These anions also affect the kinetics of the oxidative addition to ArI as well as the structure and reactivity of the arylpalladium(II) complexes produced in this reaction. The pentacoordinated anionic complexes ArPdI(Cl)L$_2$$^-$ or ArPdI(OAc)L$_2$$^-$ are formed in the reaction.

An alternative mechanism via Pd(II)/Pd(IV) cycle (Scheme 9) specially proposed for PCP complexes (pincer palladacycle with Phosphorous as donor ligands) is described in the review by Whitecomb [24]. The initiation of the catalytic cycle takes place via the oxidative addition of a vinyl C-H of CH$_2$ = CHR to the palladium (II) complex (step a). This is followed by the reductive elimination of HCl from the Pd(IV) species (step b) and is a rate determining step to generate Pd(II) species. One more oxidative addition of ArCl (step c) generates another Pd(IV) species collapsing further giving out coupled product via reductive elimination to restore the catalyst (step d). The mechanism is different than a classical Pd(0)/Pd(II) as it involves the successive oxidative addition of both aryl halide and alkene substrates and eliminates the need for the migratory insertion step. Formation of intermediate after a reversible step a, has been supported by NMR studies using deuterium.

Cavell and McGuinness [71] have reviewed the redox process involving N-heterocyclic carbene complexes and associated imidazolium salts in conjunction with Group 10 metals. The mechanistic studies with stoichiometric reaction of the [(tmiy)$_2$Pd(Ar)]$^+$ cation (where Ar = p-nitrophenyl) with butylacrylate was seen to give a complex mixture of products. It was observed that the product distribution was dependent on temperature and other reaction conditions. For example, at $-30\,^\circ$C only the reductive elimination products 2-(4-nitrophenyl)-1,3,4,5-tetramethylimidazolium ion, and traces of the 2-substituted-imidazolium ion were observed. On warming to $-20\,^\circ$C, the Heck coupling product, n-butyl (E)-4-nitrocinnamate, and a further product, 1,3,4-tetramethyl imidazolium salt, were observed. At room temperature, the main products were the Heck product and the 1,3,4,5-tetramethylimidazolium salt. This indicates that the raise in temperature leads to the Heck coupling product more rapidly, i.e., the rates of migratory insertion plus β-elimination, as required for Heck catalysis, become more competitive with reductive elimination. When the process is run under catalytic conditions, i.e., base with an excess of p-nitrophenyliodide and butylacrylate at around 120 $^\circ$C, only the Heck product is observed with high turnover frequencies (TOF) and TON and no products generated from the reductive elimination reaction were observed. This indicates the thermodynamic parameters such as activation barriers and relative exothermicities play a crucial role.

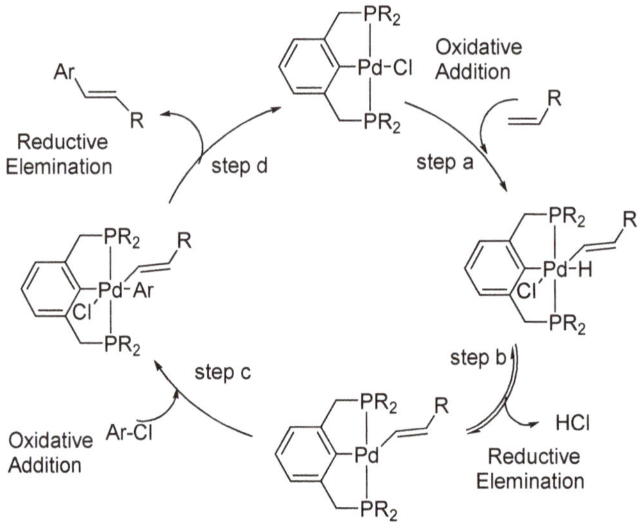

Scheme 9. An alternative mechanism for Heck reaction via Pd(II)/Pd(IV) cycle.

Trzeciak and Ziolkowski [72] have reviewed the catalytic activity of palladium in C-C bond forming processes in respect to Heck and carbonylation reactions. Catalytic systems based on palladium complexes with phosphorus ligands and phosphine-free systems and based on the Pd(0) colloid are discussed by giving a number of examples. These examples reveal that, both monomolecular Pd(0) phosphine complexes and nanosized Pd(0) colloids can act as active catalysts. This review supports the formation of conventional oxidative addition complex PhPdL$_2$X on the basis of evidences. The possibility of typical monomolecular Pd(0) complexes (e.g., [Pd(0)P4], P = phosphorus ligand) undergoing partial decomposition during the catalytic process to form nanosized phosphine-free Pd(0) particles is predicted. Palladium in the form of a colloid was observed to be a good catalyst for Heck and carbonylation reactions, especially when applied in an ammonium salt ([R$_4$N]X) or an ionic liquid (IL) medium. Ammonium salts efficiently contribute to the formation of soluble, monomolecular Pd(0) compounds of probably higher catalytic activity than bigger-sized clusters. The study reveals that carboxylate salts used as bases in the Heck reactions, can act as effective Pd(II) to Pd(0) reducing agents, and therefore they also may initiate the formation of colloidal nanoparticles.

Phan et al. [73] have a critical review on the nature of the active species in palladium catalysed Mizoroki–Heck and Suzuki–Miyaura couplings using homogeneous or heterogeneous catalysis. Regardless of what palladium precatalyst is used, it is critical for researchers to understand the nature of the true catalytic species generally used in their studies because of palladium deactivation via clustering, leaching and redeposition processes. The review summarizes each type of precatalyst of palladium, proposed to be truly active in the various form of catalysts like discrete soluble palladium complexes, solid-supported metal ligand complexes, supported palladium nano- and macroparticles, soluble palladium nanoparticles, soluble ligandfree palladium and palladium-exchanged oxides. A considerable focus is placed on solid precatalysts and on evidence for and against catalysis by solid surfaces vs. soluble species when starting with a range of precatalysts.

Based on various control experiments and tests to assess the homogeneity or heterogeneity of catalyst systems, it was concluded that, the Heck reactions using NC, SC, PC, SCS and PCP palladacycles (catalysts with N, O, P or S donor ligands) operates by Pd(0)/Pd(II) mechanism only and there is no proven example of catalysts operating by a Pd(II)/Pd(IV) catalytic cycle in Heck or Suzuki coupling reactions.

Knowles and Whiting [74] have reviewed advances in the mechanistic perspective of the Heck reaction aiming to review and compare work undertaken on the mechanism until 2007. The data provided give the evidence for a mechanism where the oxidative addition strongly depends on the conditions, particularly whether the reaction is saturated with halide. Eventually, the authors concluded that it is not possible to explain all the phenomena observed for the oxidative addition on the basis of single mechanism only and there may be several mechanisms in operation, either independently, depending on the reaction conditions, or in parallel.

Although the electron rich and chelating nature of the phosphines is required to activate aryl chlorides, it disfavors carbometallation or dissociation. It is also assumed that the rate determining step comes after initial oxidative addition; however, the nature of this step is yet unclear and has a strong dependence on the nature of the species produced in the oxidative addition step.

Kumar and co-workers [75] have reviewed the status regarding in situ generation of palladium chalcogenides phases or Pd(0) protected with organochalcogen fragments when Palladium(II) complexes of organochalcogen ligands (e.g., Pd(II) complex of an (S,C,S) pincer ligand) are used as viable alternatives to complexes of phosphine/carbene ligands for Suzuki–Miyaura and Heck C-C cross coupling reactions. These ligands are thermally stable, easy to handle and air and moisture sensitivity are not impediments with many of them.

Recently, Eremin and Ananikov [76] summarized the studies mentioning the behaviour of a "Cocktail" of catalysts and mechanistic investigations regarding the evolution of transition metal species during catalytic cycles. They have considered the transformations of molecular catalysts, leaching, aggregation and various interconversions of metal complexes, clusters and nanoparticles that occur during catalytic processes termed as "Cocktail"-type systems. These systems are considered from the perspective of the development of a new generation of efficient, selective and re-usable catalysts for synthetic applications. It also attempts to determine "What are the true active species?" in mechanism specially in coupling reactions like Heck. It has been concluded that there is nothing static in catalysis and catalytic system is always dynamic where "Cocktail"-type catalytic systems are formed when the transformation of a metal species occurs during the overall catalytic process. It was observed that the "Cocktail" picture can exist not only in homogeneous catalysis in solution but also in a heterogeneous catalytic system.

4. Reviews on Applications

The palladium-catalysed cross-coupling reactions particularly Heck reaction have great significance for both academic and industrial research and in the production of number of compounds on a large industrial scale. Fine chemicals—including pharmaceuticals, agricultural chemicals, and high-tech materials—that benefit society when designed properly in addition to the new reactivities, increased product selectivity and reduced volatile organic consumption, can lead to a great breakthrough in industrial research. Reviews targeting such chemicals are given below.

4.1. Fine Chemical Production

An overview by Blaser et al. [77] describes the industrial application of homogeneous catalysts for the chemical industry. Along with theoretical background of organometallic complexes and homogeneous catalysis, a description of the prerequisites and current problems for their industrial application mainly for manufacture of intermediates for pharmaceuticals and agrochemicals are discussed. Production processes for octyl-4-methoxycinnamate, the most common UVB sunscreen (UVB—Ultra Violet B: These rays penetrate the upper layers of the skin) developed by Dead Sea Company (Scheme 10); Naproxen, a generic analgesic, developed by Albemarle (Scheme 11) and sodium 2-(3,3,3,-trifluoropropyl)- benzenesulfonate, a key intermediate for Novartis' sulfonylurea herbicide Prosulfuron (Scheme 12) are mentioned, where, Heck reaction is one of the key step in its process, along with their yield and selectivity.

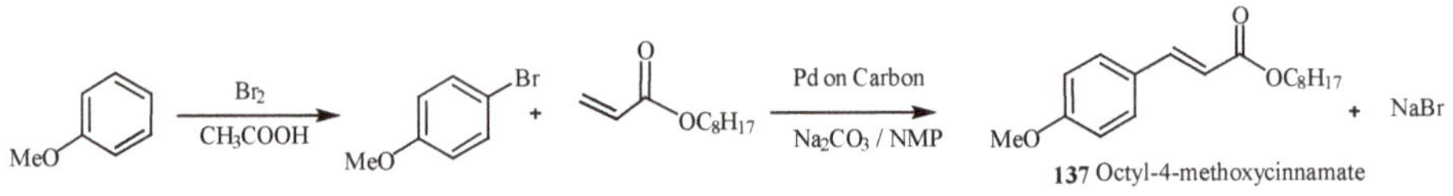

Scheme 10. Synthetic route for Octyl-4-methoxy cinnamate by Dead Sea Company.

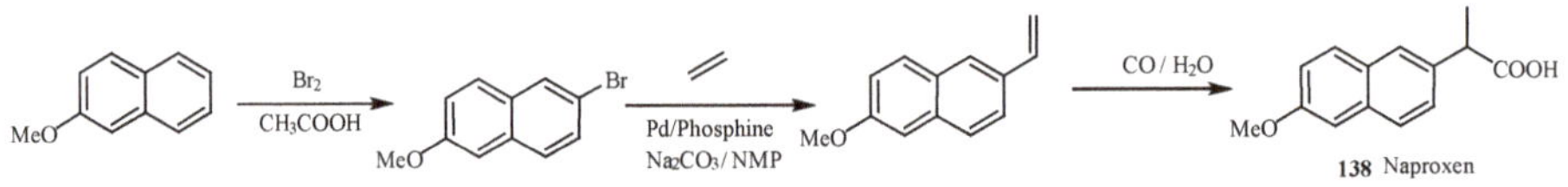

Scheme 11. Synthetic route for Naproxen by Albemarle.

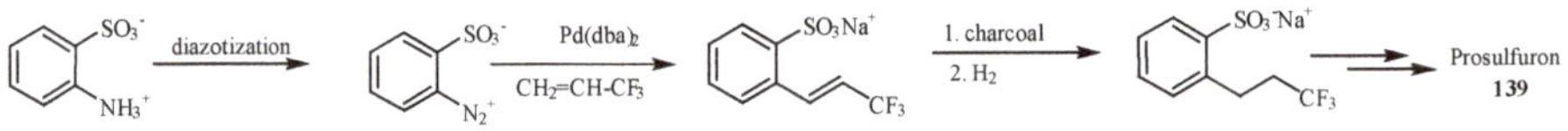

Scheme 12. Synthetic route for Prosulfuron intermediate by Ciba-Geigy/Novartis.

'The Heck reaction in the production of fine chemicals' by deVries [78], is an overview mainly focusing the commercial products produced on a scale in excess of 1 ton/year and using Heck or Heck type reaction as one of the steps during synthesis. The synthetic methods for production of a sunscreen agent, 2-ethylhexyl *p*-methoxy-cinnamate **137**; a nonsteroidal anti-inflammatory drug, Naproxen **138**; an herbicide, Prosulfuron **139**; an antiasthma agent, Singulair **140** and monomers for coatings, divinyltetramethyldisiloxanebisbenzocyclobutene **141** (Figure 20) have been discussed along with the conditions used for their synthesis. Herbicide prosulfuron is synthesized via Matsuda reaction using arenediazonium salts as an alternative to aryl halides and triflates, whereas the production of sunscreen agent 2-ethylhexyl *p*-methoxy-cinnamate uses Pd/C as catalyst. Naproxen and the monomers are synthesized via bromo derivatives, whereas Heck reaction on allylic alcohol is carried out for the production of antiasthma agent Singulair. The review also talks about the use of metals and ligands as a part of catalyst for industrial production. Although palladium has proved to be the best metal to be used for Heck reaction, from the economic standpoint its cost becomes the biggest hurdle at industrial level.

Similarly, despite the fact that palladacycles, pincers and bulky ligands are robust, they work only on few reactive substrates. Hence ligandless catalysis seems to be attractive for production. In addition, various techniques of recycling of catalyst has been discussed like immobilization of catalysts by attaching ligands to solid support, etc., however, was not found to be useful, because of leaching and reduced activity issues.

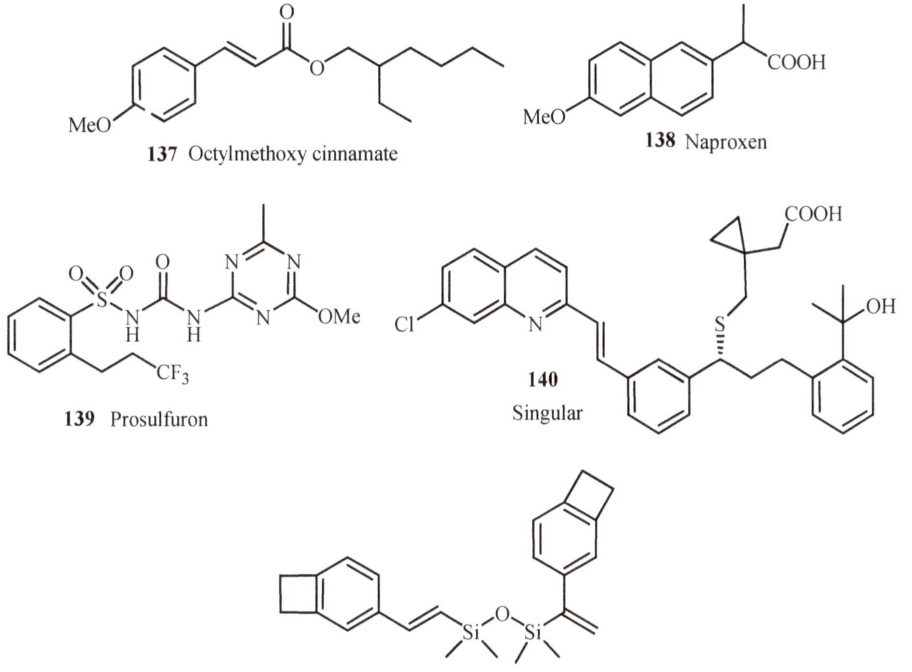

Figure 20. Fine chemical products produced on a scale in excess of 1 ton/year using Heck or Heck type reaction reviewed by deVries [78].

A review by Zapf and Beller [79] published in succeeding year gives the importance of several palladium-catalysed reactions like Heck, Suzuki, Sonogashira, telomerization [80] and carbonylation [81], developed and optimized to a stage that enables application on an industrial scale. In addition to Naproxen **138**, Prosulfuron **139** and divinyltetramethyldisiloxane-bisbenzocyclobutene **141**, use of Heck reaction in the synthesis of L-699,392 **142**, resveratrol **143**, eletriptan **144**, cincalcet **145** and montelukast sodium (Salt of Singulair) **146** (Figure 21) by different industries are discussed.

Much later, a review published by Picquet [82] edited by Beller and Blaser presents the state-of-the-art in the industrial use of organometallic or coordination complexes as catalysts for the production of fine chemicals. The review illustrates the great potential of platinum group metal (pgm) complexes as valuable tools in this field for many organic transformations, where these reactions involve platinum group metal complexes as catalysts. Heck reaction for the production of compounds Naproxen **138**, Prosulfuron **139**, Resveratrol **143**, Cinacalcet **145** and Montelukast **146** is discussed. The authors are from the industrial world so were also able to describe the technical challenges encountered in scaling up the reactions from small quantities to production amounts and also tackle issues related to it by taking numerous examples of production processes, pilot plant or bench-scale reactions.

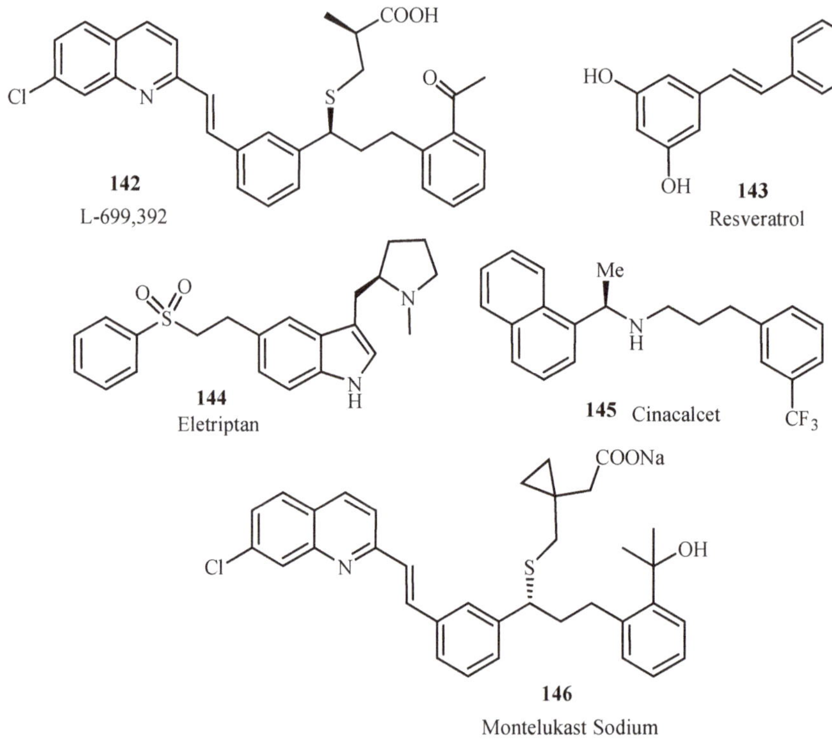

Figure 21. Additional fine chemical products produced on industrial scale using Heck reaction reviewed by Zapf and Beller [79].

4.2. Pharmaceuticals

A review by Sharma [83] shows the importance of cinnamic acid derivatives (CADs) **147**, **148** (Figure 22) for its various biological activities like antioxidant, hepatoprotective, anxiolytic, insect repellent, antidiabetic and anticholesterolemic, etc.

It is mentioned that the various methods for the preparation of cinnamic acid derivatives such as Perkin reaction [84], enzymatic method, Knoevenagel condensation, phosphorous oxychloride method and Claisen–Schmidt condensations has some or the other drawbacks or loopholes like low yield, loss of catalytic activity of enzyme, long duration of reaction time, ozone layer depletion caused by CCl_4 and tedious synthetic procedure, etc. Hence, the Heck reaction using various novel supported catalysts is by far the best for the synthesis of cinnamic acid derivatives.

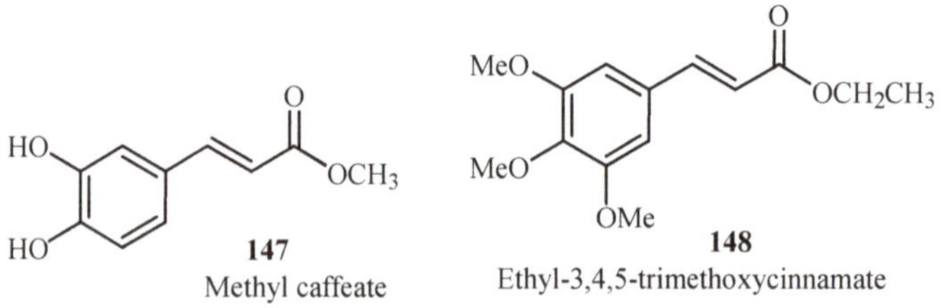

Figure 22. Cinnamic acid derivatives (CADs) reviewed by Sharma [83].

Biajoli et al. [85] have reported the applications of Pd-catalysed C-C cross-coupling reactions for the synthesis of drug components or drug candidates. The review is written in context with two earlier independent reviews by Pfizer researchers Magano and Dunetz on the large-scale applications of transition metal-catalysed coupling reactions for the manufacture of drug components in the pharmaceutical industry [86] and by Torborg and Beller [87] on the applications of Pd-catalysed coupling reactions, both covering the material till 2010. Hence, the period after that until July 2014 is covered in their review.

Cross coupling reactions like Heck, Suzuki, Negishi, Sonogashira, Stille and Kumada are discussed with number of examples. Heck reactions are intensely exploited for inter and intramolecular coupling involving double bonds for the synthesis of a idebenone **149** used for the treatment of Alzheimer's, Parkinson's diseases, free radical scavenging and action against some muscular illnesses; ginkgolic acid (13:0) **150**, a tyrosinase inhibitor; olopatadine **151** an antihistaminic drug; the vascular endothelial growth factor (VEGF) inhibitor axitinib **152**; caffeine-styryl compounds **153** possessing a dual A_{2A} antagonist/MAO-B inhibition properties and with potential application in Parkinson's disease; styryl analogues of piperidine alkaloids (+)-caulophyllumine B **154** displaying a high anti-cancer activity in vitro; (R)-tolterodine **155**, a drug employed in urinary incontinence treatment; ectenascidin 743, the anti-cancer tetrahydroisoquinoline alkaloid **156**; bioactive compounds the abamines **157**, **158** and naftifine **159** (Figure 23).

Lab scale preparation of three other drugs: cinacalcet hydrochloride **160**, alverine **161** and tolpropamine **162** are also discussed using typically Heck-Matsuda reaction (Figure 24). Cinacalcet hydrochloride is a calcimimetic drug commercialized under the trade names Sensipar and Mimpara, and is therapeutically useful for the treatment of secondary hyperthyroidism and also indicated against hypercalcemia in patients with parathyroid carcinoma. Alverine is a smooth muscle relaxant used for the treatment of gastrointestinal disorders such as diverticulitis and irritable bowel syndrome and tolpropamine is an antihistaminic drug used for the treatment of allergies.

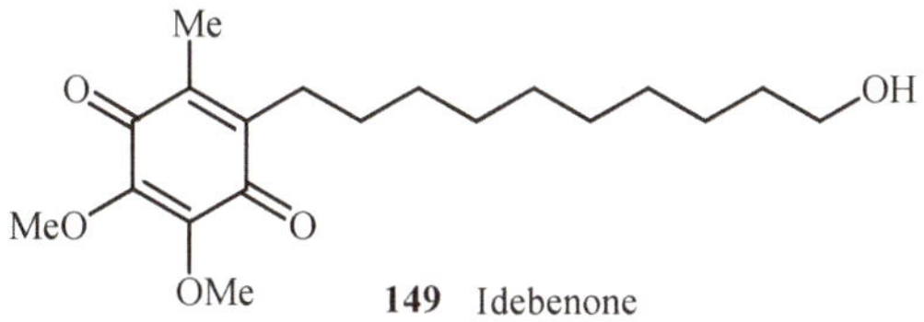

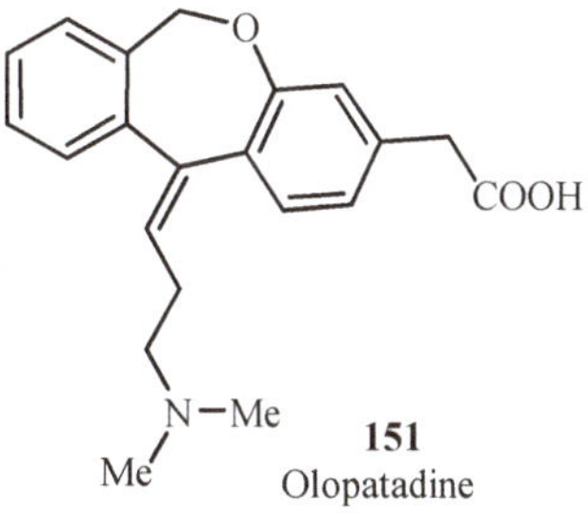

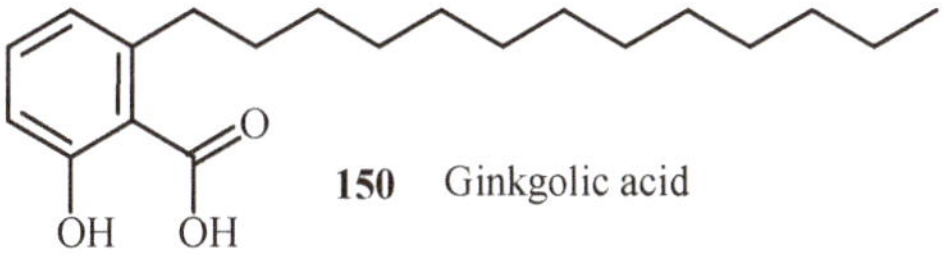

Figure 23. *Cont.*

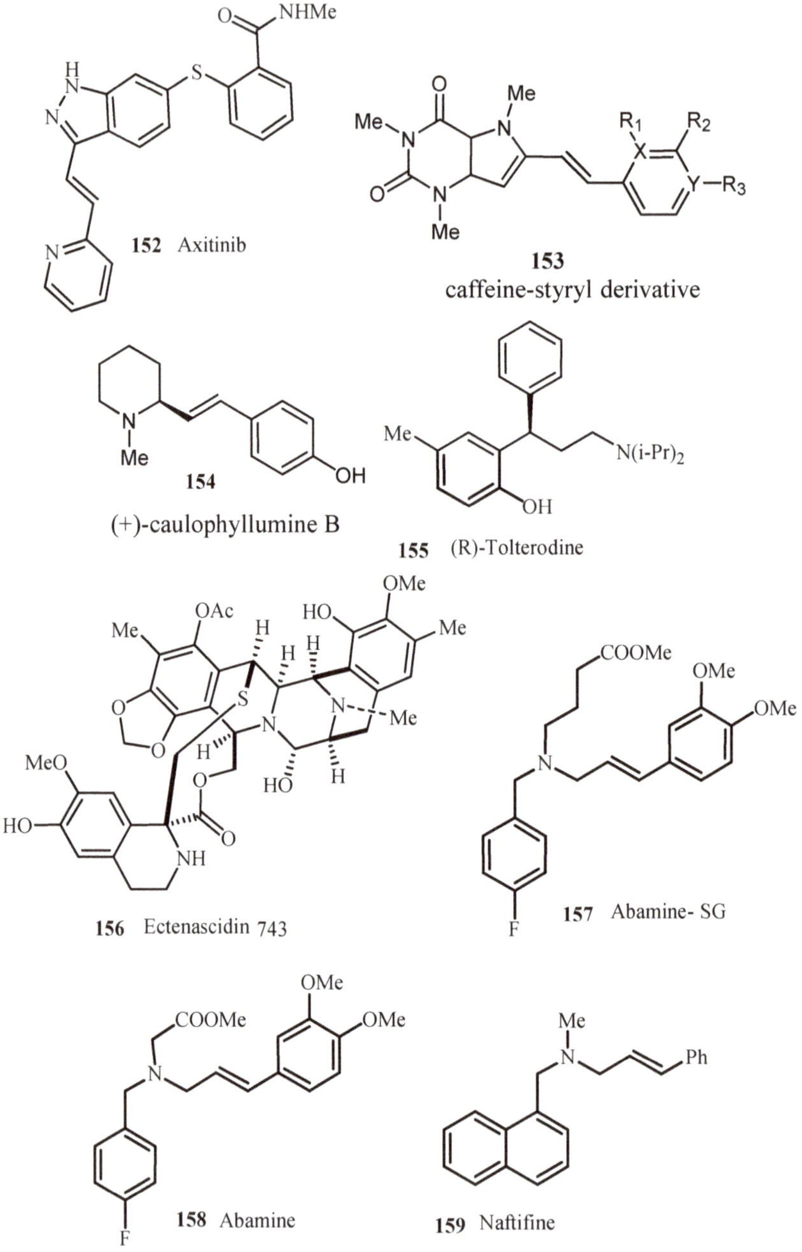

Figure 23. Drug components or drug candidates exploited for inter and intramolecular Heck coupling by Biajoli et al. [85].

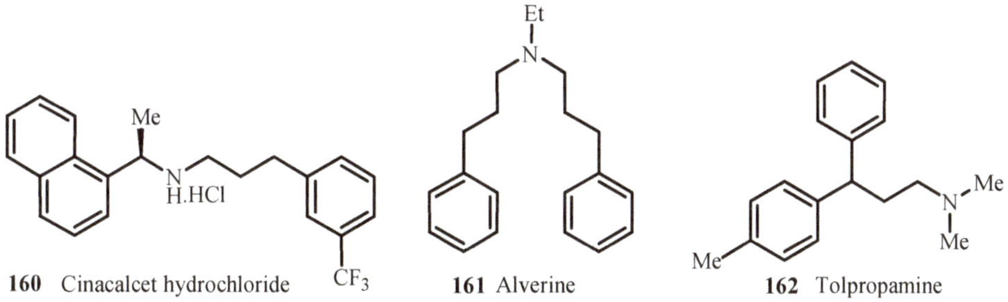

160 Cinacalcet hydrochloride

161 Alverine

162 Tolpropamine

Figure 24. Lab scale preparation drugs reviewed by Biajoli et al. [85].

4.3. Total Synthesis

The Heck reaction has been used in synthesis of more than 100 different natural products and biologically active compounds. Taxol is the first example where Heck reaction was employed for creating the eight-membered ring during its synthesis. Reviews in this section give glimpses of all such compounds.

Nicolaou et al. have corroborated the role of palladium catalyzed cross-coupling reactions in total synthesis [88]. The review highlights number of selected examples of synthesis using most commonly applied palladium-catalyzed carbon–carbon bond forming reactions including Heck reactions. The involvement of intramolecular Heck reaction in the total synthesis of (±)-dehydrotubifoline **163** with justification based on Jeffery modification [89] for the previously formed unwanted derivative has been discussed. Synthetic routes for (−)-quadrigemine C **164**, $\Delta^{9(12)}$-capnellene-3β,8β,10α-triol **165**, estrone **166**, taxol **167**, $\Delta^{9(12)}$-capnellene-3β,8β,10α,14-tetraol **168**, (+)-calcidiol **169**, Okaramine N **170**, α-tocopherol **171** (Figure 25), scopadulcic acid-B **124** and singulair **140** are given. Few examples of zipper polycyclization **172** developed by Trost group [90,91] are also mentioned.

All these processes are impressive since they do not require the preparation of reactive intermediates prior to the carbon–carbon bond forming event since they proceed by activation of stable and readily available starting materials in situ and, therefore, are both more practical and often more efficient in terms of overall yield.

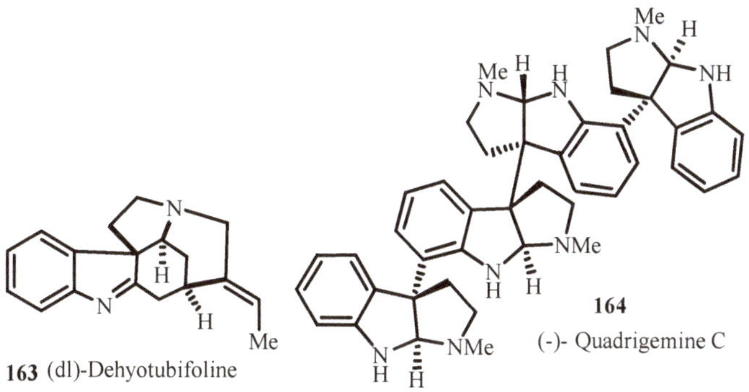

163 (dl)-Dehyotubifoline

164 (−)- Quadrigemine C

Figure 25. *Cont.*

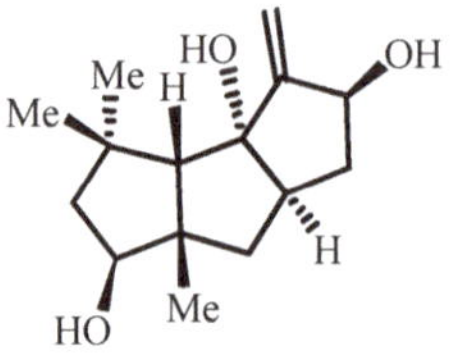

165 Δ$^{9(12)}$ – Capnellene- 3β,8β, 10α - triol

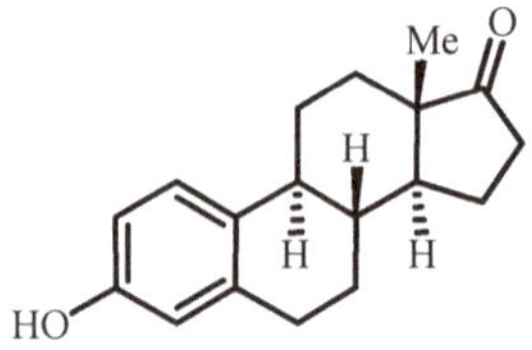

166 Estrone

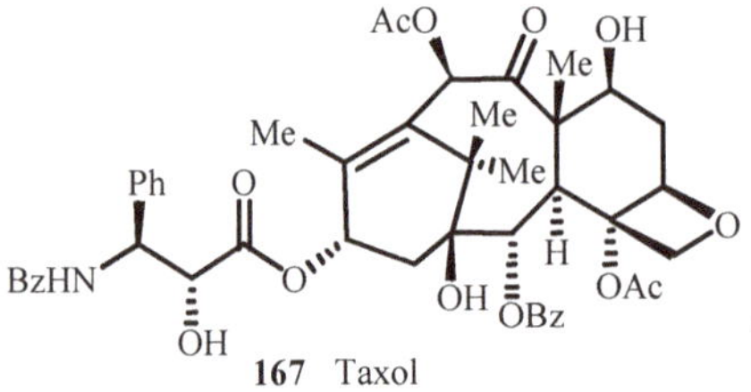

167 Taxol

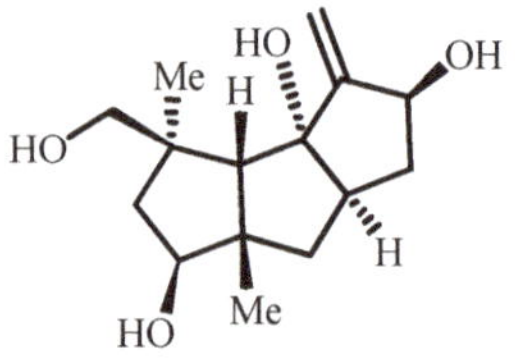

168 Δ$^{9(12)}$ – Capnellene- 3β,8β, 10α, 14- tetraol

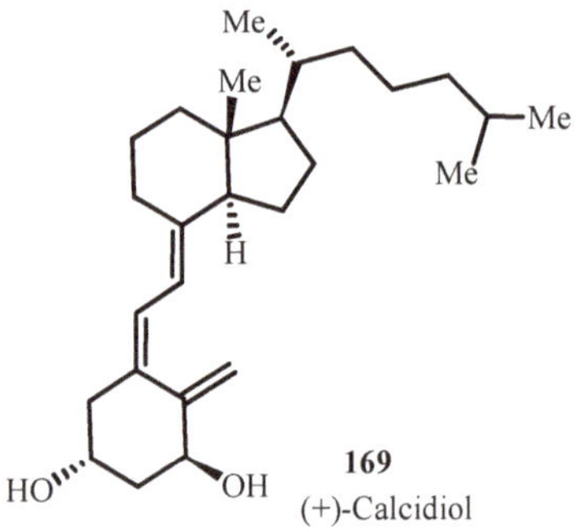

169 (+)-Calcidiol

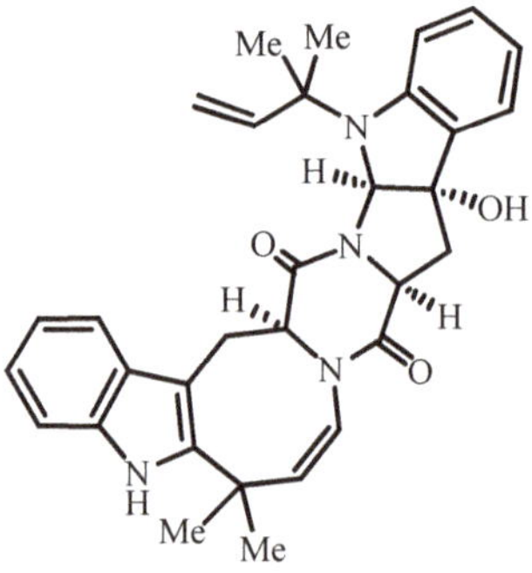

170 Okaramine N

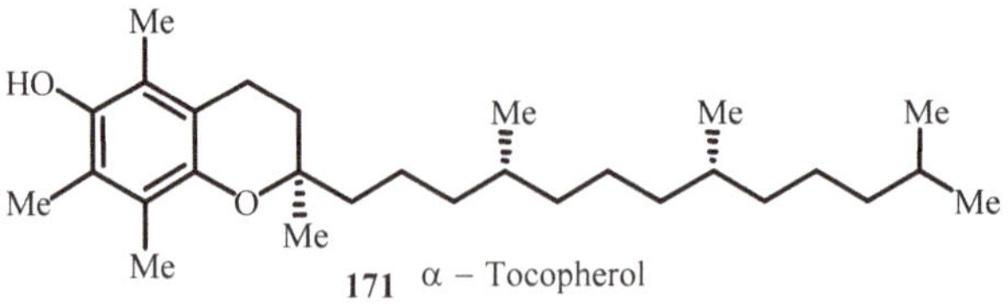

171 α – Tocopherol

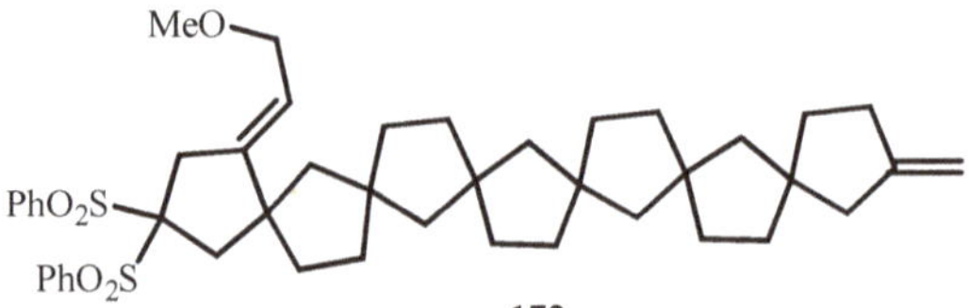

172

Figure 25. Heck reaction in Total synthesis reviewed by Nicolaou et al. [88].

4.4. Targeted Compounds

In addition to all above distinctive natural products, Heck reaction is also used in the production of some particular compounds such as aryl glycosides or heterocycles. A review by Wellington and Benner [92] is about using Heck reaction for the synthesis of aryl glycosides (also termed as C-glycosides or C-nucleosides) though there are many reports of synthesis of such compounds via non-Heck methods. Synthesis of glycosides pyrimidine c-nucleosides, purine c-nucleosides, monocyclic, bicyclic, and tetracyclic C-nucleosides **173–185** (Figure 26) via Heck reaction are discussed.

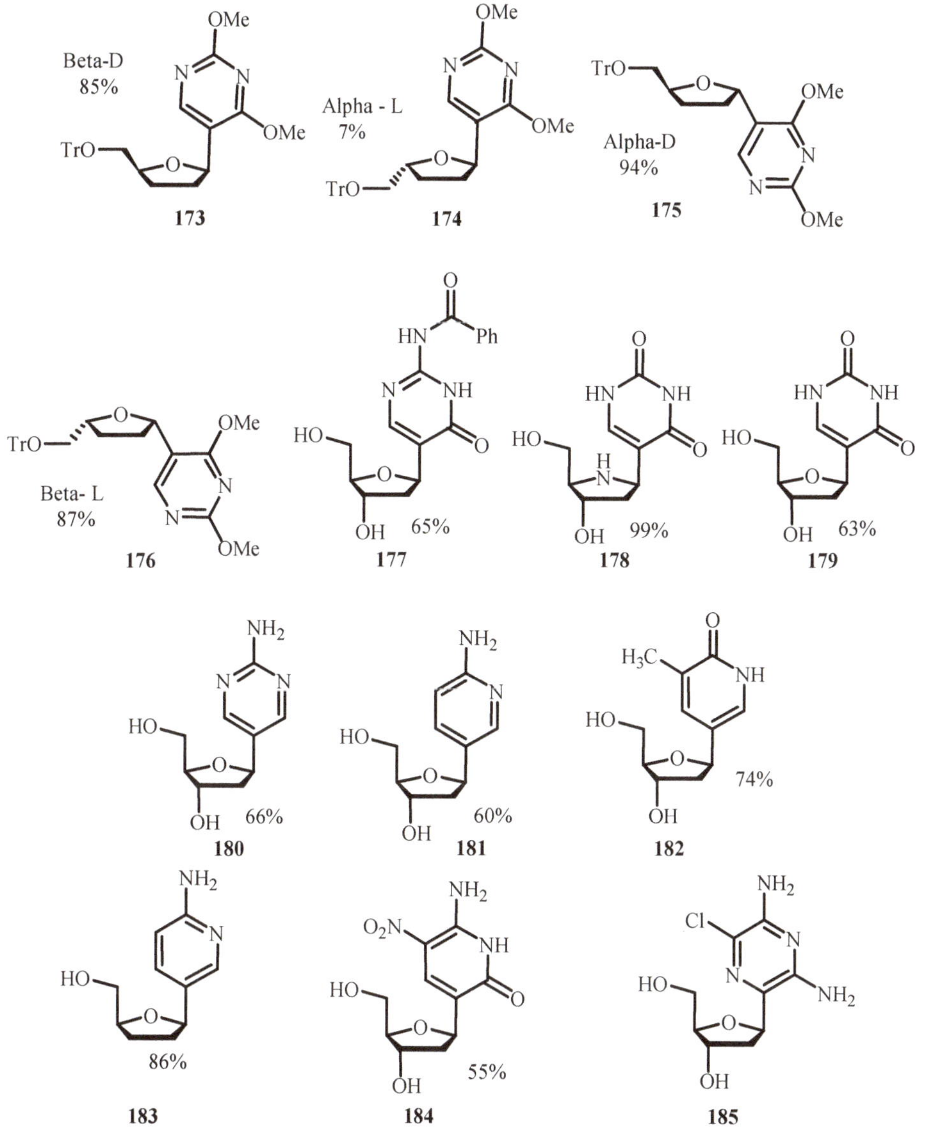

Figure 26. Glycosides synthesized via Heck reaction reviewed by Wellington and Benner [92].

The review describes the subsequent conversion with respective reagents and conditions of the Heck products corresponding to target molecules and their application. Pd(OAc)$_2$-AsPh$_3$ was observed to be serving well for most of the synthesis although few other palladium-ligand combinations have to be used in some cases. Iodides were found to be more reactive than other halides. Bidentate ligand accelerates the Heck coupling reaction and results in a higher yield of the product.

A review by Zeni and Larock [93] covers a wide range of palladium catalyzed processes involving oxidative addition- reductive elimination chemistry developed to prepare heterocycles, with the emphasis on fundamental processes used to generate the ring systems themselves. A number of examples are given for preparation of heterocycles via intramolecular Heck cyclization of aryl halides, vinylic halides, vinylic and aryl triflates. intermolecular annulation, and via asymmetric Heck cyclization.

5. Reviews on Reuse of Catalyst

Homogeneous catalysts are of great interest for synthesizing fine-chemical, specialty chemical, pharmaceutical products for their advantages of high activity and selectivity. However, the main disadvantages of traditional organic phase reactions employing homogeneous transition metal catalysts are the difficulties associated with separating the catalyst from the product and solvent albeit having less aggressive reaction conditions and increased selectivity. Hence, it is important to discuss the catalyst product separation techniques for heck reactions. This section aims to provide an overview of the current reviews on the separation/recycling methods of homogeneous transition metal catalysts.

5.1. Recapitulation of All Methods

A succinct review by Bhanage and Arai [94] describes all the various catalyst-product separation methodologies developed for Heck reactions until 2001. The review gives information about the various catalyst-product separation methodologies developed for Heck reactions until then and is illustrated with number of examples. The separation techniques are classified into two major categories: heterogeneous catalysts and heterogenized homogeneous catalysts.

Heterogeneous catalysts can easily be removed by physical separation methods. In addition, they are stable at higher temperatures and hence become interesting for the activation of less reactive but less expensive chloroaryls substrates. However, the heterogeneous catalysts have a major drawback of poor selectivity toward Heck coupling products.

Heterogeneous catalysts include

- Conventional supported metal catalysts such as Pd/C, Pd/SiO$_2$, Pd/Al$_2$O$_3$ and Pd/BaSO$_4$ and also nonpalladium based catalysts like Ni/HY zeolite, Ni/Al$_2$O$_3$, Cu/Al$_2$O$_3$ and Co/Al$_2$O$_3$. However most of them end up in leaching of metal and hence making it difficult to do effective recovery and recycling of the active metal.
- Zeolite-encapsulated catalysts include palladium-grafted mesoporous MCM-41, Pd-TMS11, etc. However, depending on zeolite structure the leaching of active Pd species was observed with this type of catalyst as well.
- Colloids–nanoparticles primarily palladium based are reported showing high activity.
- Intercalated metal compound,s e.g., palladium-graphite type catalyst, Pd-chloride- and Cu-nitrate-intercalated montmorillonite K10 clays, etc., are reported to be active for Heck reaction.

The heterogenized metal complex catalysts operate under relatively mild conditions as compared with heterogeneous catalysts, and hence can be applied to the production of pharmaceuticals and fine chemicals. The homogeneous metal complexes are heterogenized using following strategies:

- Modified silica catalysts where metal is adsorbed onto a silica support.
- Polymer-supported catalysts, where Pd anchored to phosphinated polystyrene.

- Biphasic catalysis, another exciting area of environmentally responsible catalysis. This system uses two liquid phases such that the catalyst exists in one phase while reactants and products are present in the other phase. The maximum studied catalyst of this category is sulphonated triphenylphosphine like **115, 130** complexed mainly with palladium although few other metals are also known to react. This catalyst has proven its efficiency for many other reactions too.
- Supported liquid-phase catalysts, where the catalyst-containing phase is dispersed as a thin film on a hydrophilic support like silica and is used in another immiscible solvent leading to the enhancement in reaction rate.
- NAILS also known as molten salts provide a medium in which the catalyst is generally dissolved allowing the product to be easily separated. The catalyst and ionic liquid can then be recycled.
- Perfluorinated solvents uses fluorous phase and perfluorinated ligand based organometallic catalyst and is segregated from reagents and products, either during the process or during the workup.

In addition to all above, recyclable homogeneous complexes (where they are recovered by solvent precipitation), catalysts giving High TON (as even after non-recovery of the catalyst it becomes affordable) and supercritical solvents are discussed.

5.2. Heterogeneous Catalysts

A review by Wall et al. [95] mainly focuses discussion on the Heck reaction followed by review on cinnamic acid synthesis using heterogeneous catalyst, Pd/C in particular. Progress made until then, mechanism and the conditions required for successful Heck reaction are discussed including the factors contributing to electronic control for α- or β-arylation. Limitations to the homogeneous reaction along with possible reasons and need for proceeding with heterogeneous catalysts, although the mechanism for which is not known are mentioned by giving a number of examples. Their efforts in the field of development of methods for the production of cinnamic acids, used as substrates for the synthesis of unnatural amino acids by employing phenylalanine ammonialyase are given.

Heerbeek et al. [96] have reviewed the use of dendrimers as support for recoverable catalysts and reagents. Phosphonated DAB-dendr-[N(CH$_2$PPh$_2$)$_2$]$_{16}$ dendrimer dimethylpalladium complex **186** (Figure 27) is the first such successful dendritic catalyst to recover through a precipitation procedure where no metallic palladium was observed unlike monomeric catalysts and was recycled however with lesser yields in successive runs.

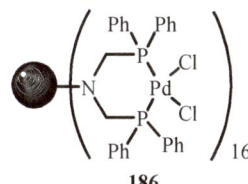

186

Figure 27. Dendrimer supported catalysts.

Solid supports have become valuable tools for catalysts immobilization nowadays for simplified product isolation and catalyst recycling. Horn et al. [97] reviewed such non-covalently solid-phase bound catalysts for organic synthesis. Immobilization of catalysts by non-covalent bonding like hydrogen bridges, or ionic, hydrophobic or fluorous interactions on solid support is found to be an alternative to covalent attachment. Such non-covalent approaches increase the flexibility in the choice of the support material, reaction conditions and work-up strategies compared to covalent attachment. Numerous catalytic reactions employing one of these non-covalent bonding strategies are given along with the examples of Heck reaction with catalyst immobilization by ionic interactions.

The problems, potential and recent advances with general approach in heterogeneously catalyzed Heck reactions especially with supported palladium catalyst such as Pd on activated carbon, oxides, polymers and in zeolites has been reviewed by Kohler et al. [98]. The advantages and limitations for practical applications are considered. A thorough literature for less activated or deactivated compounds is discussed along with the new approaches and strategies for the activation of such compounds by heterogeneous catalysts. Particular attention is given to the relation between homogeneous and heterogeneous catalysis from the mechanistic point of view. It is concluded that palladium species dissolved from the support are proven to be responsible for high activity and selectivity in Heck reactions in supported catalysis. The careful choice of optimum catalyst and reaction conditions can activate even deactivated aryl chlorides with high yields within few hours of reaction time.

Polshettiwar and Molnar [99] have reviewed the Heck coupling with silica supported catalyst and palladium redeposition. Mechanisms for these catalysts considering both less common Pd(II)/Pd(IV) catalytic cycle and the most accepted Pd(0)/Pd(II) cycle are given. A number of examples are cited for the Heck reaction in presence of silica-supported Pd complexes and the structural investigations of functionalized mesoporous silica-supported Pd catalysts. New emerging fields like sol–gel entrapped silica–Pd catalysts, nanosized Pd particles embedded in silica catalysts, colloid palladium layer–silica catalysts, silica/ionic liquid Pd catalysts and silica-supported Pd–TPPTS liquid-phase catalysts are also discussed along with testing of palladium leaching by poisoning test using pyridine and mercury.

Another review on this topic by Molnar [100] gives an account of efficient, selective, and recyclable palladium catalysts in carbon-carbon coupling reactions. The review gives a comprehensive survey, thorough analysis of the available data and discusses critically the questions related to recyclability in general. Publications reporting a stable and well characterized catalyst with high yield and minimum five recycles are cited. It is reported that, both amorphous and ordered silica materials are useful supports for palladium nanoparticles. The combined use of varied ionic liquids and related homogeneous complexes has shown interesting results. The best performance with respect to cumulative TON numbers are reported for catalyst Pd-SBA, 3.9% Pd-HS-Si(HIPE), and 4.1% Pd-grHSSi(HIPE) (HIPE = high internal phase emulsion). Thus, the catalytically active species, Pd particles and immobilized species, or the support material employed may play the decisive role in efficient catalysis.

5.3. Heterogenization of Homogeneous Catalysts

The development of new, highly efficient heterogenized catalysts is an active and important area in fine chemicals production research as it gives the maximum advantages of both homogeneous and heterogeneous catalysis. It offers several significant practical advantages for synthetic chemistry and industrial research. Reviews about various techniques used for heterogenization are given in following subsections.

5.3.1. Ionic Liquid

These days, ionic liquids are proposed as greener alternative to the classical cross-coupling procedure. They have replaced organic solvents in many metal catalyzed reactions and have gained much attention in recent times mainly because they lack the vapour pressure. However, several factors like toxicity, stability, cost, ease of processing, etc., still need to be addressed before this new technology is to be accepted by industry.

A review by Sheldon [101] is one of the finest written reviews on ionic liquids. Discovery of ionic liquids, historical development and progression for various types of reactions in them until 2001 are summarized eloquently. Scientists and industrial researchers are seriously considering ionic liquids as better replacement for toxic and/or hazardous volatile organic compounds and consider that, the use of ionic liquids as novel reaction media may offer a convenient solution to both the solvent emission and the catalyst recycling problem. Ionic liquids also provide a medium for performing clean reactions with minimum waste generation. It was observed that use of ionic liquids (Figure 28) as reaction media for catalytic transformations or, in some cases, as the catalyst itself can have a profound effect on activities and selectivities.

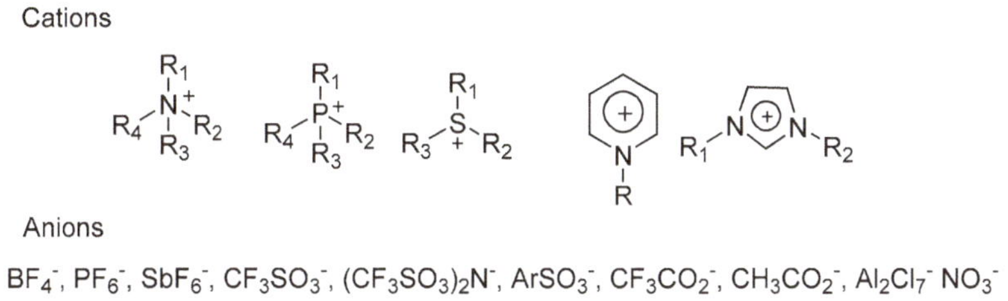

Cations

Anions

BF_4^-, PF_6^-, SbF_6^-, $CF_3SO_3^-$, $(CF_3SO_3)_2N^-$, $ArSO_3^-$, $CF_3CO_2^-$, $CH_3CO_2^-$, $Al_2Cl_7^-$ NO_3^-

Figure 28. Structure of ionic liquids used for Heck reaction reviewed by Sheldon [101].

The first example of a Heck coupling in an ionic liquid was reportedly carried out in 1996, where in ionic liquids tetraalkylammonium and tetraalkylphosphonium bromide salts, high yields for Heck reaction were obtained. No leaching of palladium was observed, the product was isolated by distillation from the ionic liquid and the ionic liquid was recycled for two more times. Similarly, ionic liquids like Bu_4NBr, $bmimPF_6$ and n-hexylpyridinium PF_6 are also reported to give improved activity for Heck reaction. In comparison to pyridinium analogues, the imidazolium ionic liquids were found to have greater catalytic activity when the reactions performed in it. The selectivity of α over β—product for the Heck arylation of electron-rich enol ethers in ionic liquids was found to be >99% in comparison with other solvents generally leading to a mixture of regio isomers owing to competition between cationic and neutral pathways.

Singh et al. [102] reviewed the use of ionic liquids medium for many palladium catalyzed reactions like Heck, Suzuki, Stille, Negishi, Trost–Tsuji and Sonogoshira coupling. A comprehensive literature about the versatility of ionic liquid in conjunction with palladium for these reactions is cited. The studies have shown that the classical transition-metal catalyzed reactions can be performed in ionic liquids. They are supposed to be a medium for clean reactions with minimal waste and efficient product extraction. In addition to the basic ionic liquids discussed by Sheldon [101] ionic liquids specially introduced after 2001 are discussed in detail. Apart from regularly studied ionic liquids, results of few specially studied ionic liquids using palladium catalyst such as functionalized ionic liquids like [bmim][TPPMS] and [bmim][OAc]; N-butyronitrile pyridinium bis(trifluoromethylsulphonyl) imide with $PdCl_2$; palladium acetate immobilized on amorphous silica with the aid of an ionic liquid [Bmim][PF_6]; palladium(II) complex from pyrazolylfunctionalized hemilabile NHC; Pd(OAc)$_2$/[PEG-mim][Cl], Pd/C-catalyzed Heck reaction in ionic liquid and Pd(0) nanoparticles (~2 nm diameter), immobilized in [bmim][PF_6] are also discussed. Most of them showed reasonable efficiency in recycle studies. Pd-nanoparticles were found to be much more efficient in TBAB than in pyridinium, phosphonium and imidazolium ionic liquids in catalyzing the Heck reactions. Few reports of reactions in ionic liquids using ultrasonic irradiation and microwaves are also mentioned to show better activity.

Bellina and Chiappe [103] have reviewed the progress and challenges for the Heck reaction in ionic liquids. Developments in the application of ionic liquids and related systems like supported ionic liquids; ionic polymers, etc., in the Heck reaction have been reviewed. Merits and achievements of ionic liquids were analysed and discussed considering the possibility of increasing the effectiveness of industrial processes. Numerous examples for homogeneous and heterogeneous conditions used for Heck reactions in ionic liquids are discussed to prove that, the ionic liquids are not only suitable solvents for Heck reaction, but their unique physico-chemical properties have ability to change the course of the reaction, activate and/or stabilize the intermediates or transition states in the reaction mechanisms. It is observed that Heck reactions performed in ionic liquids have higher reaction rates, and may be characterized by a higher control on the regio- and stereoselectivity of the coupling products. Examples with involvement of carbene complex, palladium nanoparticles and palladium anionic complexes have been discussed for the evidence. However, the study of Heck reactions in ionic liquid is still limited to the development of the ionic liquids where only simple reactions such as the coupling of aryl halides with cinnamates or styrene have been investigated. Thus, the use of ionic liquids in Heck reactions involving more complicated substrates that could give important indications about possibility of application of these alternative media on large scale need to be explored further.

Mastrorilli et al. [104] have surveyed the significant developments and the beneficial effect of the ionic liquids in terms of activity, selectivity and recyclability for palladium-catalyzed cross-coupling reactions performed in them. Insights into the reaction mechanisms reveal that, the effect of the ionic liquid on C-C bond forming reactions manifests itself not only in the energy lowering of polar transition states (or intermediates), involved in the catalytic cycles but also, depending on the cases, in the stabilization of palladium nanoparticles, in the synthesis of molecular Pd complexes with the ionic liquid anions or in the enhancement of the chemical reactivity of reactants, etc. The synergistic effect found by using appropriate mixtures of ionic liquids is also discussed.

Santos et al. [105] have reviewed Heck-Mizoroki reactions in ionic liquids, reported to be a possible green alternative to the classical cross-coupling procedure. In this review, the different approaches and achievements on the use of ionic liquids as solvents in Heck-Mizoroki coupling reactions is revisited along with a brief reference to supported ionic liquids. The results show that ionic liquids are able to modify the reaction course and activate and stabilize intermediates or transitions states. However, the examples show that, though there are large numbers of studies in this topic, most of the works involve the coupling of simple aryl halides and olefins and hence we expect more research on the use of ionic liquids in Heck reactions that include more complicated substrates affording environmentally benign chemical processes in the future.

Recently, Limberger et al. [106] have published a review on charge-tagged ligands, where there is an insertion of an ionic side chain into the molecular skeleton of a known ligand. This technique has become a useful protocol for anchoring ligands, and consequently catalysts, in polar and ionic liquid phases. The ionic modification confers a particular solubility profile making catalyst/product recovery possible, and often improving the activity of the catalytic species compared to the parent tag-free analogue. Usability of charge-tagged ligands is increasing and it has become a valuable tool in organometallic catalysis because in this, water can be used as an anchoring medium, thereby combining sustainability and efficiency. It is predicted that this technique can also be used to detect reaction intermediates in organometallic catalysis where the insertion of an ionic tag ensures the charge on the intermediates. Hence, these ligands have been used as ionic probes in mechanistic studies for several catalytic reactions. This review summarizes the selected examples on the use of charge-tagged ligands (CTLs) **187–190** (Figure 29) as immobilising agents in organometallic catalysis and as probes for studying mechanisms through ESI-MS (electrospray ionisation mass spectrometry). In these CTLs, a coordinating imidazole, pyrazole or pyridine group was attached to either C_2 or C_1 of the imidazolium moiety. These CTLs with palladium salts were found to be active for Heck reaction with the yields ranging from 70 to 99%. Moreover, it could be recycled several times and they can also be used as the solvent and ligand to stabilize the palladium species in Heck reaction.

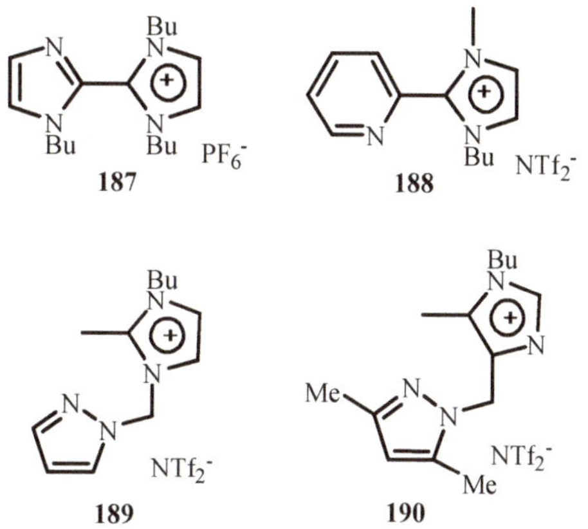

Figure 29. Charge-tagged ligands (CTLs) reviewed by Limberger et al. [106].

5.3.2. Catalysis in Multiphase Systems and Supercritical Carbon Dioxide

Bhanage et al. [107] have reviewed the multiphase Heck reactions using various types of catalysts. The multiphase Heck systems are prepared by using different types of catalyst phases, including biphasic catalysis, supported liquid phase catalysts and solvent of supercritical carbon dioxide which replaces the conventional organic solvents. These solvent systems were found to be beneficial towards easy catalyst-product separation and catalyst recycling. Supercritical carbon dioxide can serve as a better solvent system for Heck reaction under homogeneous, heterogeneous and multiphase systems provided the problem of leaching is solved.

A review by Skouta [108] presents an overview of some selected metal-catalysed chemical reactions developed in supercritical carbon dioxide, water, and ionic liquids. Heck reaction was found to give 55% yield of the coupling products in average when was performed in super critical CO_2. The water-soluble phosphine compounds m-TPPTC [tris(m-carboxyphenyl) phosphine trilithium salt], and p-TPPTC **191** (Figure 30) in addition to sulphonated phosphines like TPPTS **115**, etc., were efficient in organo-aqueous palladium-catalysed Heck reactions, and the excellent results obtained are presumably because of the steric and electronic effects of the carboxylic group in meta-position.

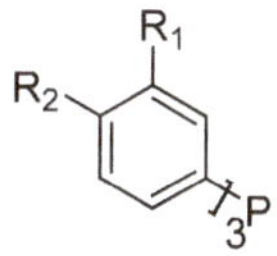

Figure 30. (tris(m-carboxyphenyl) phosphine trilithium salt (m-TPPTC)) and (p-TPPTC) (TPPTC: tris(m-carboxyphenyl) phosphine trilithium salt).

5.3.3. Catalysis in Continuous Flow

Gürsel et al. [109] gives an overview on the separation/recycling methods for homogeneous transition metal catalysts in continuous flow on the lab- and industrial scale by methods like heterogenization, scavenging, using biphasic systems and organic solvent nanofiltration. There are numerous successful demonstrations on the laboratory scale and industrial scale. Examples of Heck reaction using supported ionic liquid phase (SILP) with compressed CO_2 as the flowing medium in continuous flow; use of ionic liquid as the recycling reaction medium in an automated microreactor system catalysed by Pd catalyst immobilized in ionic liquid phase; in situ separation of the catalyst with the membrane within the reactor/separator unit, etc., are mentioned. All these methods have the advantage of ease of separation and low energy requirement compared to classical separation methods like distillation.

6. Summary

The Heck reaction has proved to be a remarkably robust and efficient method for carbon–carbon bond formation and remains a flourishing area of research. The enthusiasm and rational optimism of researchers promotes new discoveries resulting in better and efficient catalysts production and providing solutions for old problems. With the use of such reactions, many protection-deprotection procedures during organic synthesis may be reduced so as to reduce the number of steps when designed systematically in addition to the new reactivities, increasing product selectivity and reducing volatile organic consumption. The great functional group tolerance of palladium makes Heck reactions possible on even the most sensitive substrates. In addition, the intramolecular Heck reaction is an incredibly powerful method for the construction of many compounds with quaternary and/or asymmetric carbon centers. The well accepted mechanism for Heck reaction is based on intermediate species involving Pd(0)/Pd(II), however the feasibility of the Pd(II)/Pd(IV) mechanism is not ruled out by many. The Pd(II)/Pd(IV) mechanism is believed to take place when a Pd(0) to Pd(II) conversion is not available clearly or when the Pd(II) species cannot readily undergo β-hydride or reductive elimination. The prospects for application of Heck reaction for many transformations look very good via homogeneous, heterogeneous and even with heterogenized catalysis using organometallic complexes. Some of them are already scaled at an industrial level. Ironically, although a large number of metal-based catalysts are described in the literature promoting Heck reaction for the manufacture of bulk chemicals, when these reactions need to be applied for the synthesis of complex molecules like pharmaceuticals, agrochemicals or fragrances production, $Pd(OAc)_2$, $Pd_2(dba)_3$, Pd/C, $PdCl_2L_2$, $Pd(PPh_3)_4$ and $PdCl_2(dppf)_2$ still happen to be catalysts of choice.

Ample study has been made and yet a lot remains to be done. There are several challenges and expectations from the research community that need to be overcome in order to have an economic, robust and reliable industrial process, such as:

- A chase for catalyst precursor and ligands allowing high turnover numbers (TON) and turnover frequencies (TOF), making the process economically attractive, is still a challenge.
- Scaling from laboratory to production level needs a better and a cheaper catalyst precursor and ligands.
- Preparation of air and moisture stable catalyst, since the reactions easily get poisoned by molecular oxygen. In fact, preparation of such a catalyst that works better in aqueous media too.
- Standardizing the reaction conditions that are mild and also allowing lower catalyst loadings.
- Development of new generation of palladacycles promising to be excellent in future applications in industrial processes.
- Understanding of mechanism of heterogeneous catalysis is still rhetoric.
- Increased use of cheaper and easily available starting materials, especially chloroarenes and chloroalkenes, is a niche and much is needed to be done as far as finding newer protocols, along with catalyst and ligand design is concerned.
- Understanding the reasons and thereby successfully prohibiting the precipitation of metal.
- Development of the catalyst where cheaper first-row transition elements are used would truly add to the applicable research.
- Improvements in catalyst efficiency along with catalyst recycling, especially for use in pharmaceutical and other fine chemical synthetic processes.

- Solving the issue of contamination of the product at the end with palladium that needs further purification steps.

Thus, the ever-growing expansion of this coupling reaction is in the phase of new outlook that surely will enable more impressive accomplishments in the future. The useful and practical catalytic systems allowing the Heck reactions involving cheaper transition metals or the easy recovery of metal especially palladium will emerge by further research. This review aims at the stimulation of advance work in these directions.

Conflicts of Interest: The author declares no conflict of interest.

References

1. Heck, R.F. Acylation, methylation, and carboxyalkylation of olefins by group VIII metal derivatives. *J. Am. Chem. Soc.* **1968**, *90*, 5518–5526. [CrossRef]
2. Miyaura, N.; Yamada, K.; Suzuki, A. A new stereospecific cross-coupling by the palladium-catalyzed reaction of 1-alkenylboranes with 1-alkenyl or 1-alkynyl halides. *Tetrahedron Lett.* **1979**, *20*, 3437–3440. [CrossRef]
3. Sonogashira, K.; Tohda, Y.; Hagihara, N. A convenient synthesis of acetylenes: Catalytic substitutions of acetylenic hydrogen with bromoalkenes, iodoarenes and brompyridines. *Tetrahedron Lett.* **1975**, *16*, 4467–4470. [CrossRef]
4. King, A.O.; Okukado, N.; Negishi, E. Highly general stereo-, regio-, and chemo-selective synthesis of terminal and internal conjugated enynes by the Pd-catalyzed reaction of alkynylzinc reagents with alkenyl halides. *J. Chem. Soc. Chem. Commun.* **1977**, 683–684. [CrossRef]
5. Tamao, K.; Sumitani, K.; Kumada, M. Selective carbon-carbon bond formation by cross-coupling of Grignard reagents with organic halides. Catalysis by nickel-phosphine complexes. *J. Am. Chem. Soc.* **1972**, *94*, 4374–4376. [CrossRef]
6. Milstein, D.; Stille, J.K. A general, selective, and facile method for ketone synthesis from acid chlorides and organotin compounds catalyzed by palladium. *J. Am. Chem. Soc.* **1978**, *100*, 3636–3638. [CrossRef]
7. Trost, B.M.; Fullerton, T.J. New synthetic reactions. Allylic alkylation. *J. Am. Chem. Soc.* **1973**, *95*, 292–294. [CrossRef]
8. Mizoroki, T.; Mori, K.; Ozaki, A. Arylation of olefin with aryl iodide catalyzed by palladium. *Bull. Chem. Soc. Jpn.* **1971**, *44*, 581. [CrossRef]
9. Heck, R.F.; Nolley, J.P. Palladium-catalyzed vinylic hydrogen substitution reactions with aryl, benzyl, and styryl halides. *J. Org. Chem.* **1972**, *37*, 2320–2322. [CrossRef]
10. Heck, R.F. New applications of palladium in organic syntheses. *Pure Appl. Chem.* **1978**, *50*, 691–701. [CrossRef]
11. Heck, R.F. Palladium-catalyzed reactions with olefins. *Acc. Chem. Res.* **1979**, *12*, 146–151. [CrossRef]
12. Heck, R.F. Palladium catalysed vinylation of organic halides. *Org. React.* **1982**, *27*, 345–390. [CrossRef]
13. De Meijere, I.A.; Meyer, F.E. Fine feathers make fine birds: The Heck reaction in modern garb. *Angew. Chem. Int. Ed. Engl.* **1994**, *33*, 2379–2411. [CrossRef]
14. Kobetić, R.; Biliškov, N. The Heck reaction—A powerful tool in contemporary organic synthesis. *Chem. Ind.* **2007**, *56*, 391–402.
15. Sahu, M.; Sapkale, P. A review on palladium catalyzed coupling reactions. *Int. J. Pharm. Chem. Sci.* **2013**, *2*, 1159–1170.
16. Beletskaya, I.P.; Cheprakov, A.V. The Heck reaction as a sharpening stone of palladium catalysis. *Chem. Rev.* **2000**, *100*, 3009–3066. [CrossRef] [PubMed]
17. Hillier, A.C.; Grasa, G.A.; Viciu, M.S.; Lee, H.M.; Yang, C.; Nolan, S.P. Catalytic cross-coupling reactions mediated by palladium/nucleophilic carbene systems. *J. Organomet. Chem.* **2002**, *653*, 69–82. [CrossRef]
18. Herrmann, W.A. N-heterocyclic carbenes: A newconcept in organometallic catalysis. *Angew. Chem. Int. Ed.* **2002**, *41*, 1290–1309. [CrossRef]
19. Herrmann, W.A.; Öfele, K.; Preysing, D.V.; Schneider, S.K. Phospha-palladacycles and N-heterocyclic carbene palladium complexes: Efficient catalysts for C-C-coupling reactions. *J. Organomet. Chem.* **2003**, *687*, 229–248. [CrossRef]

20. Beletskaya, I.P.; Cheprakov, A.V. Palladacycles in catalysis—A critical survey. *J. Organomet. Chem.* **2004**, *689*, 4055–4088. [CrossRef]
21. Zafar, M.N.; Mohsin, M.A.; Danish, M.; Nazar, M.F.; Murtaza, S. Palladium catalyzed Heck-Mizoroki and Suzuki–Miyaura coupling reactions. *Russ. J. Coord. Chem.* **2014**, *40*, 781–800. [CrossRef]
22. Herrmann, W.A.; Böhm, V.P.; Reisinger, C.P. Application of palladacycles in Heck type reactions. *J. Organomet. Chem.* **1999**, *576*, 23–41. [CrossRef]
23. Farina, V. High-turnover palladium catalysts in cross-coupling and Heck chemistry: A critical overview. *Adv. Synth. Catal.* **2004**, *346*, 1553–1582. [CrossRef]
24. Whitecombe, N.I.; Hii, K.K.M.; Gibson, S.E. Advances in the Heck chemistry of aryl bromide and chlorides. *Tetrahedron* **2001**, *57*, 7449–7476. [CrossRef]
25. Littke, A.; Fu, G. Palladium-Catalyzed Coupling Reactions of Aryl Chlorides. *Angew. Chem. Int. Ed.* **2002**, *41*, 4176–4211. [CrossRef]
26. Zapf, A.; Beller, M. The development of efficient catalysts for palladium-catalyzed coupling reactions of aryl halides. *Chem. Commun.* **2005**, *28*, 431–440. [CrossRef] [PubMed]
27. Franzén, R. The Suzuki, the Heck, and the Stille reaction—Three versatile methods for the introduction of new C-C bonds on solid support. *Can. J. Chem.* **2000**, *78*, 957–962. [CrossRef]
28. Biffis, A.; Zecca, M.; Basato, M. Palladium metal catalysts in Heck C-C coupling reactions. *J. Mol. Catal. A Chem.* **2001**, *173*, 249–274. [CrossRef]
29. Cini, E.; Petricci, E.; Taddei, M. Pd/C Catalysis under Microwave Dielectric Heating. *Catalysts* **2017**, *7*, 89. [CrossRef]
30. Barnard, C. Palladium-catalysed C-C Coupling: Then and now. *Platinum Met. Rev.* **2008**, *52*, 38–45. [CrossRef]
31. Corbet, J.-P.; Mignani, G. Selected patented cross-coupling reaction technologies. *Chem. Rev.* **2006**, *106*, 2651–2710. [CrossRef] [PubMed]
32. Yin, L.; Liebscher, J. Carbon-carbon coupling reactions catalyzed by heterogeneous palladium catalysts. *Chem. Rev.* **2007**, *107*, 133–173. [CrossRef] [PubMed]
33. Narayanan, R. Recent advances in noble metal nanocatalysts for Suzuki and Heck cross-coupling reactions. *Molecules* **2010**, *15*, 2124–2138. [CrossRef] [PubMed]
34. Cai, S.; Wang, D.; Niu, Z.; Li, Y. Progress in organic reactions catalyzed by bimetallic nanomaterials. *Chin. J. Catal.* **2013**, *34*, 1964–1974. [CrossRef]
35. Baboo, R. Multimetallic nanomaterial based catalysis. *Int. J. Curr. Res. Acad. Rev.* **2015**, *3*, 153–157.
36. Labulo, A.H.; Martincigh, B.S.; Omondi, B.; Nyamori, V.O. Advances in carbon nanotubes as efficacious supports for palladium-catalysed carbon-carbon cross-coupling reactions. *J. Mater. Sci.* **2017**, *52*, 9225–9248. [CrossRef]
37. Moritanl, I.; Fujiwara, Y. Aromatic substitution of styrene-palladium chloride complex. *Tetrahedron Lett.* **1967**, *8*, 1119–1122. [CrossRef]
38. Gligorich, K.M.; Sigman, M.S. Recent advancements and challenges of palladiumII-catalyzed oxidation reactions with molecular oxygen as the sole oxidant. *Chem. Commun. (Camb.)* **2009**, *26*, 3854–3867. [CrossRef] [PubMed]
39. Karimi, B.; Behzadnia, H.; Elhamifar, D.; Akhavan, P.F.; Esfahani, F.K.; Zamani, A. Transition-metal-catalyzed oxidative Heck reactions. *Synthesis* **2010**, 1399–1427. [CrossRef]
40. Su, Y.; Jiao, N. Palladium-catalyzed oxidative Heck reaction. *Curr. Org. Chem.* **2011**, *15*, 3362–3388. [CrossRef]
41. Le Bras, J.; Muzart, J. Intermolecular dehydrogenative Heck reactions. *Chem. Rev.* **2011**, *111*, 1170–1214. [CrossRef] [PubMed]
42. Le Bras, J.; Muzart, J. Dehydrogenative Heck Annelations of Internal Alkynes. *Synthesis* **2014**, *46*, 1555–1572. [CrossRef]
43. Lee, A.L. Enantioselective oxidative boron Heck reactions. *Org. Biomol. Chem.* **2016**, *14*, 5357–5366. [CrossRef] [PubMed]
44. Xiaomei, Y.; Sha, M.; Yanni, D.; Yunhai, T. Progress in reductive Heck reaction. *Chin. J. Org. Chem.* **2013**, *33*, 2325–2333. [CrossRef]
45. Guiry, P.J.; Kiely, D. The Development of the intramolecular asymmetric Heck Reaction. *Curr. Org. Chem.* **2004**, *8*, 781–794. [CrossRef]
46. Oestreich, M. Breaking News on the Enantioselective Intermolecular Heck Reaction. *Angew. Chem. Int. Ed.* **2014**, *53*, 2282–2285. [CrossRef] [PubMed]

47. Kikukawa, K.; Matsuda, T. Reaction of Diazonium Salts with Transition Metals. I. Arylation of Olefins with Arenediazonium Salts Catalyzed by Zero Valent Palladium. *Chem. Lett.* **1977**, *2*, 159–162. [CrossRef]
48. Sato, Y.; Sodeoka, M.; Shibasaki, M. Catalytic asymmetric carbon-carbon bond formation: Asymmetric synthesis of cis-decalin derivatives by palladium-catalyzed cyclization of prochiral alkenyl iodides. *J. Org. Chem.* **1989**, *54*, 4738–4739. [CrossRef]
49. Carpenter, N.E.; Kucera, D.J.; Overman, L.E. Palladium-catalyzed polyene cyclizations of trienyl triflates. *J. Org. Chem.* **1989**, *54*, 5846–5848. [CrossRef]
50. Dounay, A.B.; Overman, L.E. The asymmetric intramolecular Heck reaction in natural product total synthesis. *Chem. Rev.* **2003**, *103*, 2945–2963. [CrossRef] [PubMed]
51. Diéguez, M.; Pàmies, O.; Claver, C. Ligands derived from carbohydrates for asymmetric catalysis. *Chem. Rev.* **2004**, *104*, 3189–3216. [CrossRef] [PubMed]
52. McCartney, D.; Guiry, P.J. The asymmetric Heck and related reactions. *Chem. Soc. Rev.* **2011**, *40*, 5122–5150. [CrossRef] [PubMed]
53. Daves, G.D.; Hallberg, A. 1,2-Additions to heteroatom-substituted olefins by organopalladium reagents. *Chem. Rev.* **1989**, *89*, 1433–1445. [CrossRef]
54. Bedford, R.B. Palladacyclic catalysts in C-C and C-heteroatom bond-forming reactions. *Chem. Commun.* **2003**, 1787–1796. [CrossRef]
55. Jeffery, T. On the efficiency of tetraalkylammonium salts in Heck type reactions. *Tetrahedron* **1996**, *52*, 10113–10130. [CrossRef]
56. Tucker, C.E.; de Vries, J.G. Homogeneous catalysis for the production of fine chemicals. Palladium- and nickel-catalysed aromatic carbon-carbon bond formation. *Top. Catal.* **2002**, *19*, 111–118. [CrossRef]
57. Wang, S.S.; Yang, G.Y. Recent developments in low-cost TM-catalyzed Heck-type reactions (TM = transition metal, Ni, Co, Cu, and Fe). *Catal. Sci. Technol.* **2016**, *6*, 2862–2876. [CrossRef]
58. Alonso, F.; Beletskaya, I.P.; Yus, M. Non-conventional methodologies for transition-metal catalysed carbon-carbon coupling: A critical overview. Part 1: The Heck reaction. *Tetrahedron* **2005**, *61*, 11771–11835. [CrossRef]
59. Tietze, L.F.; Ila, H.; Bell, H.P. Enantioselective palladium-catalyzed transformations. *Chem. Rev.* **2004**, *104*, 3453–3516. [CrossRef] [PubMed]
60. Shibasaki, M.; Vogl, E.M.; Ohshima, T. Asymmetric Heck reaction. *Adv. Synth. Catal.* **2004**, *346*, 1533–1552. [CrossRef]
61. Oestreich, M. Neighbouring-group effects in Heck reactions. *Eur. J. Org. Chem.* **2005**, 783–792. [CrossRef]
62. Genet, J.P.; Savignac, M. Recent developments of palladium(0) catalyzed reactions in aqueous medium. *J. Organomet. Chem.* **1999**, *576*, 305–317. [CrossRef]
63. Li, C. Organic reactions in aqueous media with a focus on carbon-carbon bond formations: A decade update. *Chem. Rev.* **2005**, *105*, 3095–3165. [CrossRef] [PubMed]
64. Velazquez, H.D.; Verpoort, F. N-heterocyclic carbene transition metal complexes for catalysis in aqueous media. *Chem. Soc. Rev.* **2012**, *41*, 7032–7060. [CrossRef] [PubMed]
65. Hervé, G.; Len, C. Heck and Sonogashira couplings in aqueous media—Application to unprotected nucleosides and nucleotides. *Sustain. Chem. Process.* **2015**, *3*, 3. [CrossRef]
66. Cabri, W.; Candiani, I. Recent developments and new perspectives in the Heck reaction. *Acc. Chem. Res.* **1995**, *28*, 2–7. [CrossRef]
67. Crisp, G.T. Variations on a theme—Recent developments on the mechanism of the Heck reaction and their implications for synthesis. *Chem. Soc. Rev.* **1998**, *27*, 427–436. [CrossRef]
68. Amatore, C.; Carre, E.; Jutand, A.; M'Barki, M.A.; Meyer, G. Evidence for the Ligation of Palladium(0) Complexes by Acetate Ions: Consequences on the Mechanism of Their Oxidative Addition with Phenyl Iodide and PhPd(OAc)(PPh$_3$)$_2$ as Intermediate in the Heck Reaction. *Organometallics* **1995**, *14*, 5605–5614. [CrossRef]
69. Amatore, C.; Jutand, A. Mechanistic and kinetic studies of palladium catalytic systems. *J. Organomet. Chem.* **1999**, *576*, 254–278. [CrossRef]
70. Amatore, C.; Jutand, A. Anionic Pd(0) and Pd(II) intermediates in palladium-catalyzed Heck and cross-coupling reactions. *Acc. Chem. Res.* **2000**, *33*, 314–321. [CrossRef] [PubMed]

71. Cavell, K.J.; McGuinness, D.S. Redox processes involving hydrocarbylmetal (N-heterocyclic carbene) complexes and associated imidazolium salts: Ramifications for catalysis. *Coord. Chem. Rev.* **2004**, *248*, 671–681. [CrossRef]
72. Trzeciak, A.M.; Ziółkowski, J.J. Structural and mechanistic studies of Pd-catalyzed C-C bond formation: The case of carbonylation and Heck reaction. *Coord. Chem. Rev.* **2005**, *249*, 2308–2322. [CrossRef]
73. Phan, N.T.S.; DerSluys, M.V.; Jones, C.W. On the Nature of the Active Species in Palladium Catalyzed Mizoroki–Heck and Suzuki–Miyaura Couplings—Homogeneous or Heterogeneous Catalysis, A Critical Review. *Adv. Synth. Catal.* **2006**, *348*, 609–679. [CrossRef]
74. Knowles, J.P.; Whiting, A. The Heck–Mizoroki cross-coupling reaction: A mechanistic perspective. *Org. Biomol. Chem.* **2007**, *5*, 31–44. [CrossRef] [PubMed]
75. Kumar, A.; Rao, G.K.; Kumar, S.; Singh, A.K. Formation and Role of Palladium Chalcogenide and Other Species in Suzuki−Miyaura and Heck C-C Coupling Reactions Catalyzed with Palladium(II) Complexes of Organochalcogen Ligands: Realities and Speculations. *Organometallics* **2014**, *33*, 2921–2943. [CrossRef]
76. Eremin, D.B.; Ananikov, V.P. Understanding active species in catalytic transformations: From molecular catalysis to nanoparticles, leaching, "Cocktails" of catalysts and dynamic systems. *Coord. Chem. Rev.* **2017**, *346*, 2–19. [CrossRef]
77. Blaser, H.U.; Indolese, A.; Schnyder, A. Applied homogeneous catalysis by organometallic complexes. *Curr. Sci.* **2000**, *78*, 1336–1344.
78. De Vries, J.G. The Heck reaction in the production of fine chemicals. *Can. J. Chem.* **2001**, *79*, 1086–1092. [CrossRef]
79. Zapf, A.; Beller, M. Fine chemical synthesis with homogeneous palladium catalysts: Examples, status and trends. *Top. Catal.* **2002**, *19*, 101–109. [CrossRef]
80. Olovnikov, A.M. A theory of marginotomy. The incomplete copying of template margin in enzymic synthesis of polynucleotides and biological significance of the phenomenon. *J. Theor. Biol.* **1973**, *41*, 181–190. [CrossRef]
81. Zoeller, J.R.; Agreda, V.H.; Cook, S.L.; Lafferty, N.L.; Polichnowski, S.W.; Pond, D.M. Eastman Chemical Company Acetic Anhydride Process. *Catal. Today* **1992**, *13*, 73–91. [CrossRef]
82. Picquet, M. Organometallics as catalysts in the fine chemical industry. *Platinum Metals Rev.* **2013**, *57*, 272–280. [CrossRef]
83. Sharma, P. Cinnamic acid derivatives: A new chapter of various pharmacological activities. *J. Chem. Pharm. Res.* **2011**, *3*, 403–423.
84. Perkin, W.H. On the artificial production of coumarin and formation of its homologues. *J. Chem. Soc.* **1868**, *21*, 53–61. [CrossRef]
85. Biajoli, A.F.P.; Schwalm, C.; Limberger, J.; Claudino, T.S.; Monteiro, A.L. Recent progress in the use of Pd-catalyzed C-C cross-coupling reactions in the synthesis of pharmaceutical compounds. *J. Braz. Chem. Soc.* **2014**, *25*, 2186–2214. [CrossRef]
86. Magano, J.; Dunetz, J.R. Large-scale applications of transition metal-catalyzed couplings for the synthesis of pharmaceuticals. *Chem. Rev.* **2011**, *111*, 2177–2250. [CrossRef] [PubMed]
87. Torborg, C.; Beller, M. Recent applications of palladium-catalyzed coupling reactions in the pharmaceutical, agrochemical, and fine chemical industries. *Adv. Synth. Catal.* **2009**, *351*, 3027–3043. [CrossRef]
88. Nicolaou, K.C.; Bulger, P.G.; Sarlah, D. Palladium-catalyzed cross-coupling reactions in total synthesis. *Angew. Chem. Int. Ed.* **2005**, *44*, 4442–4489. [CrossRef] [PubMed]
89. Jeffery, T. Heck-type reactions in water. *Tetrahedron Lett.* **1994**, *35*, 3051–3054. [CrossRef]
90. Trost, B.M.; Shi, Y. A palladium-catalyzed zipper reaction. *J. Am. Chem. Soc.* **1991**, *113*, 701–703. [CrossRef]
91. Trost, B.M. Atom economy—A challenge for organic synthesis: Homogeneous catalysis leads the way. *Angew. Chem. Int. Ed. Engl.* **1995**, *34*, 259–281. [CrossRef]
92. Wellington, K.W.; Benner, S.A. A review: Synthesis of aryl C-glycosides via the Heck coupling reaction. *Nucleotides Nucleic Acids* **2006**, *25*, 1309–1333. [CrossRef] [PubMed]
93. Zeni, G.; Larock, R.C. Synthesis of heterocycles via palladium-catalyzed oxidative addition. *Chem. Rev.* **2006**, *106*, 4644–4680. [CrossRef] [PubMed]
94. Bhanage, B.M.; Arai, M. Catalyst product separation techniques in Heck reaction. *Catal. Rev.* **2001**, *43*, 315–344. [CrossRef]
95. Wall, V.M.; Eisenstadt, A.; Ager, D.J.; Laneman, S.A. The Heck reaction and cinnamic acid synthesis by heterogeneous catalysis. *Platinum Met. Rev.* **1999**, *43*, 138–145.

96. Van Heerbeek, R.; Kamer, P.C.; van Leeuwen, P.W.; Reek, J.N. Dendrimers as support for recoverable catalysts and reagents. *Chem. Rev.* **2002**, *102*, 3717–3756. [CrossRef] [PubMed]
97. Horn, J.; Michalek, F.; Tzschucke, C.C.; Bannwarth, W. Non-covalently solid-phase bound catalysts for organic synthesis. *Top. Curr. Chem.* **2004**, *242*, 43–75. [CrossRef] [PubMed]
98. Köhler, K.; Pröckl, S.S.; Kleist, W. Supported palladium catalysts in *Heck* coupling reactions—Problems, potential and recent advances. *Curr. Org. Chem.* **2006**, *10*, 1585–1601. [CrossRef]
99. Polshettiwar, V.; Molnár, A. Silica-supported Pd catalysts for Heck coupling reactions. *Tetrahedron* **2007**, *63*, 6949–6976. [CrossRef]
100. Molnar, A. Efficient, selective, and recyclable palladium catalysts in carbon-carbon coupling reactions. *Chem. Rev.* **2011**, *111*, 2251–2320. [CrossRef] [PubMed]
101. Sheldon, R. Catalytic reactions in ionic liquids. *Chem. Commun.* **2001**, 2399–2407. [CrossRef]
102. Singh, R.; Sharma, M.; Mamgain, R.; Rawat, D.S. Ionic liquids: A versatile medium for palladium-catalyzed reactions. *J. Braz. Chem. Soc.* **2008**, *19*, 357–379. [CrossRef]
103. Bellina, F.; Chiappe, C. The Heck reaction in ionic liquids: Progress and challenges. *Molecules* **2010**, *15*, 2211–2245. [CrossRef] [PubMed]
104. Mastrorilli, P.; Monopoli, A.; Dell'Anna, M.M.; Latronico, M.; Cotugno, P.; Nacci, A. Ionic liquids in palladium-catalyzed cross-coupling reactions. *Top. Organomet. Chem.* **2013**, *51*, 237–285. [CrossRef]
105. Santos, C.I.; Barata, J.F.; Faustino, M.A.F.; Lodeiro, C.; Neves, M.G.P. Revisiting Heck–Mizoroki reactions in ionic liquids. *RSC Adv.* **2013**, *3*, 19219–19238. [CrossRef]
106. Limberger, J.; Leal, B.C.; Monteiro, A.L. Charge-tagged ligands: Useful tools for immobilising complexes and detecting reaction species during catalysis. *Chem. Sci.* **2015**, *6*, 77–94. [CrossRef] [PubMed]
107. Bhanage, B.M.; Fujita, S.I.; Arai, M. Heck reactions with various types of palladium complex catalysts: Application of multiphase catalysis and supercritical carbon dioxide. *J. Organomet. Chem.* **2003**, *687*, 211–218. [CrossRef]
108. Skouta, R. Selective chemical reactions in supercritical carbon dioxide, water, and ionic liquids. *Green Chem. Lett. Rev.* **2009**, *2*, 121–156. [CrossRef]
109. Gürsel, I.V.; Noël, T.; Wang, Q.; Hessel, V. Separation/recycling methods of homogeneous transition metal catalysts in continuous flow. *Green Chem.* **2015**, *17*, 2012–2026. [CrossRef]

© 2017 by the author. Licensee MDPI, Basel, Switzerland. This article is an open access article distributed under the terms and conditions of the Creative Commons Attribution (CC BY) license (http://creativecommons.org/licenses/by/4.0/).

Communication

Palladium-Catalyzed Regioselective Alkoxylation via C-H Bond Activation in the Dihydrobenzo[*c*]acridine Series

Benjamin Large, Flavien Bourdreux, Aurélie Damond, Anne Gaucher and Damien Prim *

Institut Lavoisier de Versailles, UVSQ, CNRS, Université Paris-Saclay, 78035 Versailles, France; benjamin.large@uvsq.fr (B.L.); flavien.bourdreux@uvsq.fr (F.B.); aurelie.damond@uvsq.fr (A.D.); anne.gaucher@uvsq.fr (A.G.)
* Correspondence: damien.prim@uvsq.fr; Tel.: +33-013-925-4455

Received: 12 March 2018; Accepted: 29 March 2018; Published: 31 March 2018

Abstract: 5,6-Dihydrobenzo[*c*]acridine belongs to the large aza-polycyclic compound family. Such molecules are not fully planar due to the presence of a partially hydrogenated ring. This paper describes the first Pd-catalyzed alkoxylation via C-H bond activation of variously substituted 5,6-dihydrobenzo[*c*]acridines. We determined suitable conditions to promote the selective formation of C-O bonds using 10% Pd(OAc)$_2$, PhI(OAc)$_2$ (2 eq.) and MeOH as the best combination of oxidant and solvent, respectively. Under these conditions, 5,6-dihydrobenzo[*c*]acridines bearing substituents at both rings A and D were successfully functionalized, giving access to polysubstitutited acridine motifs.

Keywords: C-H activation; palladium; alkoxylation; dihydrobenzo[*c*]acridine

1. Introduction

Acridines and related derivatives represent an important class of aza-polycyclic compounds that have attracted considerable interest in the last century because of their broad range of properties and applications. For example, acridines are well-known as antibacterial, antimalarial, and anticancer agents [1–3], and have also been used in pigments, dyes, and sensor devices for decades [4]. More recently, acridine motifs have found additional applications such as cell imaging probes [5], catalysis [6], Organic Light-Emitting Diodes (OLEDs) [7] and organic semiconductors [8]. The modulation and/or enhancement of these properties drove the development of synthetic methodologies to (1) construct the acridine backbone; (2) selectively install substituents; (3) modulate the substitution pattern, fusing additional rings towards extended molecules; and (4) induce distortion from planarity by including a partially saturated fragment [9–17].

In this context, 5,6-dihydrobenzo[*c*]acridine is an intriguing member of the large aza-polycyclic compound family. Indeed, 5,6-dihydrobenzo[*c*]acridine is a tetracyclic molecule comprising four fused cycles including one pyridine ring and one partially hydrogenated cycle (Figure 1). In fact, this molecule represents a mix of bicyclic quinoline and tricyclic benzo[*h*]quinoline or acridine scaffolds. Within the structure of 5,6-dihydrobenzo[*c*]acridine, the presence of a nitrogen atom, an additional condensed ring, and a cyclic "dihydro" fragment provides its originality and interest by comparison with the parent and fully aromatic quinoline or (benzo)acridine skeletons. Moreover, the joint presence of a nitrogen atom and an peri-fused aromatic ring defines an aza-bay region and the presence of the non-planar ethylene bridge induces a deviation from the planarity compared with fully aromatic analogues. This deviation from planarity has been exploited recently in the preparation of helical-shaped molecules and the evaluation of photophysical and magnetic properties of helicate-like ligands [18–20].

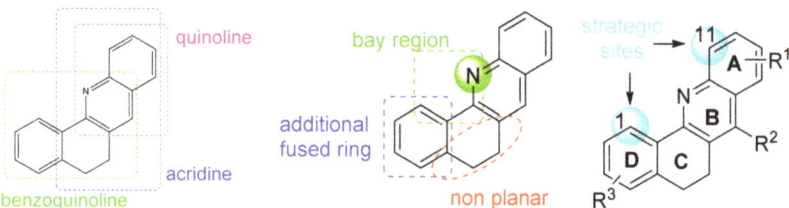

Figure 1. Molecular structure, characteristic structural features, and strategic site of 5,6-dihydrobenzo[c]acridine.

As already mentioned, installation of substitution at the 5,6-dihydrobenzo[c]acridine platform proved to be crucial to modulate both properties and shape of 3D-shaped molecular architectures. 5,6-Dihydrobenzo[h]acridine targets are usually obtained by two main routes using (1) Friedländer cyclisation between tetralone derivatives and o-aminoacetophenones [18] or (2) thermally-induced or acid-catalyzed cyclisation of 1-halovinyl-2-carboxaldehyde derivatives and anilines [14,15,20–22]. In both methodologies, substituents on rings A, C, and D usually arise from commercially available starting reactants. Post-functionalization of the acridine motif at both strategic sites (positions 1 and 11 in Figure 1) is more challenging. As already described, only the presence of bromide or iodide atoms at both positions allows metal-catalyzed installation of substituents. As examples, the formation of C-C and C-N bonds was achieved using Pd- and Cu-catalyzed strategies, respectively, from precursor bearing an iodide atom at position 11 [21] and the formation of homocoupling products was realized using the Cu-catalyzed Ullman reaction from precursor bearing a bromide atom at position 1 [18,19]. The development of methodologies that avoid the mandatory presence of halides represents a challenging alternative in the 5,6-dihydrobenzo[c]acridine series. In deep contrast to the fully aromatic benzo[h]quinoline derivatives where C-H activation and the formation of the corresponding metallacycles (Ru, Pd, Ir) at position 1 are well known and documented [23–28], C-H activation in the 5,6-dihydrobenzo[c]acridine series is scarcely reported [29]. In the dihydro analogues, the crucial point was whether the distortion from planarity due to the presence of the partially hydrogenated ring C would allow or hamper the transient palladacycle to form through C-H activation and the selective installation of substituents at position 1. In this communication, we disclose our preliminary results in the Pd-catalyzed alkoxylation via C-H bond activation within the 5,6-dihydrobenzo[c]acridine series (Figure 2).

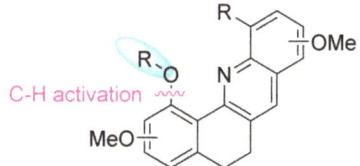

Figure 2. Polysubstituted 5,6-dihydrobenzo[c]acridines via C-H bond activation.

2. Results and Discussion

First we focused on the preparation of variously substituted acridines (Scheme 1). Based on our previous reports [15,20,21], we envisioned the synthesis of the acridine platforms from 1-chlorovinylcarboxaldehydes **1–3** (see supplementary material). The latter are readily obtained from commercially available tetralones and Vilsmeier–Haack reagent in high to quantitative yields [14,15,30]. Compounds **1–3** were reacted with 2.5 eq. of aniline derivatives in iPrOH at 90 °C for 16 h to yield acridines **4** to **7** in yields ranging from 47 to 60%. We chose a combination of various anilines including aniline, p-anisidine, and o-toluidine, and substituted 1-chlorovinylcarboxaldehydes **1–3** in order to prepare acridines which display a different substitution pattern. Indeed, as shown in Scheme 1, acridines **5** and **6** are substituted at ring D in the 2 and 3 positions, respectively. In contrast, in acridines **7** and **8**, substituents are located at ring A in the 9 and 11 positions, respectively.

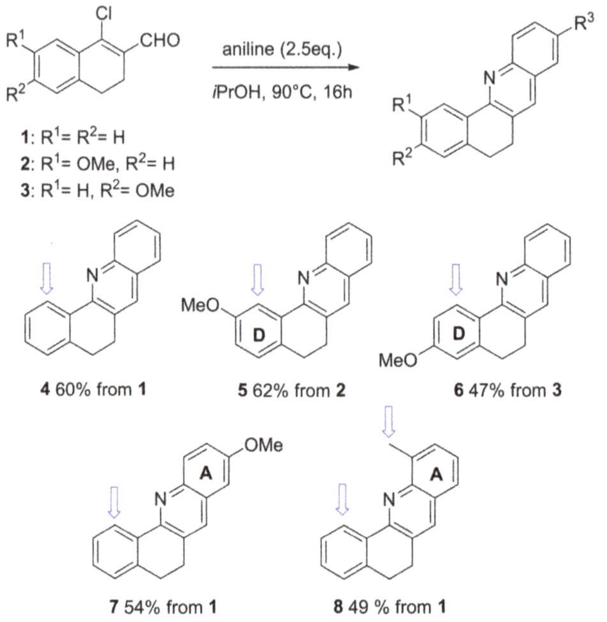

Scheme 1. General route towards variously substituted acridines **4–8**. In blue C-H activation sites.

With acridines **4** to **8** in hand, our next goal was to examine the C-H activation step. As represented in scheme 1, the acridine platform is expected to undergo cyclometallation at the 1 position in good agreement with previous reports [23–28] dealing with the fully aromatic benzo[*h*]quinoline analogues. Thus Pd-catalyzed alkoxylation should take place in the 1 position for substrates **4** to **7**. In contrast, acridine **8** is a more challenging substrate which displays two potential reaction sites: the sp^2 carbon atom located in the 1 position at ring D and the sp^3 benzylic carbon atom located at ring A. Indeed, both C-H bond might afford a five membered cyclometallated adduct [31] and thus undergo subsequent alkoxylation.

Substrate **4** was selected for initial investigation because it presents one single bond C-H(1) for directed C-H activation. Various Pd-based conditions were tested in order to determine suitable catalytic combination for the alkoxylation reaction.

We found that Pd(OAc)$_2$ (10%) was effective to obtain 1-methoxy-5,6-dihydrobenzo[*c*]acridine **9** in 82% yield (Scheme 2). Among several oxidants tested, PhI(OAc)$_2$ (2 eq.) proved superior to I$_2$ or oxone. The solvent was also a crucial parameter to ensure high conversion. Indeed, only the use of MeOH at 100 °C in a sealed tube gave the expected alkoxy acridine **9** in high yield. Decreasing the temperature even to 80 °C led to a severe decrease of conversion. Mixtures of solvents such as dioxane/MeOH similarly afforded poor conversion. The use of dichloroethane (DCE)/ MeOH as the solvent led to mixtures of 1-methoxy and 1-chloro derivatives **9** and **10**. The formation of the C-Cl bond could be unambiguously evidenced when the reaction was realized in DCE without the presence of MeOH. In this case, compound **10** was isolated in 63% yield. Moving from MeOH to EtOH and *i*PrOH led to different issues. If the use of EtOH afforded the ethoxy analogue **11** in satisfactory 61% yield, *i*PrOH failed to react.

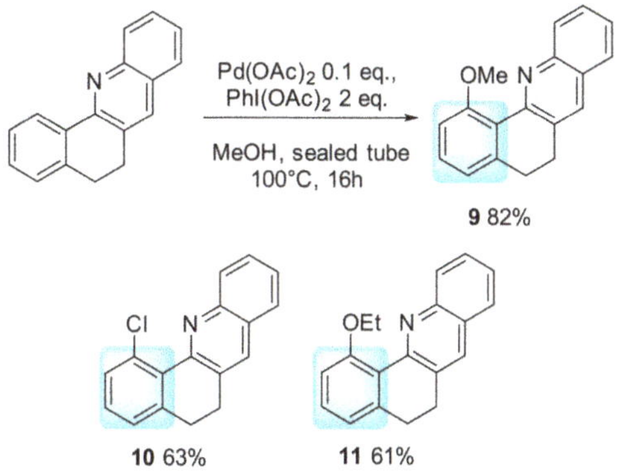

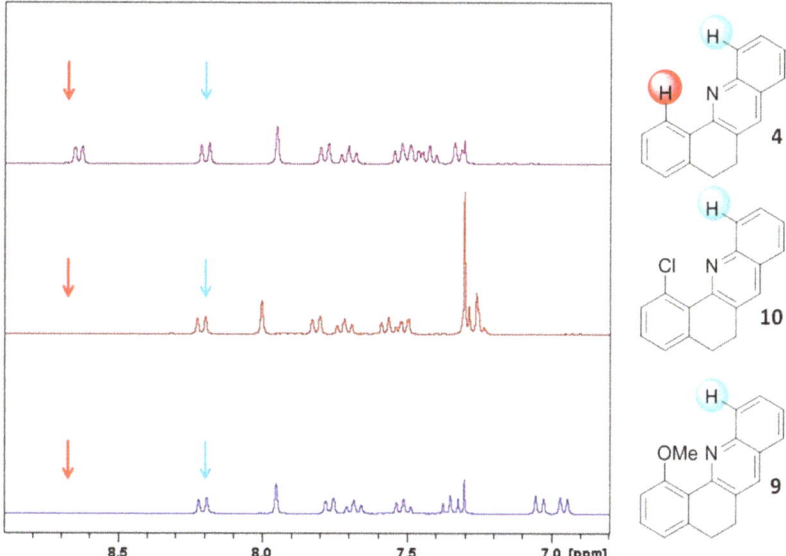

Scheme 2. Determination of best experimental conditions for C-H activation of acridine 4.

^{1}H NMR analysis of crude products allows an easy identification of both reactants and products. Indeed, except for compound 5, all other acridines 4, 6–8 display characteristic chemical shifts for H(1) ranging from 8.55 to 8.70 ppm. ^{1}H NMR of compound 4 shows two characteristic signals at 8.60 and 8.30 ppm, accounting for protons H(1) and H(11). As evidenced by Figure 3, alkoxylation or chlorination is selective at position 1 of the dihydrobenzo[c]acridine platform. Indeed, only H(11) remains unchanged in both cases.

Figure 3. ^{1}H NMR characteristic signals for H(1) – red arrow and H(11) – blue arrow acridine derivatives 4, 9, and 10.

Based on mechanistic studies reported by Sandford [31] on related 2-phenylpyridines and benzo[h]quinoline, a potential catalytic cycle is shown in Figure 4. The latter would involve successively a ligand-directed C-H activation to form a cyclometallated dimer, oxidation to generate a Pd(IV) species, and a release of the product after C-O bond-forming reductive elimination. The number and role of other ancillary ligands remain under investigation and are represented as sticks in Figure 4.

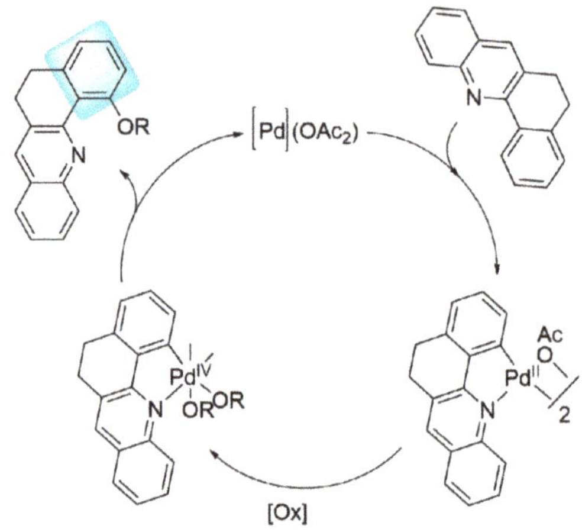

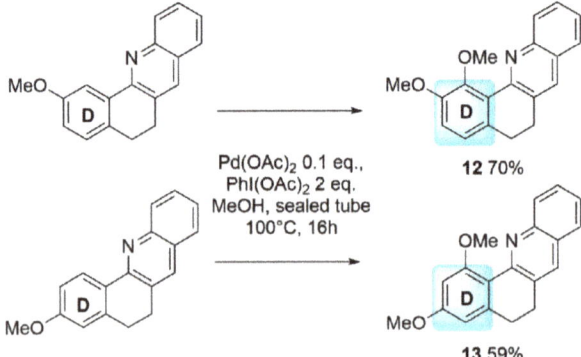

Figure 4. Possible mechanism for palladium-catalyzed regioselective alkoxylation.

The last step might proceed either by intramolecular C-OR bond elimination from the metal center or by attack of an external nucleophile in an "SN$_2$-like" reaction as suggested recently [23]. The in situ transformation of PhI(OAc)$_2$ with alcoholic solvents to afford PhI(OR)$_2$ is also suggested as a key step in alkoxylation reactions which account for the obtention C-O bonds [32].

Thus Pd(OAc)$_2$, PhI(OAc)$_2$, in methanol(or ethanol) at 100 °C afforded suitable conditions to promote alkoxylation in the dihydrobenzo[c]acridine series. With this conditions in hand, we next tried to install an additional methoxy group when ring D is already bearing a methoxy substituent, in order to prepare 1,2- and 1,3-bismethoxy acridine motifs (Scheme 3). Under the aforementioned conditions, acridines **12** and **13** were readily obtained in 70 and 59% isolated yield, respectively. Thus, the presence of a strong donating group at ring D does not hamper the C-H activation and the subsequent C-O bond formation.

Scheme 3. Preparation of 1,2- and 1,3-bismethoxy acridine motifs **12** and **13**.

Under similar catalytic conditions, acridine **7** afforded the expected bismethoxy derivative **14** in 81% yield (Scheme 4). The latter compound displays complementary substitution pattern by comparison with acridines **12** and **13**. In this case both rings A and D are independently functionalized.

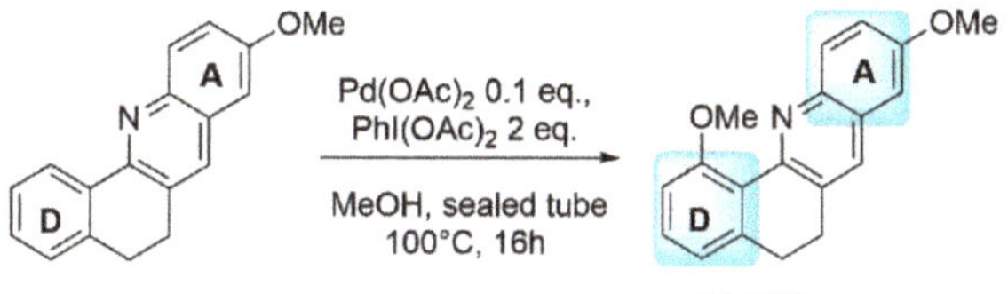

Scheme 4. Substitution at rings A and D of the acridine platform.

Finally, we decided to test our aforementioned C-H activation conditions in acridine **8**. In contrast to acridine substrates **4–7**, compound **8** displays two different C-H activation sites. Each of them might produce a transient five-membered palladacycle through C-H activation and might allow alkoxylation (Scheme 5). Unfortunately, under the aforementioned conditions in MeOH at 100 °C using one or two equivalents of oxidant, complex mixtures of alkoxylated products were obtained. In contrast, moving from MeOH to AcOH and using one equivalent of PhI(OAc)₂ allowed to isolate acridine **15** as the major product in 51% isolated yield. ¹HNMR spectra evidenced the presence of the characteristic signal of H(1), which resonates at 8.54 ppm.

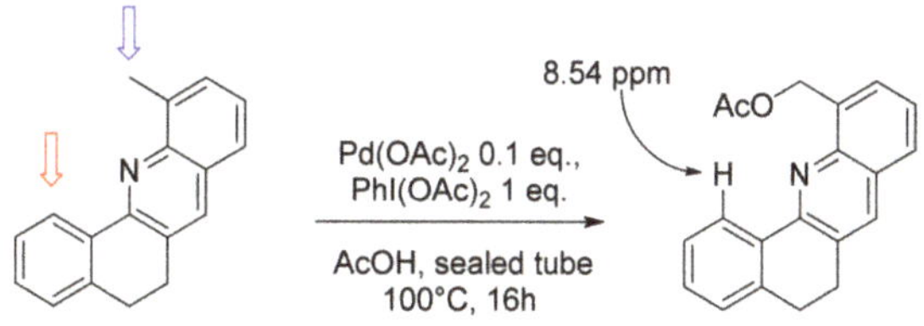

Scheme 5. Selective C-H activation at the benzylic site.

3. Conclusions

In conclusion, we succeeded in the alkoxylation of 5,6-dihydrobenzo[*h*]acridine via Pd-catalyzed C-H activation. Alkoxylation occurs selectively in position 1 of the acridine platform using 10% Pd(OAc)₂, PhI(OAc)₂ and MeOH as the best combination of catalyst, oxidant and solvent, respectively. Several bismethoxy acridine derivatives bearing all substituents at ring D or at both ring D and A have been successfully obtained. Our strategy allowed a selective functionalization of sp³ carbon atom located at the benzylic position of ring A. Current studies are focused on further exploration of the substrate scope and the extension of this methodology to the selective formation of C-C bonds via C-H activation at 5,6-dihydrobenzo[*h*]acridine architectures.

4. Materials and Methods

4.1. General Information

All reagents and solvents were obtained from commercial sources and used without further purification. Reactions were routinely carried out under nitrogen and argon atmosphere with magnetic stirring. ¹H and ¹³C NMR spectra were recorded on a Bruker AV1 300 spectrometer (Bruker BioSpin GmbH, Rheinstetten, Germany) working at 300 MHz, 75 MHz respectively for ¹H and ¹³C, with chloroform-d as solvent. Chemicals shifts were reported in δ, parts per million (ppm), relative to chloroform (δ = 7.28 ppm) as international standards unless otherwise stated for proton nuclear magnetic resonance (¹H NMR). Chemical shifts for carbon nuclear magnetic resonance (¹³C NMR) were reported in δ, parts per million (ppm), relative to the center line of the chloroform triplet (δ = 77.07 ppm). Coupling constants, *J*, were reported in Hertz (Hz) and refer to apparent peak multiplicities and not true coupling constants. The abbreviations s, d, dd, t, q, br and m stand for resonance multiplicities singlet, doublet, doublet of doublet, triplet, quartet, broad, and multiplet, respectively. Allylation diastereoselectivity was determined by ^{1}H NMR integrations of the methylene signals in the crude products. High resolution mass spectrometry data were recorded with an accuracy within 5 ppm on a quadrupole-TOF mass spectrometer (Xevo Q-Tof, Waters, Guyancourt, France) using an electrospray ionization source operating in positive mode. Thin-layer chromatography (TLC) was carried out on aluminum sheets pre-coated with silica gel plates (Fluka Kiesel gel 60 F254, Merck, Bucharest, Romania) and visualized by a 254 nm UV lamp and potassium permanganate. Melting points (Mp) were determined on a System Kofler type WME apparatus (Fisher Scientific SAS, Illkirch, France).

4.2. General Procedure for Pd-Catalyzed Alkoxylation

The acridine derivative (1 eq.), PhI(OAc)$_2$ (2 eq.), and Pd(OAc)$_2$ (0.1 eq.) were place in screw-capped tube. MeOH (3 mL) was next added and the reaction mixture was stirred for 15 min. The tube was sealed and the suspension was heated with stirring to 100 °C for 16 h. The crude mixture was filtered through Celite and the solvent evaporated. The solid residue was extracted between ethyl acetate and successively water and brine. The organic layers were dried over sodium sulfate and the solvent was removed under vacuum. In all cases the residue was purified by flash column chromatography on silica gel (petroleum ether/dichloromethane, 6:4) to afford the expected alkoxyacridine derivative.

Supplementary Materials: The following are available online at http://www.mdpi.com/2073-4344/8/4/139/s1, experimental procedure for the alkoxylation reaction couplings as well as analytical data for new compounds.

Acknowledgments: University of Versailles St Quentin, MENRT-France, and LabEx CHARMMMAT (ANR-11-LABEX-0039) are gratefully acknowledged for financial supports and grant (BL).

Author Contributions: Benjamin Large and Damien Prim performed the experiments, Anne Gaucher, Aurélie Damond and Flavien Bourdreux analyzed the data. Damien Prim wrote the paper.

Conflicts of Interest: The authors declare no conflict of interest.

References

1. Zhou, Y.-J.; Chen, D.-S.; Li, Y.-L.; Liu, Y.; Wang, X.-S. Combinatorial synthesis of pyrrolo[3,2-f]quinoline and pyrrolo[3,2-a]acridine derivatives via a three-component reaction under catalyst-free conditions. *ACS Comb. Sci.* **2013**, *15*, 498–502. [CrossRef] [PubMed]
2. Yu, X.-M.; Ramiandrasoa, F.; Guetzoyan, L.; Pradines, B.; Quintino, E.; Gadelle, D.; Forterre, P.; Cresteil, T.; Mahy, J.-M.; Pethe, S. Synthesis and biological evaluation of acridine derivatives as antimalarial agents. *ChemMedChem* **2012**, *7*, 587–605. [CrossRef] [PubMed]
3. Liu, F.; Suryadi, J.; Bierbach, U. Cellular recognition and repair of monofunctional-interactive platinum-DNA adducts. *Chem. Res. Toxicol.* **2015**, *28*, 2170–2178. [CrossRef] [PubMed]
4. Geddes, C.D. Optical thin film polymeric sensors for the determination of aqueous chloride, bromide and iodide ions at high pH, based on the quenching of fluorescence of two acridinium dyes. *Dyes Pigment.* **2000**, *45*, 243–251. [CrossRef]
5. Warther, D.; Bolze, F.; Leonard, J.; Gug, S.; Specht, A.; Puliti, D.; Sun, X.-H.; Kessler, P.; Lutz, Y.; Vonesch, J.-L.; et al. Live-cell one- and two-photon uncaging of a far-red Emitting acridinone fluorophore. *J. Am. Chem. Soc.* **2009**, *132*, 2585–2590. [CrossRef] [PubMed]
6. Sahoo, D.; Yoo, C.; Lee, Y. Direct CO$_2$ addition to a Ni(0)–CO species allows the selective generation of a Nickel(II) carboxylate with expulsion of CO. *J. Am. Chem. Soc.* **2018**, *140*, 2179–2185. [CrossRef]
7. Dos Santos, P.L.; Ward, J.S.; Bryce, M.R.; Monkman, A.P. Using guest–host interactions to optimize the efficiency of TADF Oleds. *J. Phys. Chem. Lett.* **2016**, *7*, 3341–3346. [CrossRef] [PubMed]
8. Goel, A.; Kumar, V.; Singh, S.P.; Sharma, A.; Prakash, S.; Singh, C.; Anand, R.S. Non-aggregating solvatochromic bipolar benzo[f]quinolines and benzo[a]acridines for organic electronics. *J. Mater. Chem.* **2012**, *22*, 14880–14888. [CrossRef]

9. Martins, A.P.; Frizzo, C.P.; Moreira, D.N.; Buriol, L.; Machado, P. Solvent-free heterocyclic synthesis. *Chem. Rev.* **2009**, *109*, 4140–4182. [CrossRef] [PubMed]
10. Wang, M.; Fan, Q.; Jiang, X. Nitrogen-iodine exchange of diaryliodonium salts: Access to acridine and carbazole. *Org. Lett.* **2018**, *20*, 216–219. [CrossRef] [PubMed]
11. Su, Q.; Li, P.; He, M.; Wu, Q.; Ye, L.; Mu, Y. Facile synthesis of acridine derivatives by $ZnCl_2$-promoted intramolecular cyclization of o-arylaminophenyl Schiff bases. *Org. Lett.* **2014**, *16*, 18–21. [CrossRef] [PubMed]
12. Wang, T.-J.; Chen, W.-W.; Li, Y.; Xu, M.-H. Facile synthesis of acridines via Pd(0)-diphosphine complex-catalyzed tandem coupling/cyclization protocol. *Org. Biomol. Chem.* **2015**, *13*, 6580–6586. [CrossRef] [PubMed]
13. Li, A.; Kindelin, P.J.; Klumpp, D.A. Charge migration in dicationic electrophiles and its application to the synthesis of aza-polycyclic aromatic compounds. *Org. Lett.* **2006**, *8*, 1233–1236. [CrossRef] [PubMed]
14. Karthikeyan, P.; Meena Rani, A.; Saiganesh, R.; Balasubramanian, K.K.; Kabilan, S. Synthesis of 5,6-dihydrobenz[c]acridines: A comparative study. *Tetrahedron* **2009**, *65*, 811–821. [CrossRef]
15. Souibgui, A.; Gaucher, A.; Marrot, J.; Bourdreux, F.; Aloui, F.; Ben Hassine, B.; Prim, D. New series of acridines and phenanthrolines: Synthesis and characterization. *Tetrahedron* **2014**, *70*, 3042–3048. [CrossRef]
16. Gogoi, S.; Shekarrao, K.; Duarah, A.; Bora, T.C.; Gogoi, S.; Boruah, R.C. A microwave promoted solvent-free approach to steroidal quinolines and their in vitro evaluation for antimicrobial activities. *Steroids* **2012**, *77*, 1438–1445. [CrossRef] [PubMed]
17. De, S.; Mishra, S.; Kakde, B.N.; Dey, D.; Bisai, A. Expeditious approach to pyrrolophenanthridones, phenanthridines, and benzo[c]phenanthridines via organocatalytic direct biaryl-coupling promoted by potassium tert-butoxide. *J. Org. Chem.* **2013**, *78*, 7823–7844. [CrossRef] [PubMed]
18. Jierry, L.; Harthong, S.; Aronica, C.; Mulatier, J.-C.; Guy, L.; Guy, S. Efficient dibenzo[c]acridine helicene-like synthesis and resolution: Scale up, structural control, and high chiroptical properties. *Org. Lett.* **2012**, *14*, 288–291. [CrossRef] [PubMed]
19. Speed, S.; Pointillart, F.; Mulatier, J.-C.; Guy, L.; Golhen, S.; Cador, O.; Le Guennic, B.; Riobé, F.; Maury, O.; Ouahab, L. Photophysical and magnetic properties in complexes containing 3d/4f elements and chiral phenanthroline-based helicate-like ligands. *Eur. J. Inorg. Chem.* **2017**, *14*, 2100–2111. [CrossRef]
20. Souibgui, A.; Gaucher, A.; Marrot, J.; Aloui, F.; Mahuteau-Betzer, F.; Ben Hassine, B.; Prim, D. A Flexible strategy towards thienyl-, oxazolyl- and pyridyl-fused fluorenones. *Eur. J. Org. Chem.* **2013**, *21*, 4515–4522. [CrossRef]
21. Solmont, K.; Boufroura, H.; Souibgui, A.; Fornarelli, P.; Gaucher, A.; Mahuteau-Betzer, F.; Ben Hassine, B.; Prim, D. Divergent strategy in the synthesis of original dihydro benzo- and dihydronaphtho-acridines. *Org. Biomol. Chem.* **2015**, *13*, 6269–6277. [CrossRef] [PubMed]
22. Some, S.; Ray, J.K. Chemoselective arylamination of β-bromovinylaldehydes followed by acid catalyzed cyclization: A general method for polycyclic quinolines. *Tetrahedron Lett.* **2007**, *48*, 5013–5016. [CrossRef]
23. Dick, A.R.; Hull, K.L.; Sanford, M.S. A highly selective catalytic method for the oxidative functionalization of C-H bonds. *J. Am. Chem. Soc.* **2004**, *126*, 2300–2301. [CrossRef] [PubMed]
24. Seki, B. Arylation using ruthenium catalyst. In *Catalytic Transformations via C-H Activation*; Yu, J.-Q., Ed.; Georg thieme Verlag KG: Stuttgart, Germany, 2016; Volume 1, pp. 119–153, ISBN 978-3-13-171141-0.
25. Powers, D.C.; Benitez, D.; Tkatchouk, E.; Goddard, W.A., III; Ritter, T. Bimetallic reductive elimination from dinuclear Pd(III)complexes. *J. Am. Chem. Soc.* **2010**, *132*, 14092–14103. [CrossRef] [PubMed]
26. Aiello, I.; Crispini, A.; Ghedini, M.; La Deda, M.; Barigelletti, F. Synthesis and characterization of a homologous series of mononuclear palladium complexes containing different cyclometalated ligands. *Inorg. Chim. Acta* **2000**, *308*, 121–128. [CrossRef]
27. Selbin, J.; Gutierrez, M.A. Cyclometallation IV. Palladium(II) compounds with benzo[h]quinoline and substituted 2,6-diarylpyridines. *J. Organomet. Chem.* **1983**, *246*, 95–104. [CrossRef]
28. Zhang, X.; Wang, H.; Yuan, J.; Guo, S. Palladacycles incorporating a carboxylate-functionalized phosphine ligand: Syntheses, characterization and their catalytic applications toward Suzuki couplings in water. *Transit. Met. Chem.* **2017**, *42*, 727–738. [CrossRef]
29. Li, C.; Sun, P.; Yan, L.; Pan, Y.; Cheng, C.-H. Synthesis and electroluminescent properties of Ir complexes with benzo[c]acridine or 5,6-dihydro-benzo[c]acridine ligands. *Thin Solid Films* **2008**, *516*, 6186–6190. [CrossRef]

30. Sako, M.; Takeuchi, Y.; Tsujihara, T.; Kodera, J.; Kawano, T.; Takizawa, S.; Sasai, H. Efficient enantioselective synthesis of oxahelicenes using redox/acid cooperative catalysts. *J. Am. Chem. Soc.* **2016**, *138*, 11481–11484. [CrossRef] [PubMed]
31. Lyons, T.W.; Sanford, M.S. Palladium-catalyzed ligand-directed C-H functionalization reactions. *Chem. Rev.* **2010**, *110*, 1147–1169. [CrossRef] [PubMed]
32. Schardt, B.C.; Hill, C.L. Preparation of iodobenzene dimethoxide. A new synthesis of [18O]iodosylbenzene and a reexamination of its infrared spectrum. *Inorg. Chem.* **1983**, *22*, 1563–1565. [CrossRef]

© 2018 by the authors. Licensee MDPI, Basel, Switzerland. This article is an open access article distributed under the terms and conditions of the Creative Commons Attribution (CC BY) license (http://creativecommons.org/licenses/by/4.0/).

Article

The First Catalytic Direct C–H Arylation on C2 and C3 of Thiophene Ring Applied to Thieno-Pyridines, -Pyrimidines and -Pyrazines

Joana F. Campos [1,2], Maria-João R. P. Queiroz [2] and Sabine Berteina-Raboin [1,*]

1 Institut de Chimie Organique et Analytique (ICOA), Université d'Orléans UMR-CNRS 7311, BP 6759, rue de Chartres, 45067 Orléans CEDEX 2, France; joana-filomena.mimoso-silva-de-campos@univ-orleans.fr
2 Departmento/Centro de Química, Escola de Ciências, Universidade do Minho, Campus de Gualtar, 4710-057 Braga, Portugal; mjrpq@quimica.uminho.pt
* Correspondence: sabine.berteina-raboin@univ-orleans.fr; Tel.: +33-238-494-856

Received: 13 March 2018; Accepted: 28 March 2018; Published: 30 March 2018

Abstract: A practical one-pot procedure for the preparation of diverse thieno[3,2-*d*]pyrimidines is reported here for the first time. This two-step process via C–H activation in position C-2 of thiophene led to the development of an improved methodology for the synthesis of numerous compounds. This new methodology is an efficient alternative to the conventional methods currently applied. The C–H activation of the thiophene C-3 position was also achieved and can be selective. The optimized conditions can also be applied to thienopyridines and thienopyrazines.

Keywords: C–H activation; regioselectivity; thienopyridines; thienopyrimidines; thienopyrazines

1. Introduction

The direct functionalization of C–H bonds in catalytic coupling reactions is considered to be a significant step to achieve molecular diversity and great progress has already been made in this field. Furthermore, environmentally benign, operationally simple, and robust reactions, particularly those employing heterogeneous catalysts, are of significant interest to the chemical industry. The direct palladium-catalyzed C–H activation of heteroaromatic compounds has recently been extensively studied. The main challenge of this approach is the control of regioselectivity with heterocyclic substrates containing multiple C–H bonds that may have similar reactivities [1–6]. Thienopyridines, thienopyrimidines, and thienopyrazines are sulfur-containing heterocyclic molecules that, when functionalized, are often incorporated into important molecular scaffolds used in materials science and in particular in biology and medicine [7]. To the best of our knowledge, however, no studies have yet been reported on the C–H activation of the thiophene ring in thieno[3,2-*b*]pyridines, thieno[3,2-*d*]pyrimidines, and thieno[2,3-*b*]pyrazines. The present study explores this topic.

A study of the literature revealed that only three teams have worked on the C–H activation of thieno[3,4-*b*]pyrazine, 2,3-thienoisoquinolines, and 3,4-thienoisoquinolines (Schemes 1–3).

McNamara et al. [8] presented the synthesis of a series of thieno[3,4-*b*]pyrazine derivatives as fluorescent compounds through the direct palladium-catalyzed activation of the C–H bonds of thiophene using Pd(OAc)$_2$ with X-Phos or PtBu$_3$ as ligands. Moderate yields were achieved, with some variation depending on the reagents (Scheme 1).

In 2014, Chen et al. [9] developed a synthetic route on 2,3-thienoisoquinoline-phenylsulfamide which was successfully functionalized with phenyl, 3-EtO$_2$CC$_6$H$_4$, 4-EtO$_2$CC$_6$H$_4$, 3-CH$_3$OC$_6$H$_4$, and 4-CH$_3$OC$_6$H$_4$ bromides employing a C–H activation reaction on position 5 of the thiophene moiety. Yields ranged from 25% to 86% (Scheme 2).

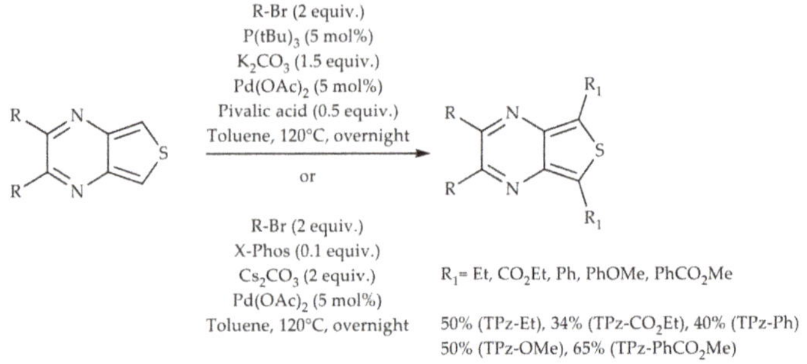

Scheme 1. McNamara et al.'s work (2016).

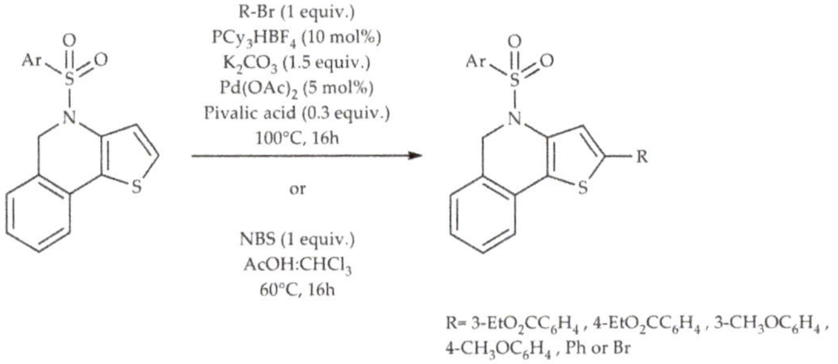

Scheme 2. Chen et al.'s work (2014).

Wong and Forngione [10] reported the synthesis of a unique class of highly functionalized 3,4-thienoisoquinolines via an efficient double C-H palladium-catalyzed one-pot activation using Pd(OAc)$_2$, PCy$_3$ HBF$_4$, PivOH, and K$_2$CO$_3$ in DMF at 100 °C for 6 h in good yields (Scheme 3).

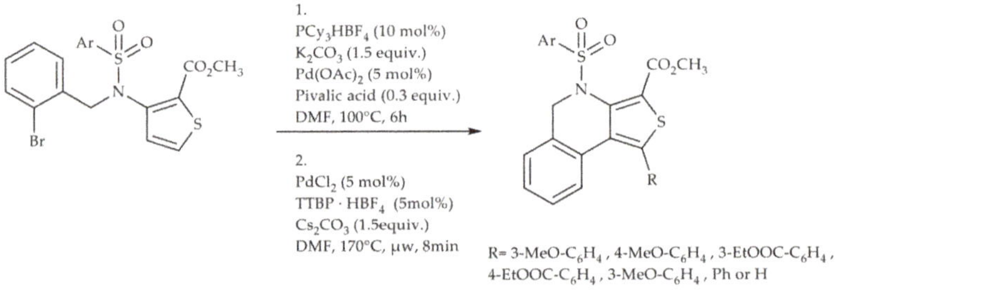

Scheme 3. Wong and Forngione's work (2012).

2. Results and Discussion

Pursuing our previous work [11] on various annelated sulfur-containing heterocycles, where we investigated the functionalization of the C3 position of the thiophene ring in several thieno-pyridines, -pyrimidines, and -pyrazines (Figure 1) using C3-bromoderivatives and various boronic acids in optimized Suzuki conditions, we worked on a Pd-catalyzed direct arylation to generate diversity.

Herein we report our investigation of the development of efficient Pd-catalyzed direct arylation on position C-3 of the thiophene ring from the thieno-pyridines, -pyrimidines, and -pyrazines presented in Figure 1. In a first attempt, the thienoderivatives 1 and 4 were reacted using the conditions already described by our team for the Pd-catalyzed direct arylation of pyrazolo[1,5-*a*]pyrimidine [12].

Unfortunately, these conditions did not yield the expected products (Table 1). The thienoderivatives 1 and 4 in presence of 1.5 equivalent of 1-iodo-4-methoxybenzene, Pd(OAc)$_2$ (10% mol), P(tBu)$_3$ (20% mol), and cesium carbonate (2 equiv.) in toluene after 48 h at 110 °C led only to recovery of the starting material and no trace of the expected product was detected. The use of P(Cy)$_3$HBF$_4$ as ligand did not lead to any improvement even when the reaction time was increased up to 70 h (Scheme 4).

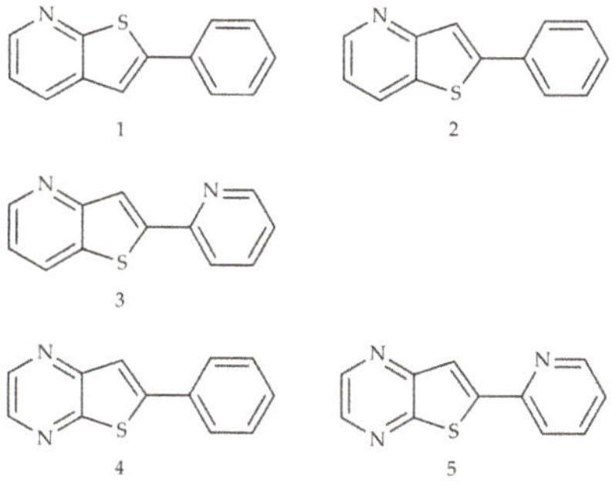

Figure 1. Thienopyridines and thienopyrazines used in first attempts of Pd-catalyzed direct arylation.

Table 1. Optimization of direct C–H arylation in position 3 of the 2-phenylthieno[3,2-b]pyridine 2.

Entry	Pd(OAc)$_2$ (mol %)	Ligand (mol %)	Cu (equiv.)	Base (equiv.)	Solvent	Br–X (1 equiv.)	Time (h)	T (°C)	Yield [a] (%)
1	(10)	-	CuI (20)	K$_2$CO$_3$ (2)	Toluene	Bromo benzene	24	120	0
2	(10)	-	Cu(OAc)$_2$ (20)	K$_2$CO$_3$ (2)	Toluene	Bromo benzene	24	120	0
3	(10)	-	CuI (20)	KOAc (2)	Toluene	Bromo benzene	24	120	0
4	(10)	-	Cu(OAc)$_2$ (20)	KOAc (2)	Toluene	Bromo benzene	24	120	0
5	(5)	P(t-Bu)$_2$MeHBF$_4$ (10)	-	K$_2$CO$_3$ (1)	DMA	Bromo benzene	24	120	7
6	(1)	P(t-Bu)$_2$MeHBF$_4$ (3)	-	K$_2$CO$_3$ + AgOTf (0.2) + (0.1)	DMA	Bromo benzene	24	145	0
7	(5)	P(t-Bu)$_2$MeHBF$_4$ (10)	-	K$_2$CO$_3$ (1)	DMA	1-bromo-4-methyl benzene	96	120	26
8	(5)	P(t-Bu)$_2$MeHBF$_4$ (10)	-	K$_2$CO$_3$ (1)	DMA	1-bromo-4-(trifluoro methyl) benzene	96	120	20
9	(1)	-	-	KOAc (2)	DMA	Bromo aceto phenone	24	150	0

[a] Isolated yield after column chromatography.

Applying the conditions reported by Hull and Sanford [13] on thienoderivatives 3 and 5 was likewise unsuccessful. This methodology involves the direct oxidative coupling of two arene C–H substrates with a very large excess of one of them. The use of Pd(OAc)$_2$ in presence of Ag$_2$CO$_3$ as base and benzoquinone Bzq as oxidant in DMSO at 130 °C with 100 equivalents of 1,2-dichlorobenzene did not yield the expected compound after 16 h, and once again only the starting material was recovered. After 24 h of reaction time 20% of the expected compound was obtained from 2-(pyridin-2-yl)thieno[3,2-b]pyridine **3** and 10% from 6-(pyridin-2-yl)thieno[2,3-b]pyrazine **5**. When 100 equivalents of 1,2-dimethoxybenzene, nitrobenzene, methoxybenzene, or 1,3-dimethoxy-2-nitrobenzene were used in the same conditions, only the starting material was recovered and no product was detected. Although the product was obtained with a very low yield in two cases, given the large quantity of one of the reagents that had to be used in this method, we made no further attempts to optimize the conditions in terms of time or other factors (Scheme 5).

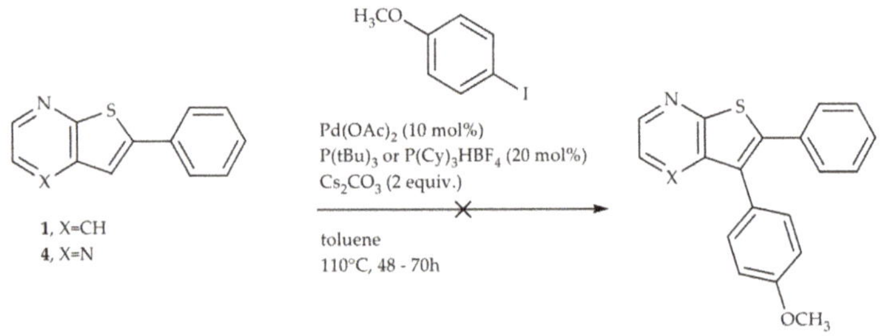

Scheme 4. Pd-catalyzed direct arylation on position C-3 of the thiophene ring from the thieno-pyridines and -pyrazines.

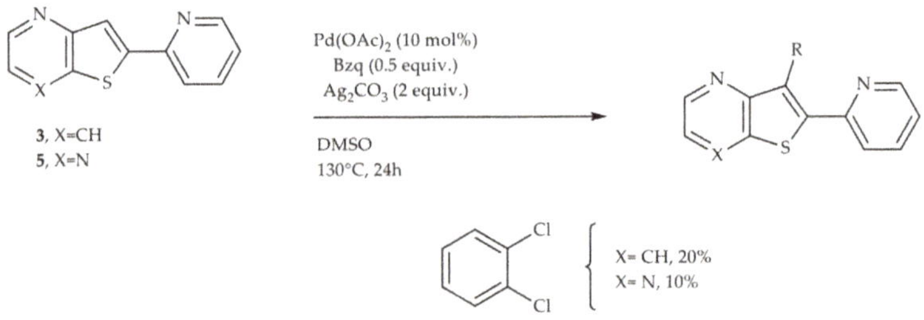

Scheme 5. Results of conditions reported by Hull and Sanford on thienoderivatives.

The work by Wei et al. [14] and Yang et al. [15] reported procedures for Pd-catalyzed direct arylation with aryl boronic acids which were applied to our thienoderivatives **2** and **3** using phenyl boronic acid but yet again, the reaction was unsuccessful (Scheme 6).

One of the other possible options was to use aryl bromides [16,17]. The 2-phenylthieno[3,2-b]pyridine **2** was chosen for these experiments and was reacted with various amounts of Pd(OAc)$_2$, K$_2$CO$_3$, or KOAc as base, with or without ligands in different solvents. The results are summarized in Table 1. In entries 1 to 4 copper was added as co-catalyst without any improvement, while in entries 5 to 8 P(t-Bu)$_2$MeHBF$_4$ was used as ligand with which the best yield was obtained (Table 1, Entry 7, yield 26%).

Despite several attempts, however, we did not manage to exceed a 26% yield. Using a different strategy, with 4-chlorothieno[3,2-d]pyrimidine **6** as starting material in the conditions reported by Hull and Sanford [13], also proved unsuccessful (Scheme 7).

To avoid a possible interaction of the chlorine atom in the Pd-catalyzed direct arylation (Scheme 7), it was substituted by various amines, generating a new series of potentially biologically active molecules. This new strategy is shown in Scheme 8. After adding one chain by SnAr, the scope of the reactivity and regioselectivity of Pd-catalyzed direct arylation in positions 2 or 3 of the thiophene moiety was evaluated.

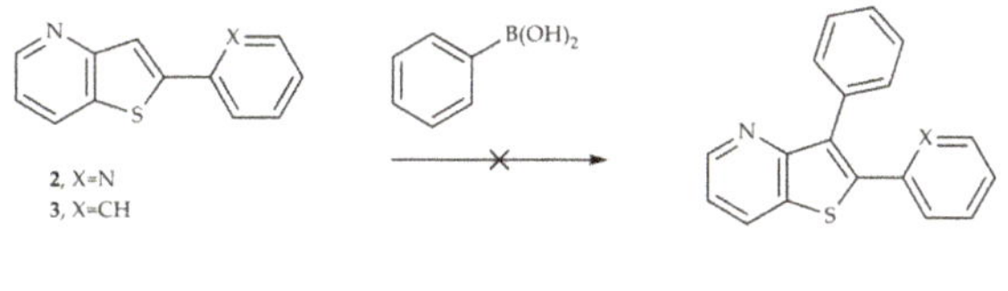

1) Pd(OAc)$_2$ (5 mol%), Ag$_2$CO$_3$ (2 equiv.), K$_2$CO$_3$ (0.5 equiv.), p-toluic acid (0.3 equiv.), DMA, 16h, 110 °C: no reaction
2) Pd(OAc)$_2$ (5 mol%), Cu(OAc)$_2$ (1 mol%), TFA, 63h, r.t.: no reaction
3) Pd(OAc)$_2$ (10 mol%), AcOH, 24h, r.t.: no reaction

Scheme 6. Results of conditions reported by Wei et al. and Yang et al. on thienoderivatives.

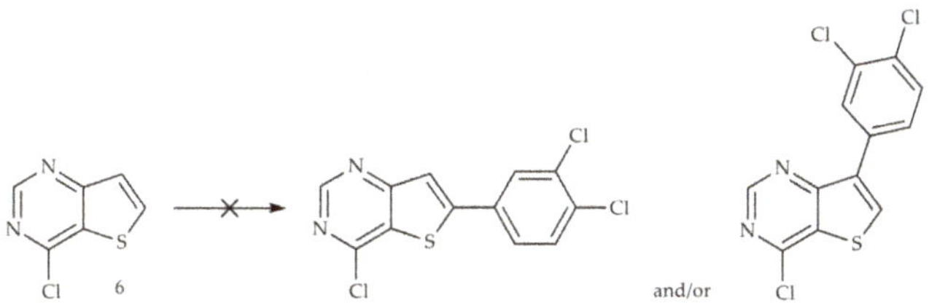

Conditions: Pd(OAc)$_2$ (10 mol%), Benzoquinone (0.5 equiv.), Ag$_2$CO$_3$ (2 equiv.), DMSO, 16h, 130°C.

Scheme 7. C–H activation from 4-chlorothieno[3,2-d]pyrimidine using conditions reported by Hull and Sanford.

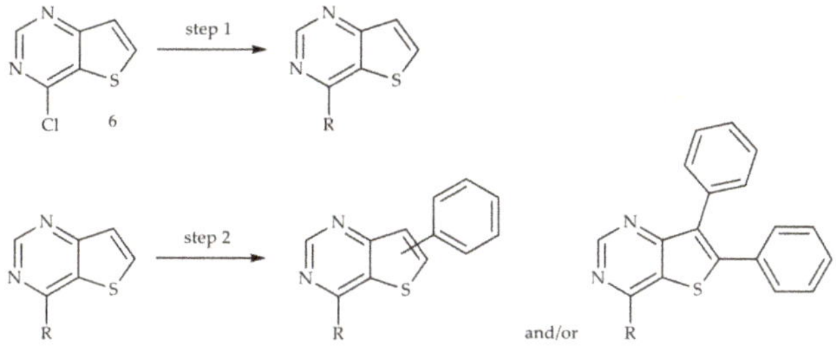

Scheme 8. New synthesis strategy.

Several strategies to conduct an SnAr reaction of 4-chlorothieno[3,2-d]pyrimidine using amino derivatives are reported in the literature [18–22]. With a view to developing procedures with the lowest possible environmental impact, we tested the use of PEG 400 as solvent. This compound, like the other heterocyclic scaffolds, underwent the SnAr reaction using amines and produced good yields in only 5 min [23], but our goal was to diversify our core structure by C–H activation which tried using Polyethylene glycol 400 as solvent without success in a one-pot process. We therefore decided, in the present work, to use toluene with various amines to generate precursors of C–H catalyzed cross-coupling. In these conditions all the expected substituted 4-amino-thieno[3,2-d]pyrimidine compounds were successfully synthesized in good yields of 60% to 96%. The lowest yield was obtained for the deactivated aniline substituted in the *ortho* position by the electro-withdrawing trifluoromethyl group (Table 2).

Table 2. SnAr substitution of chlorine on the six-membered ring by various amines.

Entry	Amine Reagent	Product	Yield [a]
1	morpholine	morpholine-substituted thienopyrimidine	7, 86%
2	dibutylamine	dibutylamino-substituted thienopyrimidine	8, 96%
3	piperidine	piperidine-substituted thienopyrimidine	9, 82%
4	2-(trifluoromethyl)aniline	2-(trifluoromethyl)anilino-substituted thienopyrimidine	10, 60%
5	4-methoxyaniline	4-methoxyanilino-substituted thienopyrimidine	11, 86%
6	methyl 4-aminobenzoate	methyl 4-(thienopyrimidinylamino)benzoate	12, 67%
7	1-adamantylamine	1-adamantylamino-substituted thienopyrimidine	13, 68%

[a] Isolated yield after column chromatography.

Optimization of the Pd-catalyzed direct arylation and its regioselectivity in position C-2 was achieved starting from 4-(thieno[3,2-d]pyrimidin-4-yl)morpholine **7** using bromobenzene and by varying the amount of Pd(OAc)$_2$ with or without ligand, and the type and amount of K$_2$CO$_3$ or KOAc as base. Various solvents and temperatures were also tested, as summarized in Table 3.

Table 3. Optimization of the C–H activation with regioselectivity in position C-2 of 4-(thieno[3,2-d]pyrimidin-4-yl)morpholine **7**.

Entry	Pd Catalyst (equiv.)	Ligand (mol %)	Base (equiv.)	Solvent	Time (h)	T (°C)	Yield [a] (%)	
							14	15
1	Pd(OAc)$_2$ (20%)	-	K$_2$CO$_3$ (4)	Toluene	46	140	64	31
2	Pd(OAc)$_2$ (10%)	PCy$_3$ (20)	K$_2$CO$_3$ (2)	Dioxane	46	130	50	44
3	Pd(OAc)$_2$ (10%)	TTBP · HBF$_4$ (20)	K$_2$CO$_3$ (2)	Toluene	46	130	81	9
4	Pd(OAc)$_2$ (10%)	Phenantroline (20)	K$_3$PO$_4$/K$_2$CO$_3$ (1)/(1)	DMA	46	140	22	0
5	Pd(OAc)$_2$/Bu$_4$NBr (20%)/(2)	-	KOAc (6)	DMF	24	80	26	0
6	Pd(OAc)$_2$/Bu$_4$NBr (20%)/(2)	-	KOAc (6)	Water	24	80	0	0
7	Pd(OAc)$_2$ (10%)	TTBP · HBF$_4$ (20)	K$_2$CO$_3$ (2)	Toluene	46	100	72	0

[a] Isolated yield after column chromatography.

The desired compound was obtained in moderate to good yields in presence of Pd(OAc)$_2$ with K$_2$CO$_3$ in toluene (Table 3, Entries 1 and 3) and the presence of ligand increased the regioselectivity (Table 3, Entries 3–4). When the reaction was performed in water no results were obtained and only the starting material was recovered (Table 3, entry 6). The temperature had a pronounced effect on regioselectivity: when 4-(thieno[3,2-d]pyrimidin-4-yl)morpholine was stirred using Pd(OAc)$_2$, TTBP · HBF$_4$ as ligand, K$_2$CO$_3$ in toluene at 100 °C, complete regioselectivity was obtained.

This process proved to be a good alternative to the most commonly used synthetic routes [24–29] (Figure 2).

Based on our results (Table 3, Entry 7) the scope and limitations of the one pot SnAr–Pd-catalyzed direct arylation on 4-chlorothieno[3,2-d]pyrimidine **6** were assessed using several bromo-benzenes (Table 4).

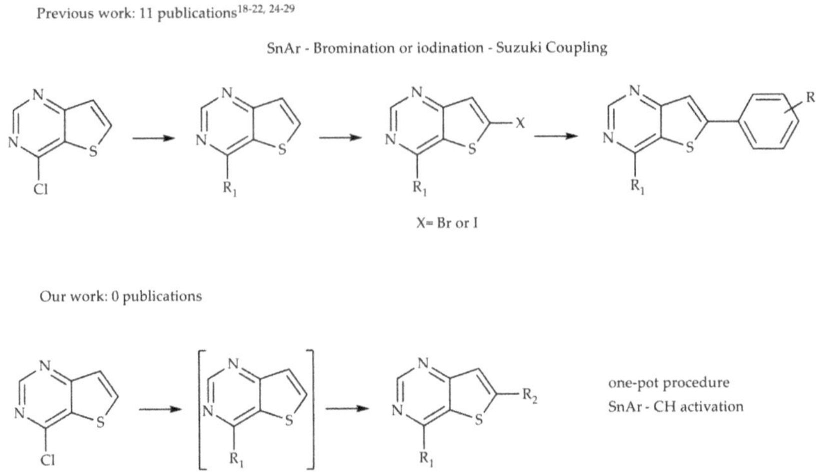

Previous work: 11 publications[18-22, 24-29]

SnAr - Bromination or iodination - Suzuki Coupling

X = Br or I

Our work: 0 publications

one-pot procedure
SnAr - CH activation

Figure 2. Literature and previous work.

Table 4. Scope and limitations of the one pot SnAr-Pd-catalyzed direct arylation from 4-chlorothieno[3,2-d]pyrimidine **6**.

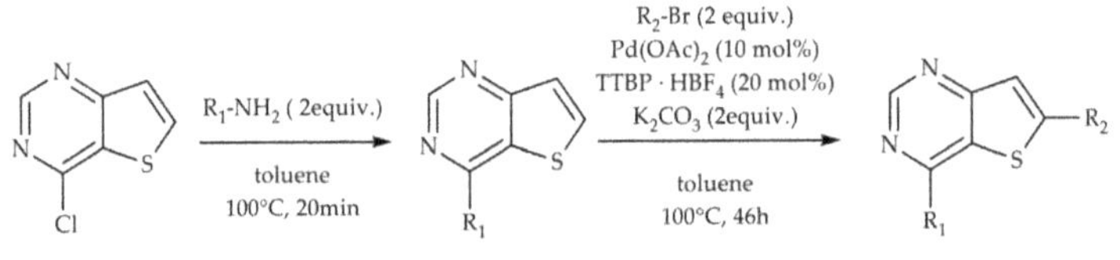

Reaction conditions: R_1-NH_2 (2 equiv.), toluene, 100°C, 20min; then R_2-Br (2 equiv.), Pd(OAc)$_2$ (10 mol%), TTBP·HBF$_4$ (20 mol%), K$_2$CO$_3$ (2 equiv.), toluene, 100°C, 46h.

Entry	Amine Reagent	R_2-Br	Product	Yield [a]
1	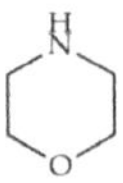 morpholine	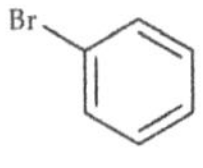 Br-phenyl	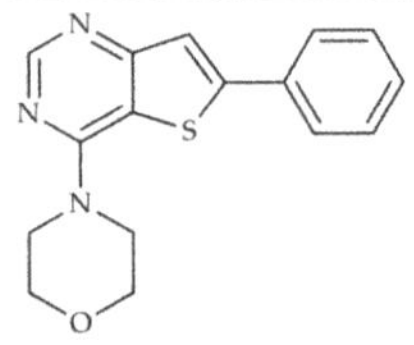 (product)	**14**, 70%
2	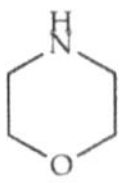 morpholine	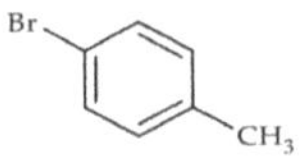 Br-C$_6$H$_4$-CH$_3$	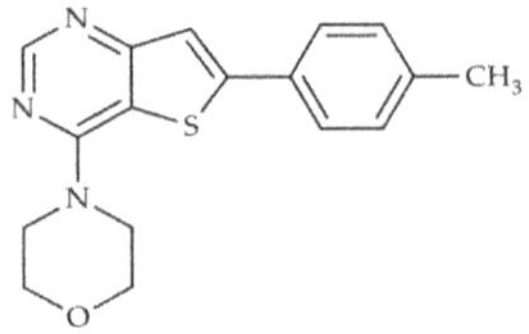 (product)	**16**, 58%
3	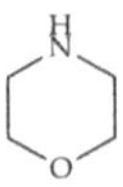 morpholine	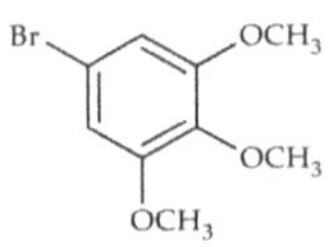 Br-C$_6$H$_2$(OCH$_3$)$_3$	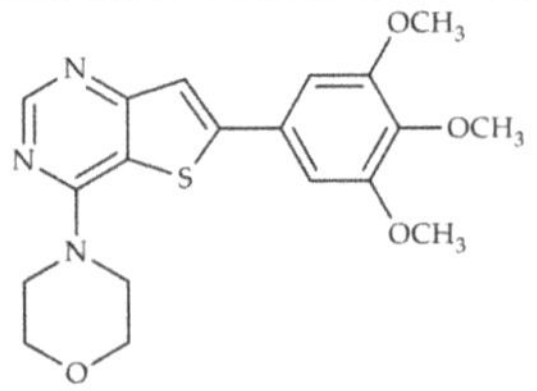 (product)	**17**, 63%

Table 4. Cont.

Entry	Amine Reagent	R$_2$-Br	Product	Yield [a]
4	morpholine	Br-C$_6$H$_4$-CO$_2$CH$_3$ (para)	thienopyrimidine-morpholine with 4-CO$_2$CH$_3$ phenyl	18, 61%
5	morpholine	Br-C$_6$H$_4$-CO$_2$CH$_3$ (meta)	thienopyrimidine-morpholine with 3-CO$_2$CH$_3$ phenyl	19, 54%
6	morpholine	2-bromobenzothiophene	thienopyrimidine-morpholine with benzothiophene	20, 43%
7	morpholine	1-bromonaphthalene	thienopyrimidine-morpholine with naphthyl	21, 84%
8	morpholine	Br-C$_6$H$_4$-CN	thienopyrimidine-morpholine with 4-CN phenyl	22, 55%
9	1-adamantylamine	bromobenzene	thienopyrimidine-adamantylamine with phenyl	23, 67%

[a] Isolated yield after column chromatography.

Several 2-aryl-thieno[3,2-*d*]pyrimidine compounds amino-substituted in position 4 were synthesized in moderate to excellent yields, demonstrating the generalizability of this method. From these results, we started to develop the Pd-catalyzed direct C–H activation on position C-3 of the thiophene moiety. This was studied from compound **14** using bromobenzene (2 equiv.), different amounts of Pd(OAc)$_2$ and K$_2$CO$_3$ with or without ligand, in toluene at different temperatures (Table 5). The desired product was obtained in moderate yields (Table 5, Entries 1–3). A temperature between

130–140 °C induced activation in the C-3 position, but when the reaction was performed at 100 °C, direct C–H activation did not take place (Table 5, Entry 4).

Table 5. Optimization of the C–H activation at position C-3 of 4-(6-phenylthieno[3,2-*d*]pyrimidin-4-yl)morpholine **14**.

Entry	Pd Catalyst (mol %)	Ligand (mol %)	Base (equiv.)	T (°C)	Yield [a] (%)
1	Pd(OAc)$_2$ (20)	-	K$_2$CO$_3$ (4)	140	55
2	Pd(OAc)$_2$ (10)	TTBP·HBF$_4$ (20)	K$_2$CO$_3$ (2)	130	49
3	Pd(OAc)$_2$ (10)	PCy$_3$ (20)	K$_2$CO$_3$ (2)	130	34
4	Pd(OAc)$_2$ (10)	TTBP·HBF$_4$ (20)	K$_2$CO$_3$ (2)	100	0

[a] Isolated yield after column chromatography.

The scope and limitations of the one-pot reaction were then explored. We started with morpholine; then bromobenzene was used for the Pd-catalyzed C–H activation on the C-2 position, and for the last step, the Pd-catalyzed C–H activation on the C-3 position, bromobenzene and bromobenzonitrile were tested. The conditions of entry 2 (Table 5) were chosen even though they required a ligand because the reaction can be carried out with a smaller amount of base and at a lower temperature, which may be important for some sensitive reagents. These conditions appeared to be a good compromise. The results obtained for the one-pot three-step activation are summarized in Table 6.

For these two examples of one-pot three-step SnAr, C–H activation in C-2 then C–H activation in C-3 positions, we were able to obtain the expected compounds with reasonable yields. This method is also a valid alternative to standard synthetic strategies. To validate our Pd-catalyzed C–H arylation conditions on a large panel of heterocycles possessing a thiophene moiety we chose two compounds used at the beginning of this work, and investigated the scope and limitations first on the 2-phenylthieno[3,2-*b*]pyridine core structure (Table 7) and then on the 6-phenyl thieno[2,3-*b*]pyrazine, a heterocycle known to have a particularly low reactivity (Table 8) [11].

Six different 3-aryl-2-phenylthienopyridines (**25–29**) were synthesized in moderate to good yields (41 to 91%) by direct Pd-catalyzed C–H arylation in C-3 position.

Several 5-aryl-6-phenylthieno[2,3-*b*]pyrazines (**30–32**) were synthesized in moderate yields, which demonstrated the generality and the feasibility of this method. As expected from the latest results of our team, the pyrazine showed lower yields than the pyridine. These results (Tables 7 and 8) represented an improvement and a good alternative to the most commonly used synthetic routes [11,24–29].

Table 6. One pot three-step selective SnAr, Pd-catalyzed C–H activation on the C-2 then on the C-3 positions of the thiophene moiety.

Entry	Amine	R$_2$-Br	R$_3$-Br	Product	Yield [a]
1	morpholine	bromobenzene	bromobenzene		15, 48%
2	morpholine	bromobenzene	4-bromobenzonitrile		24, 36%

[a] Isolated yield after column chromatography.

Table 7. Pd-catalyzed C–H arylation in position C-3 of the 2-phenylthieno[3,2-b]pyridine **2**.

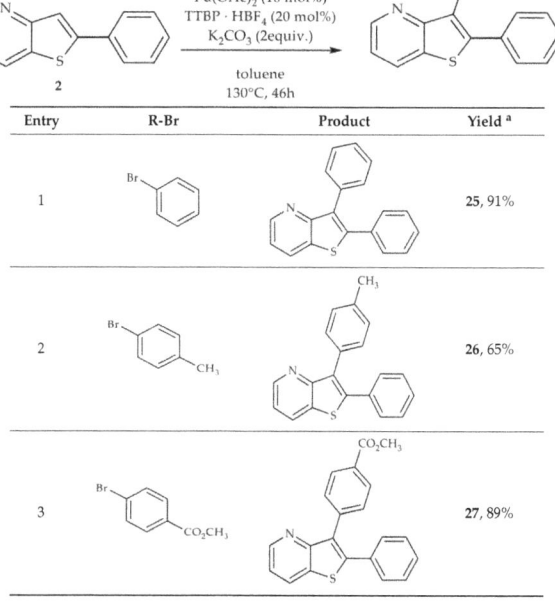

Entry	R-Br	Product	Yield [a]
1	bromobenzene		25, 91%
2	4-bromotoluene		26, 65%
3	methyl 4-bromobenzoate		27, 89%

Table 7. Cont.

Entry	R-Br	Product	Yield [a]
4	Br-C6H4-CO2CH3	(structure with CO2CH3)	28, 41%
5	Br-C6H4-CN	(structure with CN)	29, 84%

[a] Isolated yield after column chromatography.

Table 8. Pd-catalyzed C–H arylation in C-3 of the 6-phenyl thieno[2,3-b]pyrazine 4.

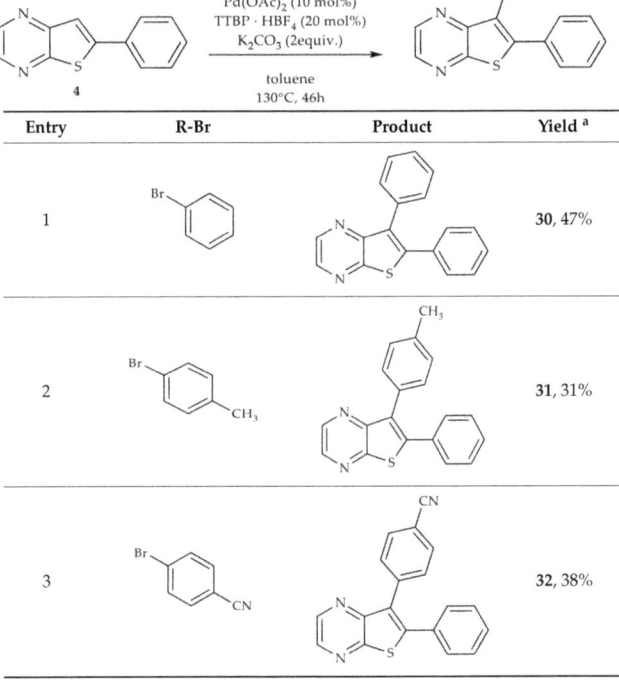

Entry	R-Br	Product	Yield [a]
1	Br-C6H5	(structure)	30, 47%
2	Br-C6H4-CH3	(structure)	31, 31%
3	Br-C6H4-CN	(structure)	32, 38%

[a] Isolated yield after column chromatography.

3. Conclusions

In summary, we have disclosed a convenient one-pot synthesis of thieno-pyridine, -pyrimidine, and -pyrazine scaffolds. The conditions reported make this methodology an interesting alternative to conventional routes, as it avoids the bromination, iodination, or chlorination processes generally used to synthesize various annelated sulfur-containing heterocycles.

Supplementary Materials: Supplementary materials are available online at http://www.mdpi.com/2073-4344/8/4/137/s1.

Acknowledgments: We acknowledge the Region Centre for financial support.

Author Contributions: Joana F. Campos and Sabine Berteina-Raboin conceived and designed the experiments. Joana F. Campos performed the experiments. Maria-João R. P. Queiroz participated in analyzing the data. Joana F. Campos and Sabine Berteina-Raboin analyzed the data and wrote the paper.

Conflicts of Interest: The authors declare no conflict of interest.

References

1. Crabtree, R.H.; Lei, A. Introduction: CH Activation. *Chem. Rev.* **2017**, *117*, 8481–8482. [CrossRef] [PubMed]
2. Tzouras, N.V.; Stamatopoulos, I.K.; Papastavrou, A.T.; Liori, A.A.; Vougioukalakis, G.C. Sustainable metal catalysis in C–H activation. *Coord. Chem. Rev.* **2017**, *343*, 25–138. [CrossRef]
3. Liu, C.; Yuan, J.; Gao, M.; Tang, S.; Li, W.; Shi, R.; Lei, A. Oxidative Coupling between Two Hydrocarbons: An Update of Recent C–H Functionalizations. *Chem. Rev.* **2015**, *115*, 12138–12204. [CrossRef] [PubMed]
4. Kuhl, N.; Schroder, N.; Glorius, F. Formal S_N-Type Reactions in Rhodium (III)-Catalyzed C-H Bond Activation. *Adv. Synth. Catal.* **2014**, *356*, 1443–1460. [CrossRef]
5. Girard, S.A.; Knauber, T.; Li, C.J. The Cross-Dehydrogenative Coupling of C_{sp}^3-H Bonds: A Versatile Strategy for C-C Bond Formations. *Angew. Chem. Int. Ed.* **2014**, *53*, 74–100. [CrossRef] [PubMed]
6. Kuhl, N.; Hopkinson, M.N.; Wencel-Delord, J.; Glorius, F. Beyond Directing Groups: Transition-Metal-Catalyzed C-H Activation of Simple Arenes. *Angew. Chem. Int. Ed.* **2012**, *51*, 10236–10254. [CrossRef] [PubMed]
7. Elrazaz, E.Z.; Serya, R.A.T.; Ismail, N.S.M.; Ella, D.A.A.E.; Abouzid, K.A.M. Thieno[2,3-d]pyrimidine based derivatives as kinase inhibitors and anticancer agents. *Future J. Pharm. Sci.* **2015**, *1*, 33–41. [CrossRef]
8. McNamara, L.E.; Liyanage, N.; Peddapuram, A.; Murphy, J.S.; Delcamp, J.H.; Hammer, N.I. Donor–Acceptor–Donor Thienopyrazines via Pd-Catalyzed C–H Activation as NIR Fluorescent Materials. *J. Org. Chem.* **2016**, *81*, 32–42. [CrossRef] [PubMed]
9. Chen, F.; Wong, N.W.Y.; Forgione, P. One-Pot Tandem Palladium-Catalyzed Decarboxylative Cross-Coupling and C-H Activation Route to Thienoisoquinolines. *Adv. Synth. Catal.* **2014**, *356*, 1725–1730. [CrossRef]
10. Wong, N.W.Y.; Forgione, P. A one-pot double C-H activation palladium catalyzed route to a unique class of highly functionalized thienoisoquinolines. *Org. Lett.* **2012**, *14*, 2738–2741. [CrossRef] [PubMed]
11. Campos, J.F.; Queiroz, M.J.R.P.; Berteina-Raboin, S. Synthesis of New Thieno[3,2-b]pyridines and Thieno[2,3-b]pyrazines by Palladium Cross-Coupling. *ChemistrySelect* **2017**, *2*, 6945–6948. [CrossRef]
12. Bassoude, I.; Berteina-Raboin, S.; Massip, S.; Leger, J.M.; Jarry, C.; Essassi, E.; Guillaumet, G. Catalyst- and Base-Controlled Site-Selective sp^2 and sp^3 Direct Arylation of 5,7-Dimethyl-2-phenylpyrazolo[1,5-a]pyrimidine Using Aryl Bromides. *Eur. J. Org. Chem.* **2012**, 2572–2578. [CrossRef]
13. Hull, K.L.; Sanford, M.S. Catalytic and Highly Regioselective Cross-Coupling of Aromatic C−H Substrates. *J. Am. Chem. Soc.* **2007**, *129*, 11904–11905. [CrossRef] [PubMed]
14. Wei, Y.; Kan, J.; Wang, M.; Su, W.; Hong, M. Palladium-Catalyzed Direct Arylation of Electron-Deficient Polyfluoroarenes with Arylboronic Acids. *Org. Lett.* **2009**, *11*, 3346–3349. [CrossRef] [PubMed]
15. Yang, S.D.; Sun, C.L.; Fang, Z.; Li, B.J.; Li, Y.Z.; Shi, Z.J. Palladium-catalyzed direct arylation of (hetero)arenes with aryl boronic acids. *Angew. Chem. Int. Ed. Engl.* **2008**, *47*, 1473–1476. [CrossRef] [PubMed]
16. Campeau, L.-C.; Parisien, M.; Jean, A.; Fagnou, K. Catalytic Direct Arylation with Aryl Chlorides, Bromides, and Iodides: Intramolecular Studies Leading to New Intermolecular Reactions. *J. Am. Chem. Soc.* **2006**, *128*, 581–590. [CrossRef] [PubMed]
17. Bensaid, S.; Doucet, H. Influence of the solvent and of the reaction concentration for palladium-catalysed direct arylation of heteroaromatics with 4-bromoacetophenone. *Comptes Rendus Chim.* **2014**, *17*, 1184–1189. [CrossRef]
18. Wang, J.; Su, M.; Li, T.; Gao, A.; Yang, W.; Sheng, L.; Zang, Y.; Li, J.; Liu, H. Design, synthesis and biological evaluation of thienopyrimidine hydroxamic acid based derivatives as structurally novel histone deacetylase (HDAC) inhibitors. *Eur. J. Med. Chem.* **2017**, *128*, 293–299. [CrossRef] [PubMed]

19. Desroches, J.; Kieffer, C.; Primas, N.; Hutter, S.; Gellis, A.; El-Kashef, H.; Rathelot, P.; Verhaeghe, P.; Azas, N.; Vanelle, P. Discovery of new hit-molecules targeting Plasmodium falciparum through a global SAR study of the 4-substituted-2-trichloromethylquinazoline antiplasmodial scaffold. *Eur. J. Med. Chem.* **2017**, *125*, 68–86. [CrossRef] [PubMed]
20. Baugh, S.D.P.; Pabba, P.K.; Barbosa, J.; Coulter, E.; Desai, U.; Gay, J.P.; Gopinathan, S.; Han, Q.; Hari, R.; Kimball, S.D.; et al. Design, synthesis, and in vivo activity of novel inhibitors of delta-5 desaturase for the treatment of metabolic syndrome. *Bioorg. Med. Chem. Lett.* **2015**, *25*, 3836–3839. [CrossRef] [PubMed]
21. Kemnitzer, W.; Sirisoma, N.; May, C.; Tseng, B.; Drewe, J.; Cai, S.X. Discovery of 4-anilino-*N*-methylthieno[3,2-*d*]pyrimidines and 4-anilino-*N*-methylthieno[2,3-*d*]pyrimidines as potent apoptosis inducers. *Bioorg. Med. Chem. Lett.* **2009**, *19*, 3536–3540. [CrossRef] [PubMed]
22. Song, Y.H. A facile synthesis of new 4-(phenylamino) thieno[3,2-*d*]pyrimidines using 3-aminothiophene-2-carboxamide. *Heterocycl. Commun.* **2007**, *13*, 33–34. [CrossRef]
23. Campos, J.F.; Loubidi, M.; Scherrmann, M.C.; Berteina-Raboin, S. A Greener and Efficient Method for Nucleophilic Aromatic Substitution of Nitrogen-Containing Fused Heterocycles. *Molecules* **2018**, *23*, 684. [CrossRef] [PubMed]
24. Woodring, J.L.; Patel, G.; Erath, J.; Behera, R.; Lee, P.J.; Leed, S.E.; Rodriguez, A.; Sciotti, R.J.; Mensa-Wilmot, K.; Pollastri, M.P. Evaluation of aromatic 6-substituted thienopyrimidines as scaffolds against parasites that cause trypanosomiasis, leishmaniasis, and malaria. *Med. Chem. Commun.* **2015**, *6*, 339–346. [CrossRef] [PubMed]
25. Devine, W.; Woodring, J.L.; Swaminathan, U.; Amata, E.; Patel, G.; Erath, J.; Roncal, N.E.; Lee, P.J.; Leed, S.E.; Rodriguez, A.; et al. Protozoan Parasite Growth Inhibitors Discovered by Cross-Screening Yield Potent Scaffolds for Lead Discovery. *J. Med. Chem.* **2015**, *58*, 5522–5537. [CrossRef] [PubMed]
26. Ni, Y.; Gopalsamy, A.; Cole, D.; Hu, Y.; Denny, R.; Ipek, M.; Liu, J.; Lee, J.; Hall, J.P.; Luong, M.; et al. Identification and SAR of a new series of thieno[3,2-*d*]pyrimidines as Tpl2 kinase inhibitors. *Bioorg. Med. Chem. Lett.* **2011**, *21*, 5952–5956. [CrossRef] [PubMed]
27. Cabrera, D.G.; Douelle, F.; Manach, C.L.; Han, Z.; Paquet, T.; Taylor, D.; Njoroge, M.; Lawrence, N.; Wiesner, L.; Waterson, D.; et al. Structure–Activity Relationship Studies of Orally Active Antimalarial 2,4-Diamino-thienopyrimidines. *J. Med. Chem.* **2015**, *58*, 7572–7579. [CrossRef] [PubMed]
28. Gopalsamy, A.; Shi, M.; Hu, Y.; Lee, F.; Feldberg, L.; Frommer, E.; Kim, S.; Collins, K.; Wojciechowicz, D.; Mallon, R. B-Raf kinase inhibitors: Hit enrichment through scaffold hopping. *Bioorg. Med. Chem. Lett.* **2010**, *20*, 2431–2434. [CrossRef]
29. Snégaroff, K.; Lassagne, F.; Bentabed-Ababsa, G.; Nassar, E.; Ely, S.C.S.; Hesse, S.; Perspicace, E.; Derdour, A.; Mongin, F. Direct metallation of thienopyrimidines using a mixed lithium–cadmium base and antitumor activity of functionalized derivatives. *Org. Biomol. Chem.* **2009**, *7*, 4782–4788. [CrossRef] [PubMed]

© 2018 by the authors. Licensee MDPI, Basel, Switzerland. This article is an open access article distributed under the terms and conditions of the Creative Commons Attribution (CC BY) license (http://creativecommons.org/licenses/by/4.0/).

Review

Advances in Enantioselective C–H Activation/Mizoroki-Heck Reaction and Suzuki Reaction

Shuai Shi [1], Khan Shah Nawaz [1], Muhammad Kashif Zaman [1] and Zhankui Sun [1,2,*]

[1] School of Pharmacy, Shanghai Jiao Tong University, No. 800 Dongchuan Rd., Shanghai 200240, China; shishuai@sjtu.edu.cn (S.S.); snkhan@sjtu.edu.cn (K.S.N.); kashifzaman43@sjtu.edu.cn (M.K.Z.)
[2] Huzhou Research and Industrialization Center for Technology, Chinese Academy of Sciences, 1366 Hongfeng Road, Huzhou 313000, China
* Correspondence: zksun@sjtu.edu.cn; Tel.: +86-21-3420-8590

Received: 31 January 2018; Accepted: 15 February 2018; Published: 23 February 2018

Abstract: Traditional cross-coupling reactions, like Mizoroki-Heck Reaction and Suzuki Reaction, have revolutionized organic chemistry and are widely applied in modern organic synthesis. With the rapid development of C–H activation and asymmetric catalysis in recent years, enantioselective C–H activation/cross-coupling reactions have drawn much attention from researchers. This review summarizes recent advances in enantioselective C–H activation/Mizoroki-Heck Reaction and Suzuki Reaction, with emphasis on the structures and functions of chiral ligands utilized in different reactions.

Keywords: enantioselective C–H activation; C–C cross-coupling; Suzuki reaction; Mizoroki-Heck reaction

1. Introduction

Over the past few decades, C–H bond activation was established as a credible and viable strategy in organic synthesis [1–18]. On the basis of previous work, C–H activation/functionalization is employed as a dynamic strategy for the synthesis of a great deal of highly valuable natural products and other classes of compounds for pharmaceutics or research interests [19–22]. Chemists have long been captivated by such synthetic techniques, owing to the obvious merits. From a philosophical point of view, chemists regard C–H bonds as dormant equivalents of various pre-functionalized groups. Moreover, the direct modification of ubiquitously-existing C–H bonds successfully live up to the criteria of one perfect catalytic reaction that ought to be atom-economic and environmental-friendly [23]. Thus, it provides new synthetic disconnections in retrosynthetic analysis [24–28].

It is widely acknowledged that carbon-carbon (C–C) bond formation is fundamental and essential in organic chemistry. Numerous methods have been well developed to enable such C–C bond formation to proceed smoothly. Among these reactions, transition-metals (Pd, Rh, Ru, Cu, Zn, Sn, Mg, etc.) catalyzed cross-coupling reactions are efficient techniques to realize C–C bond formation, which can be well exemplified by Suzuki [29], Mizoroki-Heck [30], Sonogashira [31], Negishi [32], Stille [33], Kumada [34] coupling reaction, etc. These coupling reactions are renowned for their extraordinary utility, practicality, and reliability, and have been broadly utilized in many syntheses, involving pharmaceuticals, fine chemicals, agrochemicals, etc. Consequently, Richard Mizoroki-Heck, Ei-ichi Negishi, and Akira Suzuki jointly won the Nobel Prize in Chemistry 2010 for their excellent work of "palladium-catalyzed cross-coupling reactions in organic synthesis", which furnished a novel way to achieve C–C bond formation that substantially accelerated the development of pharmaceutics and electronics industries [35]. Although lots of outstanding achievements in this field have been made, from an atom-economic and environmental-harmoniously perspective, there are still shortcomings in these known and powerful traditional cross-coupling reactions. For instance, the substrates must be pre-functionalized, such as using organic (pseudo) halides or organometallic reagents, to gain reactivity during cross-coupling processes. Subsequently, it would result in metal (pseudo) halide wastes, which are supposed to be evaded for atom-economic purpose. Therefore, it is a brilliant notion to develop new methods replacing the pre-functionalized substrates with raw arenes or hydrocarbons, which can achieve the C–H activation and C–C cross-coupling simultaneously. When compared with traditional cross-coupling reactions, the C–H activation/C–C cross coupling reactions have obvious superiority in the aspect of atom-economy and environmental benignity.

With the rapid development of C–H activation/C–C cross-coupling reactions and the emerging of asymmetric catalysis, the enantioselective C–H activation/C–C cross-coupling reactions drew much more attention from researchers. It is a dynamic research frontier to achieve C–H activation/C–C cross-coupling reactions in a stereoselective manner. However, the direct asymmetric functionalizations of inert C–H bonds still remain challenging in current organic synthesis because of the poor reaction selectivity (regioselectivity and enantioselectivity). This is not odd due to the properties of C–H bond: (1) Substrates commonly contain diverse C–H bonds, which usually have high but comparable bond dissociation energy (BDE of C–H bonds are typically 90–110 kcal·mol^{-1}) within one molecular, rather than bearing a single targeted C–H bond [36–40]; (2) Higher reaction temperature is required for most of the reactions that are related to C–H activation in order to meet the high energy for the cleavage of C–H bonds, which would undoubtedly impose a detrimental effect on the asymmetric induction or coordination between the chiral ligands and the transition metals. Despite the difficulty in controlling stereoselectivity of C–H cleavage, tremendous progresses have been made in transition-metal catalyzed enantioselective C–H activation in the past few decades [41–46]. Among these highly efficient synthetic methodologies, complex chiral ligands or neighboring directing groups are usually needed. For instance, the Pd-catalyzed desymmetrization of prochiral C–H bonds has emerged as a promising strategy that can lead to a wide range of corresponding coupling products.

To our knowledge, numerous articles and reviews have been reported previously [30,36,39–41,46–52], particularly in the field of enantioselective coupling via asymmetric cross-dehydrogenative-coupling (CDC) [40,51] and enantioselective metal carbenoid/nitrenoid insertion into unactivated C–H bonds [40,46,53,54]. As such, this review is meant to briefly highlight, discuss, and illustrate the latest progresses and encountered challenges on enantioselective C–H activation/Mizoroki-Heck reaction and Suzuki reaction with a focus on the origin of chirality. The corresponding mechanism and characteristic features of each part will be discussed in detail, meanwhile, emphasizing the structures and functions of chiral ligands utilized in different reactions.

2. Enantioselective C–H Activation/Suzuki Reaction

Among different cross-coupling strategies, the Suzuki reaction between a $C(sp^2/sp^3)$-halide or $C(sp^2/sp^3)$-triflate with a $C(sp^2/sp^3)$-boronic acid or ester manifests remarkable performance in realizing C-C bond formation (Scheme 1). It has fascinated enormous attention since its first disclosure in 1979 and become a dynamic and powerful tool [29].

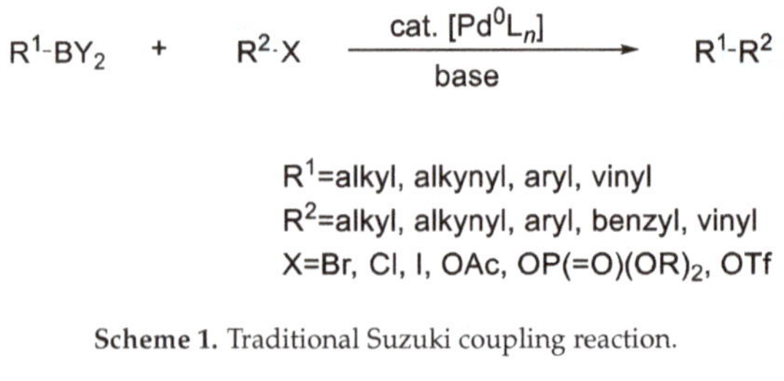

$$R^1\text{-BY}_2 \;+\; R^2\text{-X} \xrightarrow[\text{base}]{\text{cat. }[Pd^0L_n]} R^1\text{-}R^2$$

R^1=alkyl, alkynyl, aryl, vinyl
R^2=alkyl, alkynyl, aryl, benzyl, vinyl
X=Br, Cl, I, OAc, OP(=O)(OR)$_2$, OTf

Scheme 1. Traditional Suzuki coupling reaction.

The great influences of Suzuki reaction are attributed to its high tolerance of different functional groups and extraordinary efficiency under facile reaction conditions in various media. In addition, its prominent compatibility to diverse processes, including microwave [55] and continuous flow conditions [56,57], and the easy accessibility of organic boronic acid or ester coupling partners, which are usually stable to oxygen, water, as well as harsh reaction conditions, are also attractive to scientists.

In recent years, the direct asymmetric C–H activation/Suzuki cross-coupling reactions have substantially attracted the attention of synthetic chemists. Herein, we review the recent progresses on enantioselective C–H activation/Suzuki cross-coupling reactions of prochiral substrates.

In 2008, Li et al., developed the asymmetric C(sp^3)–H arylation reactions of tetrahydroisoquinolines with aryl boronic acids [58]. The reaction proceeded smoothly with CuBr as the catalyst and T-HYDRO as the oxidant in DME. Interestingly, when chiral PhPyBox was added as the ligand, the desired product was obtained with 30% *ee* (Scheme 2). The addition of CuOTf instead of CuBr further increased the enantioselectivity to 44%.

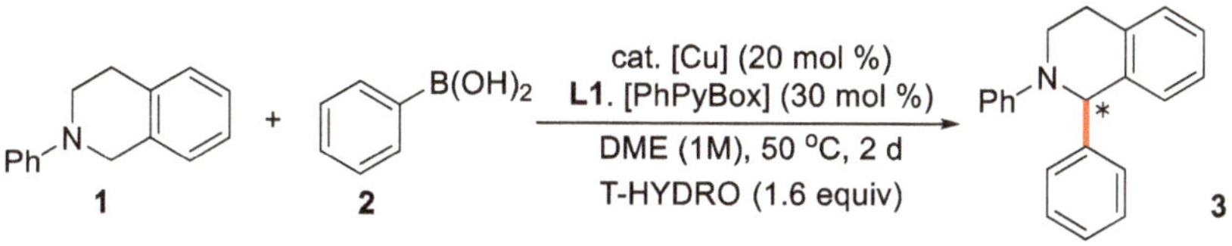

Copper salt	*ee* (%)
CuBr	30
CuOTf	44

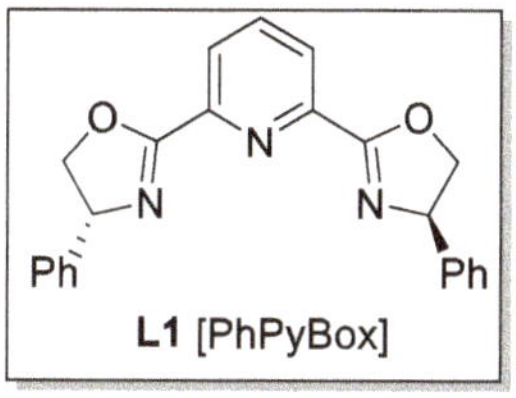

Scheme 2. Copper-catalyzed oxidative C(sp^3)–H bond arylation with aryl boronic acids. Reproduced from Reference [58].

There were two proposed mechanism pathways (Scheme 3) and path B is preferred. In pathway B, the copper-iminium ion intermediate is generated and the subsequent coupling with aryl boronic acids gives access to the desired product. Meanwhile, the introduction of chiral ligands could induce the enantioselective C–H activation process. However, the enantioselectivity still remains to be further improved for practical applications.

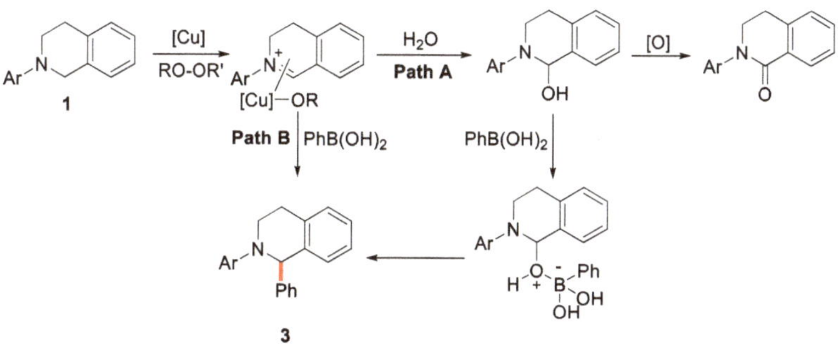

Scheme 3. Proposed mechanism of the oxidative arylation reaction. Reproduced from Reference [58].

Since 2009, the direct arylation of inert C–H of heteroarenes catalyzed by Pd-catalysts has rapidly evolved as a reliable and practical method. In 2011, Itami et al., developed the Pd-catalyzed oxidative C4-selective C–H arylation of thiophenes and thiazoles enabled by boronic acids (Scheme 4) [59].

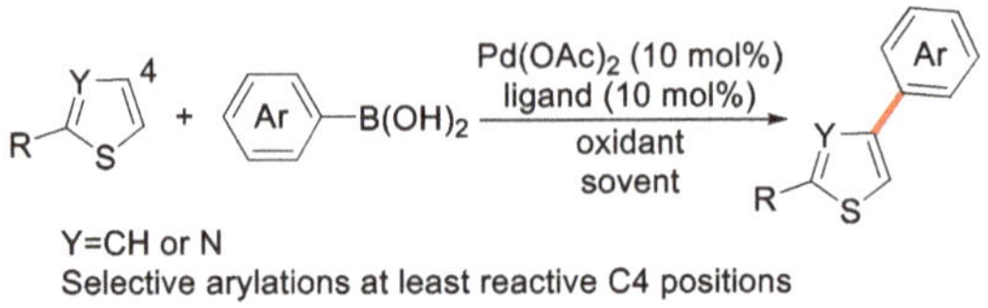

Scheme 4. Selective arylation of heteroarenes via C–H activation [59].

This approach was further improved by Itami group in 2012 [60]. In order to access the hindered heterobiaryls, the screening of ligands was executed employing **4a** and **5a** (2,3-dimethylthiophene and 2-methylnaphthalenyl-1-boronic acid respectively) as substrates and Pd^{II} as the catalyst. Consequently, the desired coupling product was obtained with bisoxazolines ligands in the presence of Pd(OAc)$_2$, and L3 was proved to be the most effective ligand (Scheme 5). Under the optimized reaction conditions, the scope of the substrates (thiophenes and arylboronic acids) was examined. A variety of different substrates were amenable for this system, delivering moderate yields of up to 84% and excellent C4 regioselectivities up to 99%. The successful synthesis of the tetra-*ortho*-substituted heterobiaryl **6aa** manifested the high efficiency of this bisoxazoline-Pd catalytic system.

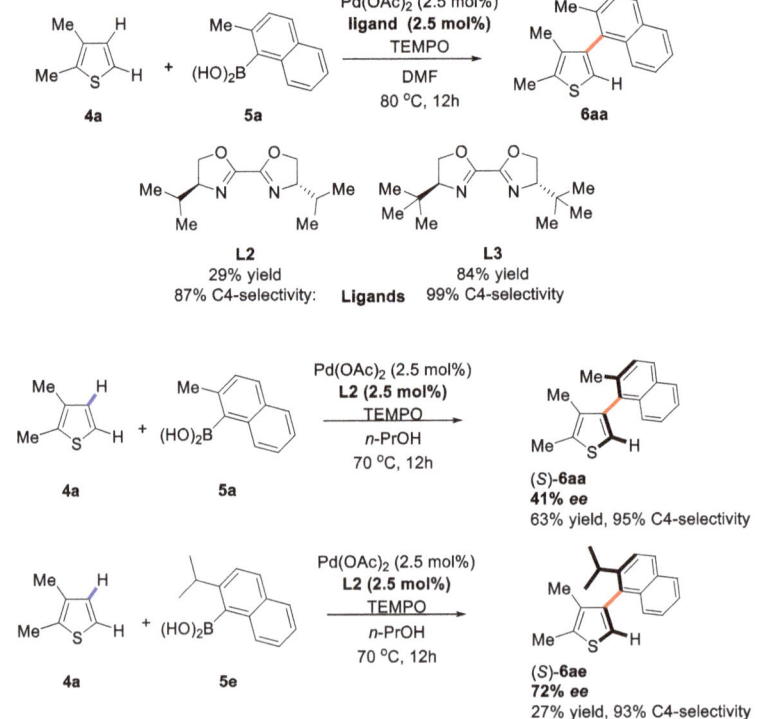

Scheme 5. Generation of axially chiral biaryls via C–H activation/Suzuki reaction [60].

As the ligands delivering excellent performance (**L2** and **L3**) are chiral, the enantioselective synthesis of axially chiral heterobiaryls was predicted to be possible. To have a better understanding of the axial chirality of the substituted heterobiaryls, the rotation energy of 3-methyl-4-(2-methylnaphthalen-1-yl)-thiophene (**6ba**) between two conformations was investigated by DFT calculations (Figure 1). It was found that the rotation energy was high enough for the two atropisomers to exist stably at room temperature.

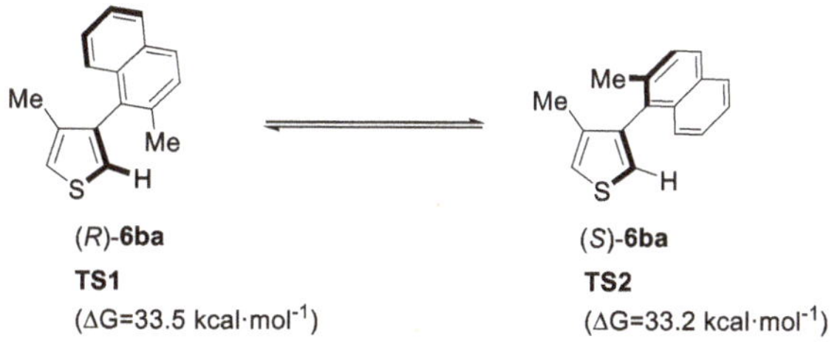

Figure 1. Rotation energy of **6ba** calculated at B3LYP/6-31G(d) level. Reproduced from Reference [60].

After investigating various conditions, impressively, the asymmetric induction truly occurred (Scheme 5). When the *n*-PrOH solution of **4a** and **5a** was exposed to Pd(OAc)$_2$/**L2** and TEMPO at 70 °C for 12 h under air, the product (*S*)-**6aa** was obtained with 41% *ee* and 63% yield. When a more sterically encumbered arylboronic acid **5e** was used, the enantiometric excess of the corresponding product was improved to 72%, albeit with a lower yield. The absolute stereo-configuration of the asymmetric products was determined by X-ray crystallography.

In 2013, the same group revealed another version of this reaction [61]. In this paper, they utilized a Pd^{II}–sulfoxide–oxazoline/iron–phthalocyanine (**FePc**) dual catalyst system for the syntheses of sterically hindered heterobiaryls with air as the oxidant instead of using TEMPO as the stoichiometric co-oxidant (Scheme 6). It was proposed that the ligands (e.g., **L4**) took effect in the form of a **Pd-sox** complex, which exhibited higher reactivity in coupling hindered partners.

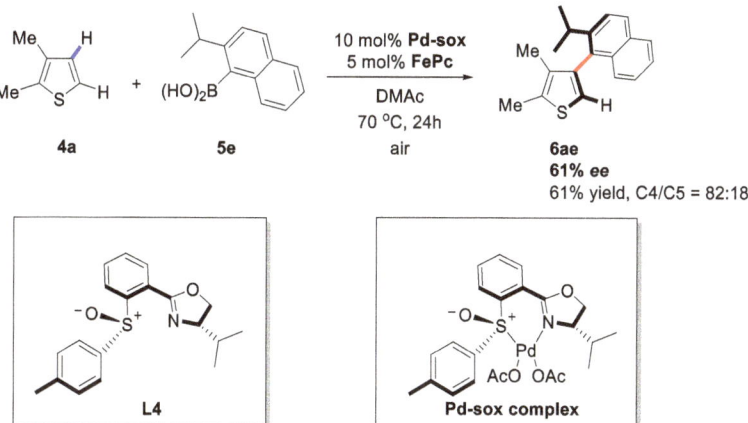

Scheme 6. Enantioselective C–H activation/C–C coupling between heteroarenes and arylboronic acids [61].

In 2008, a huge advance was made by Yu and coworkers [62]. They developed the Pd^{II}/chiral mono-*N*-protected amino acid (**MPAA**) system and applied it in the desymmetrization reaction of prochiral substrates (Scheme 7).

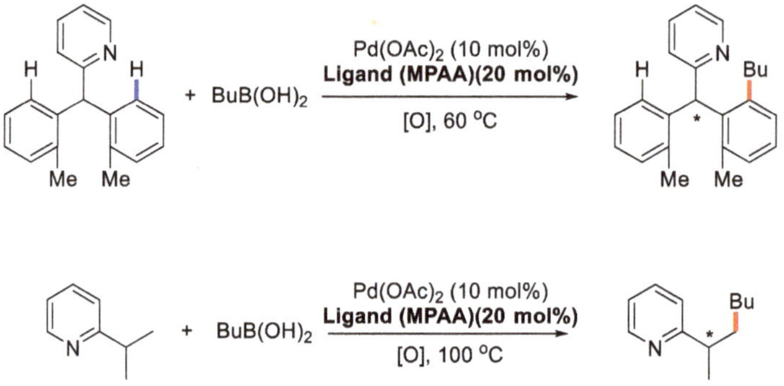

Scheme 7. Desymmetrization of prochiral C(sp²)–H and C(sp³)–H bonds [62].

Detailed mechanism studies by Yu and coworkers suggested that using conformationally rigid chiral carboxylic acids as ligands as well as Pd^{II} catalyst might induce the enantioselective C–H activation process. Indeed, when Boc-protected chiral amino acids were used, the asymmetric induction took place. With compound 7 and butylboronic acid as model substrates, Boc-L-leucine afforded the corresponding product with 63% yield and 90% *ee* (Scheme 8). Intriguingly, when the Boc protecting group was removed or substituted by methyl group, the reactions failed to occur, which indicates that an electron-withdrawing group on the nitrogen atom is essential in maintaining the electrophilicity of Pd^{II} towards the C–H bond. Moreover, the esterification of the amino acid or the decrease of the nitrogen protecting group size resulted in significant declines of enantioselectivities. Thus, the bulkier menthoxycarbonyl protecting groups were introduced and ligand **L13** was found to give the best results.

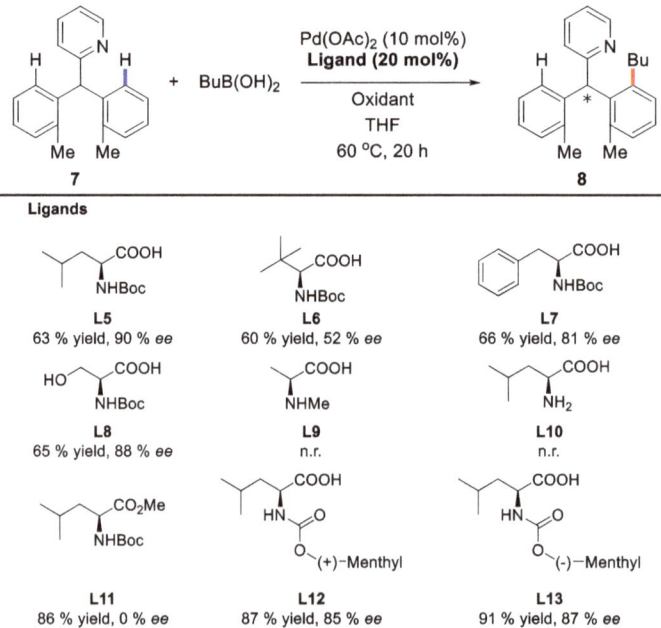

Scheme 8. Influence of ligands on the enantioselectivities [62].

With the optimized condition in hand, different substrates and boronic acids were investigated and the products were obtained in good yields and moderate to excellent *ee* (Table 1). Furthermore, enantioselective alkylation of C(sp^3)–H bonds, such as substrate **9**, was also executed and the desired product was obtained with 38% yield and 37% *ee* (Scheme 9).

Table 1. Investigation on substrate scope [62].

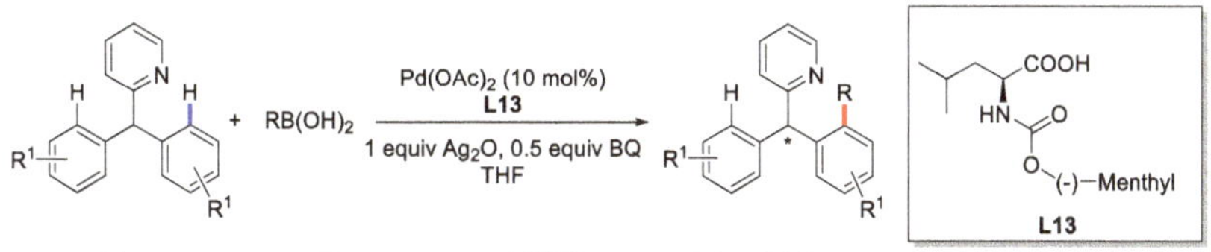

R^1	R	T [°C]	L [mol %]	Yield (%)	*ee* (%)
o-Me	*n*-Bu	50	20	50	95
o-Me	*n*-Bu	60	10	96	88
H	*n*-Bu	80	20	47	79
o-Me	Et	60	10	81	84
o-Me	Cy	60	10	61	89
m-Me	*n*-Bu	60	10	58	84
m-OMe	*n*-Bu	80	10	55	54
m-OAc	*n*-Bu	80	10	43	72
p-Me	*n*-Bu	80	10	61	78

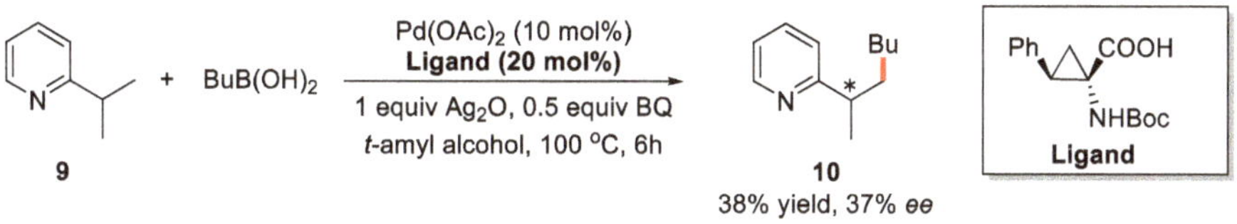

Scheme 9. Enantioselective alkylation of C(sp^3)–H bond [62].

In the proposed transition state, both the nitrogen atom and the carboxylate group of the amino acid ligand coordinate with the Pd^{II} center in a bidentate way, providing a chiral environment. Transition state **11a** rather than **11b** is preferred, in that the steric repulsion between the substituent on the newly generated chiral center (*o*-Tol) and the Boc group on the nitrogen center is minimized (Figure 2).

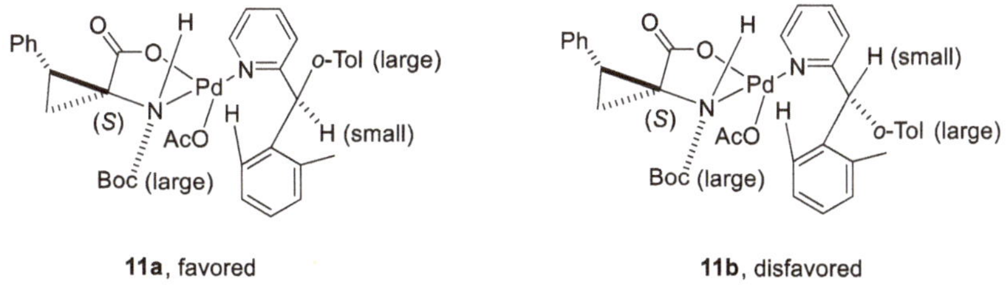

Figure 2. Key intermediates in the proposed mechanistic model. Reproduced from Reference [62].

In 2011, Yu and coworkers reported the enantioselective C(sp^3)–H activation of cyclopropanes catalyzed by Pd^{II}/**MPAA** (Scheme 10) [44].

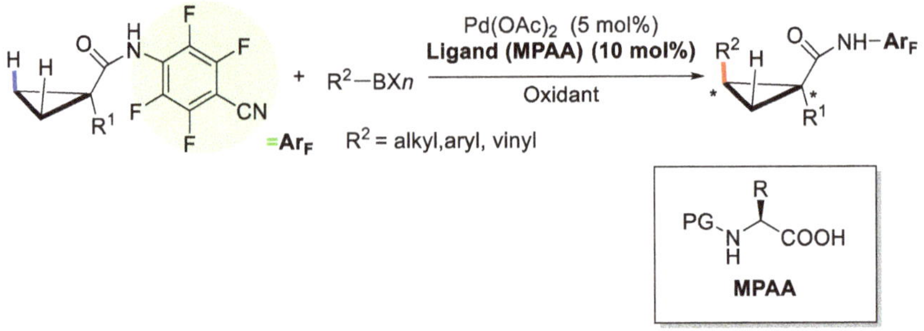

Scheme 10. Asymmetric C–H activation/C–C coupling reaction of cyclopropane [44].

In this reaction, the amide derivative of 1-methylcyclopropanecarboxylic acid was utilized as the substrate and the electron-deficient arylamide group plays as a directing group. In the process of screening the chiral **MPAA** ligands, it was discovered that both the protecting group of the amine and the backbone of the amino acid were crucial for the enantioseectivities (Scheme 11). When the protecting group Boc was changed into TcBoc, the enantioselectivity increased dramatically to 78% from 31%, indicating that CCl_3 might serve as a bulkier group and an electron-withdrawing group (EWG) simultaneously. Further screening revealed that phenylalanine derivative **L23** was the best ligand, and up to 93% ee was achieved.

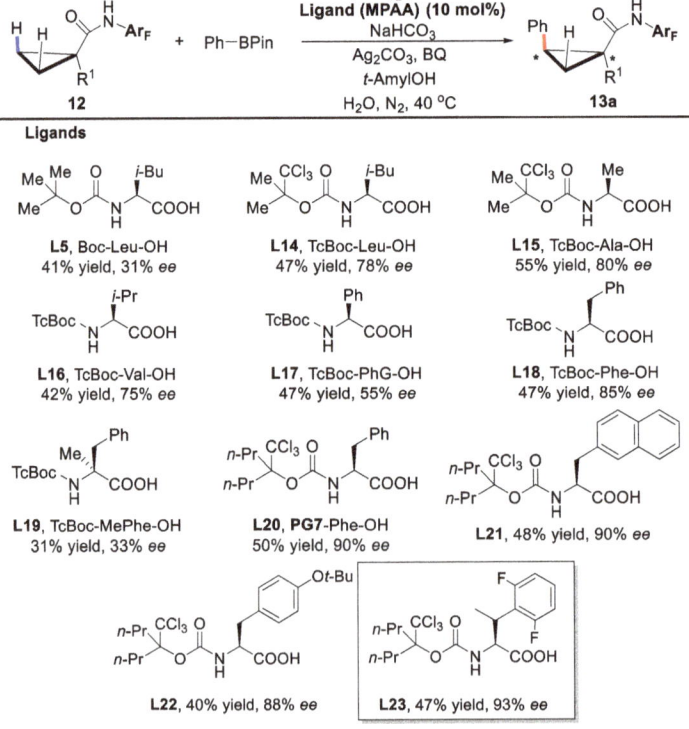

Scheme 11. Screening of chiral mono-protected amino acid ligands [44].

With the optimized conditions, different cyclopropanes and organoboronic compounds were investigated and the products were obtained in good yields and good to excellent *ee* (Figure 3).

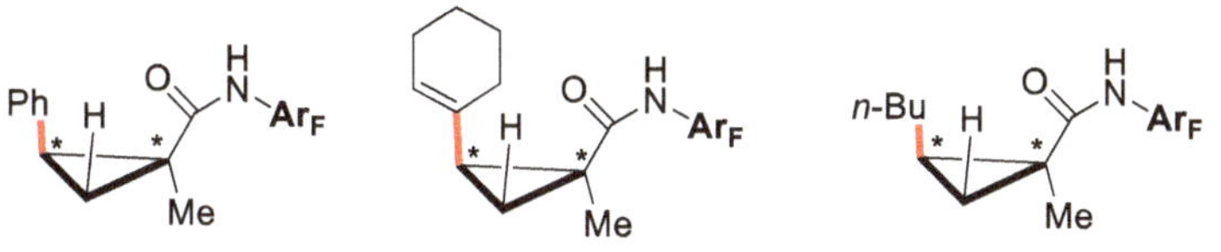

13a, 81% yield, 91% ee   **13b**, 60% yield, 82% ee   **13c**, 49% yield, 62% ee

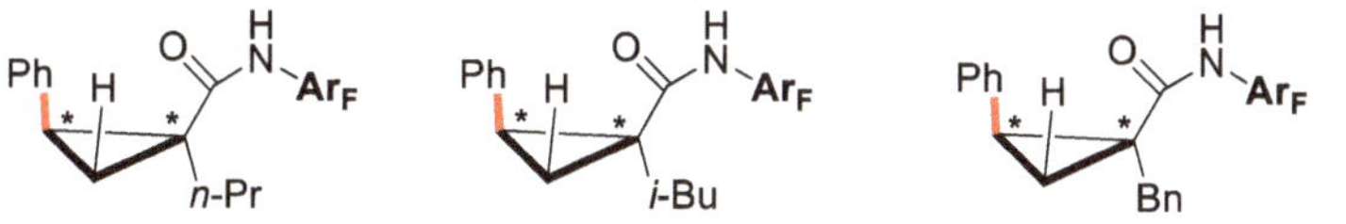

14a, 70% yield, 92% ee   **15a**, 70% yield, 91% ee   **16a**, 80% yield, 92% ee

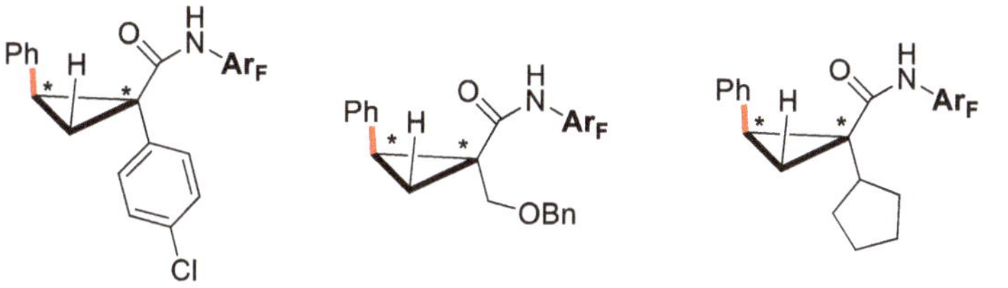

17a, 51% yield, 89% ee   **18a**, 62% yield, 92% ee   **19a**, 66% yield, 87% ee

Figure 3. The substrate scope of asymmetric cyclopropane C–H functionalization [44].

In 2014, the same group reported a further research for the arylation of methylene β-C(sp^3)–H bonds of cyclobutanecarboxylic acid derivatives with arylboron reagents using palladium(II) catalyst with chiral mono-N-protected α-amino-O-methylhydroxamic acid (**MPAHA**) as the ligand (Scheme 12) [63]. This method provided a complementary protocol for the syntheses of enantioenriched cyclobutanes containing chiral quaternary stereocenters [64,65].

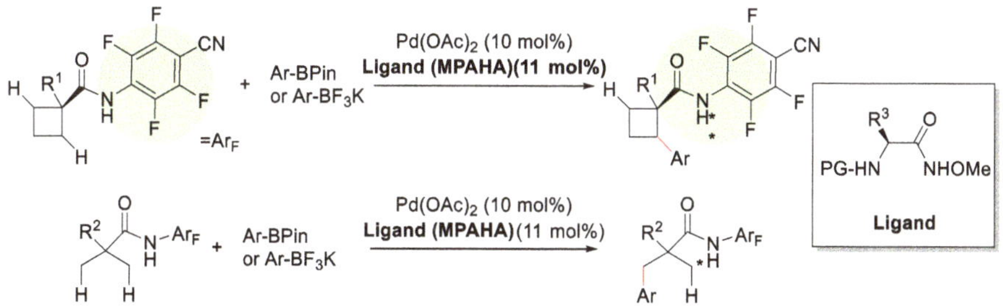

Scheme 12. Enantioselective C(sp^3)–H activation of cyclobutanes [63].

Similar to the precedent studies, **MPAHA** can generate a chiral complex with Pd^{II} catalyst and is the key to obtain appreciable yield and enantioselectivity. When O-methylhydroxamic acids were used as ligands instead of the previously used mono-protected amino acids, a significant boost of enantioselectivity was observed, which might derive from the stronger coordination between the ligand and the Pd^{II} center. Further evaluation revealed that the Boc protecting group and an aromatic side chain within the ligand were prone to elevate the enantioselectivities. Of the various ligands that were tested, **L32** gave the best results (Scheme 13). Further optimization of the solvents, bases and catalysts eventually led to the desired product with 75% yield and 92% ee.

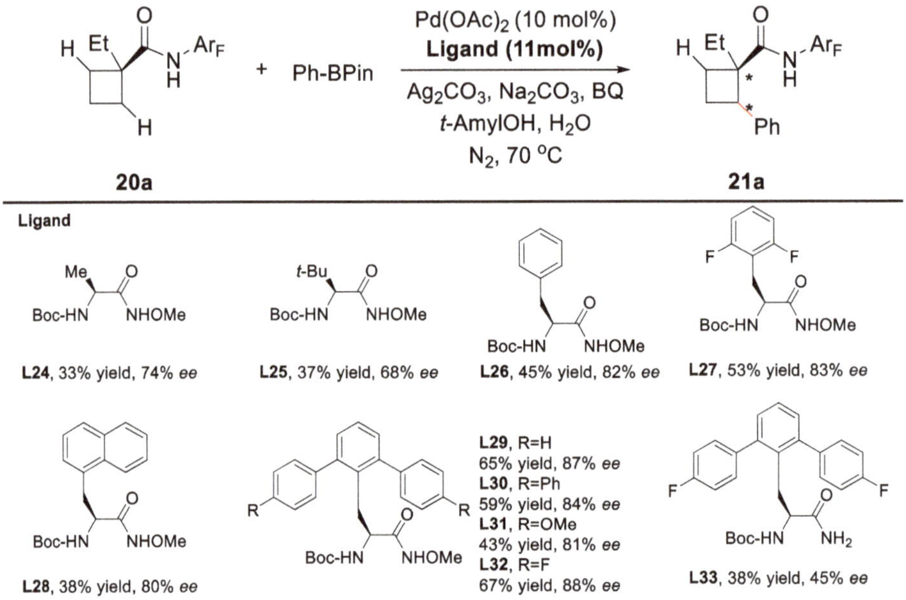

Scheme 13. Designs of chiral ligands [63].

Under the optimized condition, the reaction was found to work well between a variety of arylboronic acid pinacol esters and various 1-substituted 1-cyclobutanecarboxylic acid derivatives (Scheme 14).

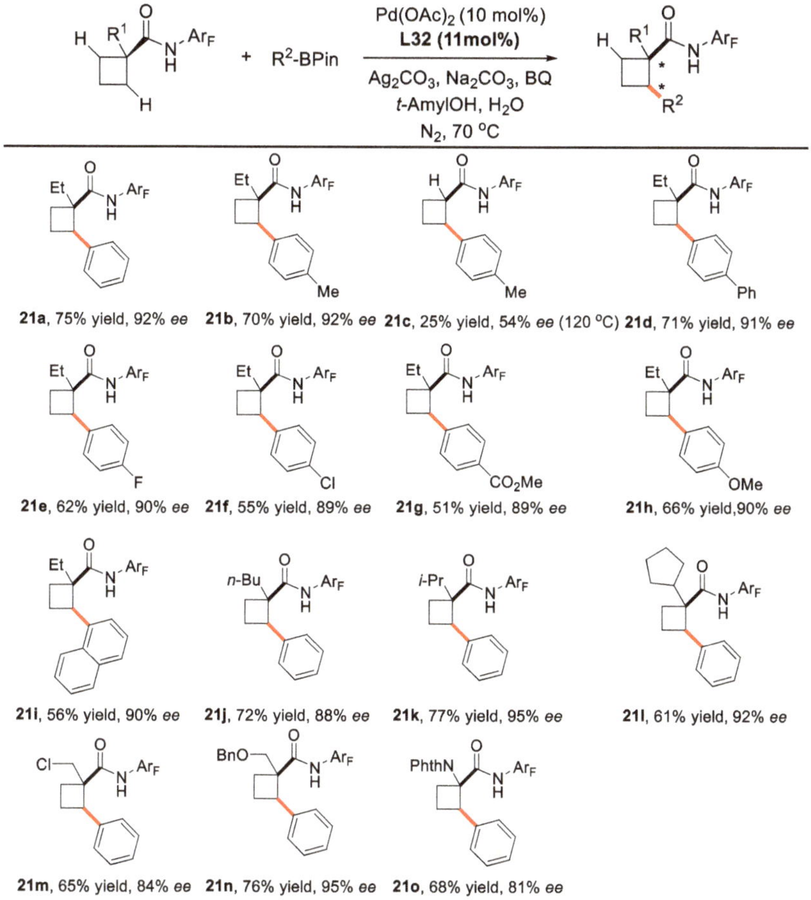

Scheme 14. Substrates scope for C(sp^3)–H activation of cyclobutanes [63].

It is well known that planar chiral ferrocenes are frequently applied as highly efficient catalysts or ligands in asymmetric synthesis [66–71]. Inspired by previous studies from Yu group [44,62], You et al., developed an enantioselective syntheses of planar chiral ferrocenes via palladium-catalyzed direct coupling with arylboronic acids in 2013 (Scheme 15) [72].

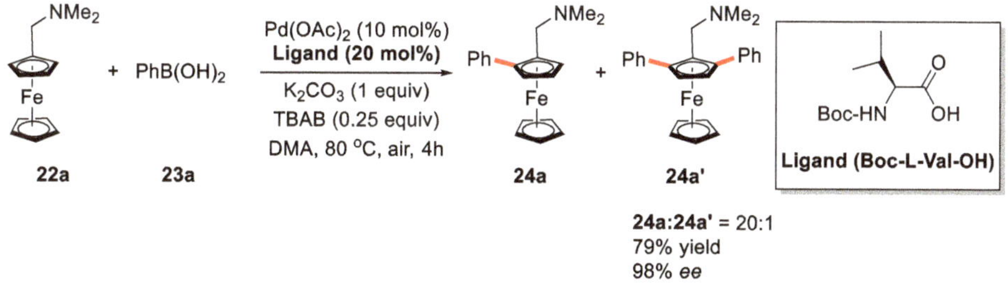

Scheme 15. Synthesis of planar-Chiral Ferrocenes via enantioselective C–H activation [72].

In this work, dimethylaminomethylferrocene **22a** and phenylboronic acid **23a** were chosen as model substrates. The reaction proceeded smoothly in the presence of 10 mol % Pd(OAc)$_2$, 20 mol % Boc-L-Val-OH, and 1 equiv of K$_2$CO$_3$ in DMA (Dimethylacetamide) at 80 °C under air, providing the desired product with 58% yield and 97% ee. The yield was further improved to 79% with 25 mol % TBAB as the additive when the reaction was performed at 60 °C.

With the optimized condition, various aminomethylferrocene derivatives and boronic acids were examined. Substituted arylboronic acids bearing either an electron-donating group or an electron-withdrawing group were well-tolerated and afforded the corresponding products in good yields and excellent enantioselectivities. Moreover, the reaction was also general for aminomethylferrocenes with different alkyl groups on the nitrogen atom (Scheme 16). In addition, a large scale reaction was executed smoothly, which further confirmed the practicality of this method.

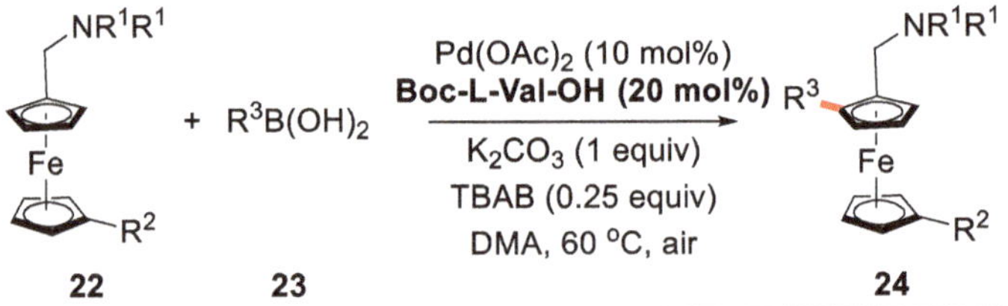

R^1	R^2	R^3	Yield (%)	ee (%)
Me	H	C$_6$H$_5$	79	98
Me	H	3-MeC$_6$H$_4$	81	99
Me	H	4-MeOC$_6$H$_4$	59	96
Me	H	2-naphthyl	75	96
Me	H	4-FC$_6$H$_4$	55	97
Me	H	4-CF$_3$C$_6$H$_4$	61	94
Me	H	4-EtOCOC$_6$H$_4$	72	95
Me	H	Me	14	ND
Et	H	C$_6$H$_5$	67	90
-(CH$_2$)$_4$-	H	C$_6$H$_5$	71	98
Me	Br	C$_6$H$_5$	69	97

Scheme 16. Investigation of substrate scope [72].

At last, the planar-chiral P,N-ligand **L34** prepared from compound **24** was successfully utilized in the palladium-catalyzed allylic alkylation reaction (Scheme 17), which fully demonstrated the potential application of this novel protocol.

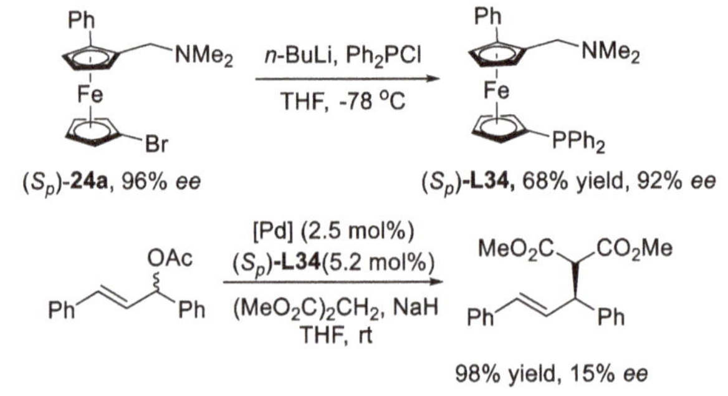

Scheme 17. Application of the synthesized planar-chiral ferrocenes. Reproduced from Reference [72].

It is well known that chiral phosphorus compounds play important roles as ligands or organocatalysts in asymmetric synthesis [73–76]. In 2015, Han group reported the asymmetric syntheses of traditionally inaccessible P-stereogenic phosphinamides via Pd-catalyzed enantioselective C(sp^2)–H functionalization (Scheme 18) [77].

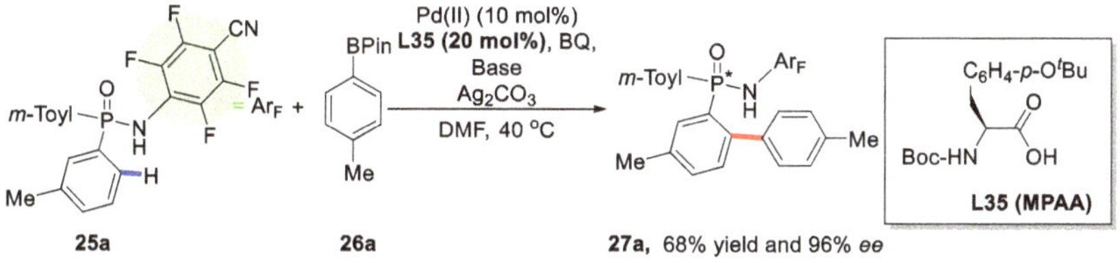

Scheme 18. Enantioselective synthesis of P-stereogenic phosphinamides via asymmetric C–H arylation [77].

Similarly, chiral mono-N-protected amino acids (**MPAA**) were used as ligands in this reaction. The presence of the carbamate moiety employed as the N-protecting group and the carboxylic acid group within the ligand were essential to deliver the desired products with good enantioselectivities. Of the various chiral ligands tested, ligand **L35** was found to be the optimal ligand, affording 68% yield and 96% *ee* (Scheme 18). The best reaction condition was found to be 10 mol % Pd(OAc)$_2$, 20 mol % **L35**, 0.5 equiv of BQ, 1.5 equiv of Ag$_2$CO$_3$, 3.0 equiv of Li$_2$CO$_3$, and 40.0 equiv of H$_2$O in anhydrous DMF at 40 °C under air. In addition, an array of substrates, including arylboronic esters decorated with different groups and different diarylphosphinamides, were subjected to this protocol and most of the reactions occurred efficiently (Scheme 19). Practically, this novel approach could be carried out in the gram scale with consistent efficiency.

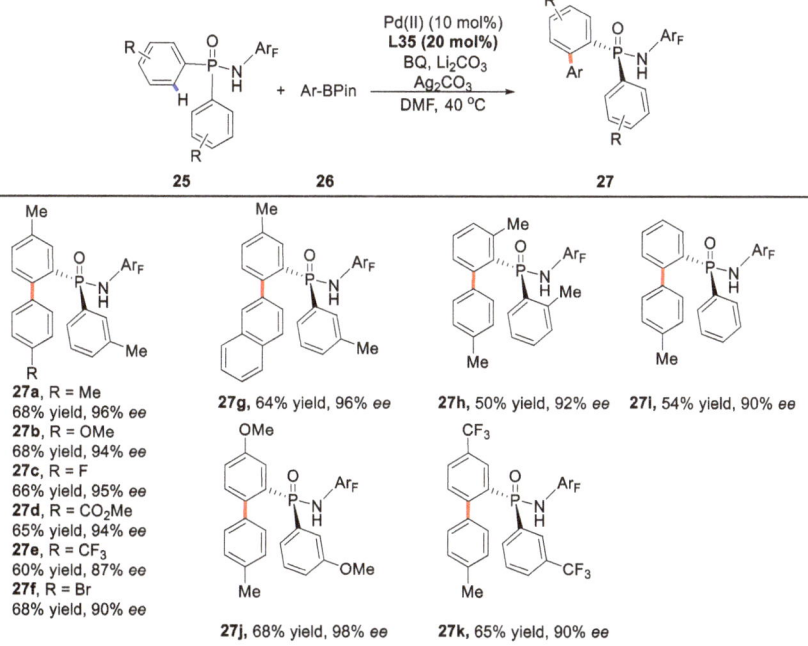

Scheme 19. Evaluation of substrate scope [77].

Recently, the enantioselective *ortho*-C(sp^2)–H coupling between *para*-nitrobenzenesulfonyl (nosyl) protected diarylmethylamines and arylboronic acid pinacol esters was established by Yu and coworkers (Scheme 20) [78].

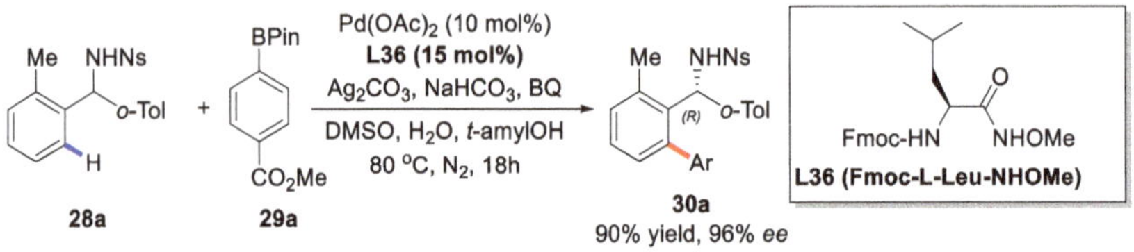

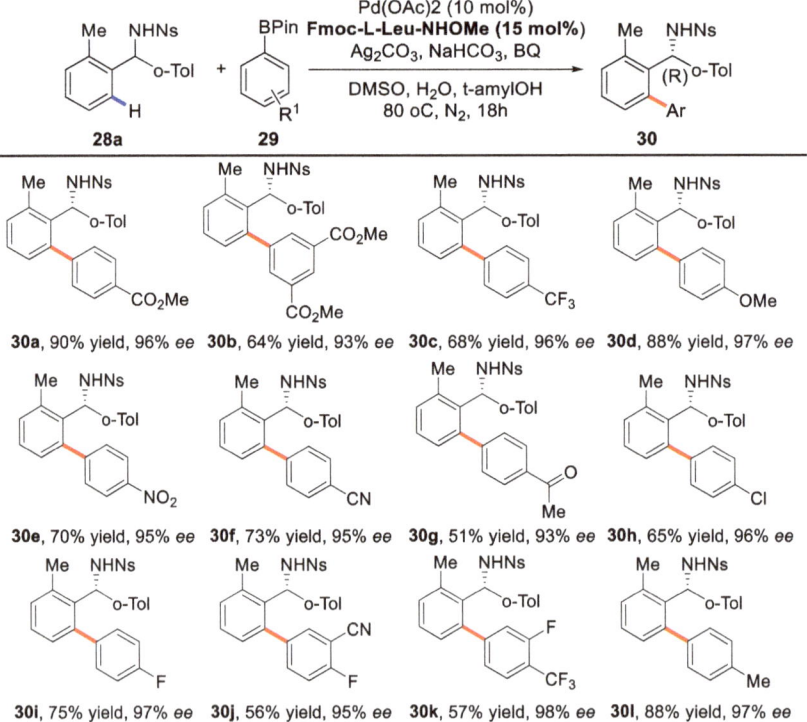

Scheme 20. Enantioselective *ortho*-C–H activation of diarylmethylamines [78].

Herein, chiral mono-*N*-protected amino acids (**MPAA**) were adopted as chiral ligands at first. Further screening revealed that carbamate-protected, *N*-methoxyamide-substituted aliphatic amino acids were the best choice. When Fmoc-L-Leu-NHOMe was used, the desired product could be obtained in 90% yield and 96% *ee*. This method was further applied to a variety of different diarymethylamines with arylboronic acid pinacol esters as coupling partners. Under the optimized condition, most of the reactions proceeded smoothly with good yields and excellent enantioselectivities (Schemes 21 and 22).

Scheme 21. Reactions with different arylboronic acid pinacol esters [78].

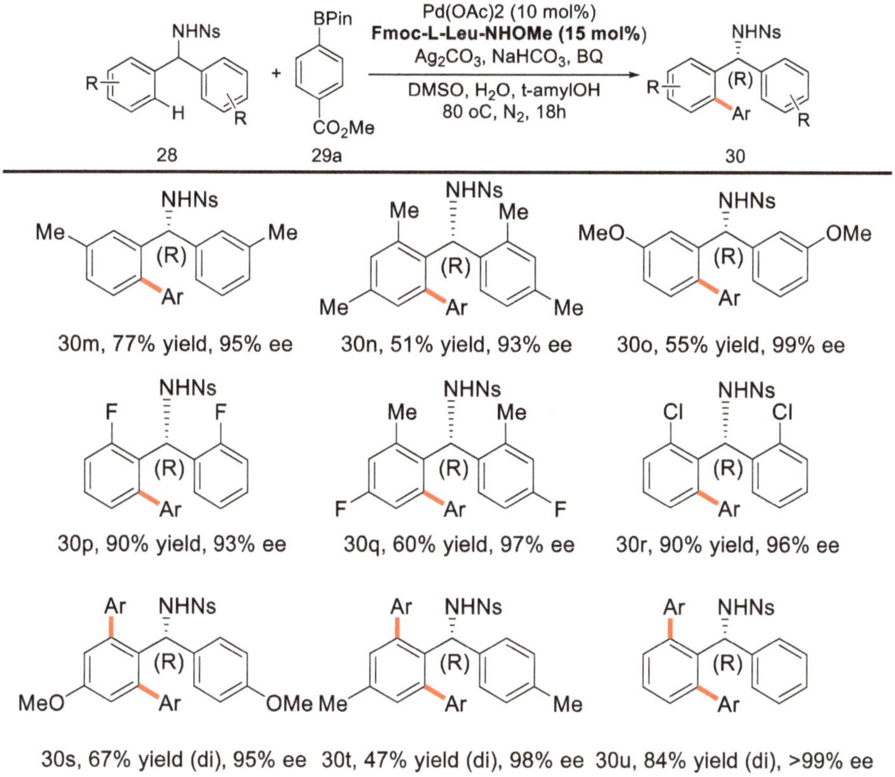

Scheme 22. Reactions with different diarymethylamines [78].

A stereochemical model was proposed (Figure 4). It was assumed that the coordination between the imine moiety of the deprotonated anionic sulfonamide and Pd^{II} center promoted the stereoselective C–H activation followed by arylation.

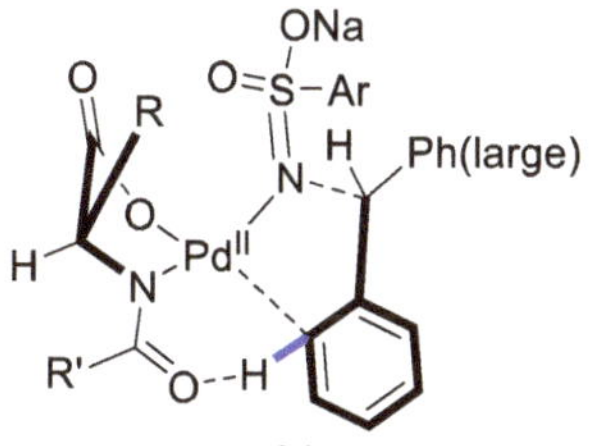

Figure 4. Proposed transition state. Reproduced from Reference [78].

3. Enantioselective C–H Activation/Mizoroki-Heck Type Reaction

Another important type of cross coupling reaction is the Mizoroki-Heck reaction (Scheme 23), which exhibits extraordinary performance with high efficiency in assembling C–C bonds [79,80]. Herein, we will highlight the recent progresses in the field of enantioselective C–H activation concerning Mizoroki-Heck type reaction.

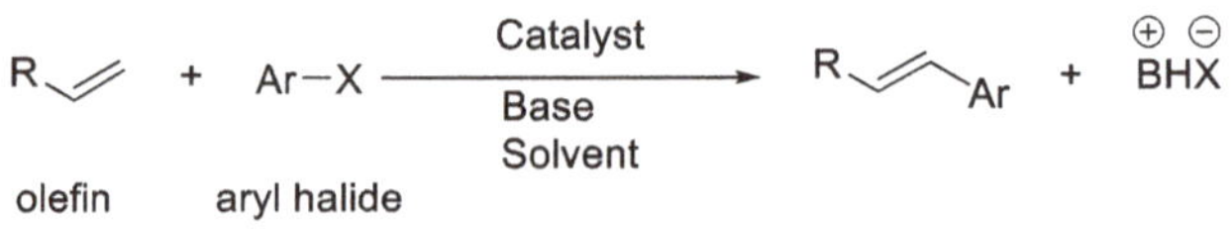

Scheme 23. Mizoroki-Heck reaction.

Inspired by the excellent performances of monoprotected amino acids (**MPAA**) as ligands for enantioselective C–H activation [62], Yu and coworkers developed an enantioselective C–H olefination reaction of diphenylacetic acids using **MPAA** as chiral ligands [81].

Among the various chiral monoprotected α-amino acids examined, Boc-Ile-OH proved to be the best one. The yield could be improved to 73% (97% ee) with the preformed sodium salt of the starting material and KHCO$_3$ as the base (Scheme 24).

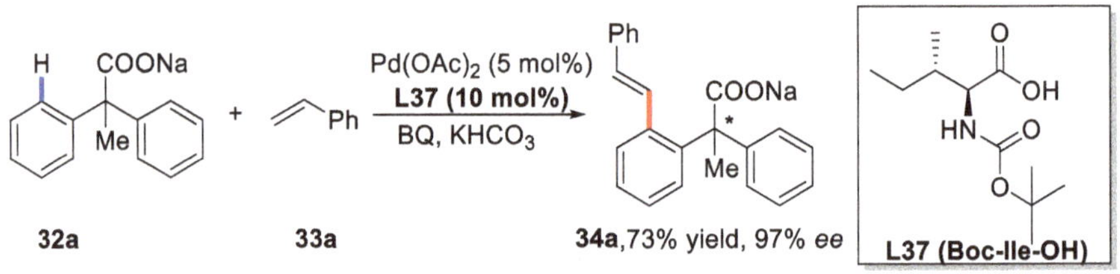

Scheme 24. Enantioselective C–H olefination of diphenylacetic acids [81].

A broad range of styrenes with different substituents were inspected and it was found that styrenes with *para* and *meta* alkyl substituents gave higher enantioselectivities (92–97% ee). In addition, acrylate coupling partners were also tolerant to such condition, affording 99% ee. However, a mixture of the desired olefination product and the corresponding conjugated addition product was isolated (Scheme 25).

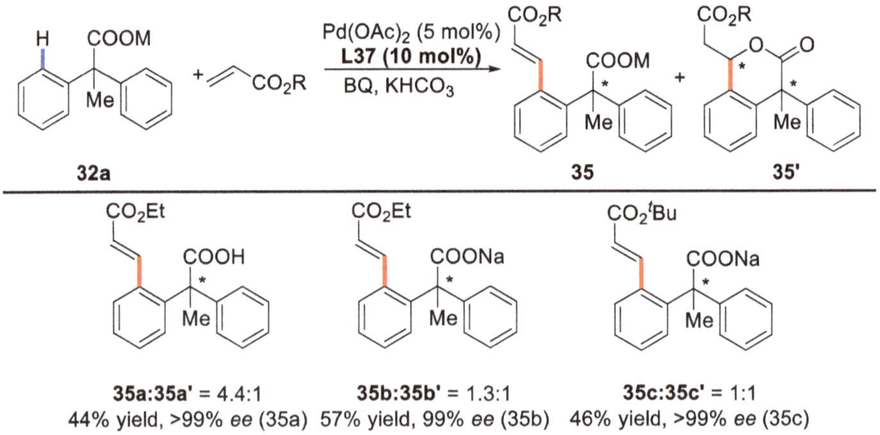

Scheme 25. Acrylates employed as the coupling partners [81].

Different carboxylic acids were also treated with this strategy and most of them proceeded efficiently except for α-hydrogen containing substrate (58% ee). Besides, substrates containing electron-donating groups and moderately electron-withdrawing groups were well compatible to this procedure, although olefination of the latter gave lower yields (Scheme 26).

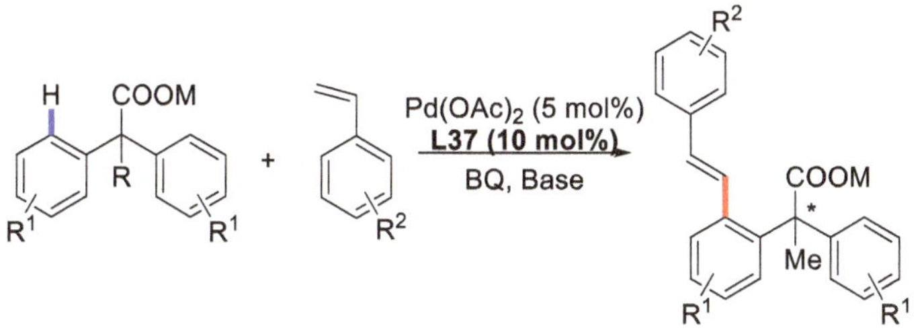

Scheme 26. Enantioselective C–H olefination with substituted styrenes as coupling partners [81].

R	R^1	R^2	Yield (%)	ee (%)
Me	H	H	73	97
Me	H	p-Me	71	97
Me	H	m-Me	63	92
Me	H	o-Me	51	80
Me	H	p-Cl	74	96
Me	H	p-F	51	89
Me	H	p-^{t}Bu	51	95
Me	p-Me	H	63	90
Me	m-Me	H	58	92
Me	3,4-dimethyl	H	63	82
Me	p-^{t}Bu	H	45	88
Me	p-OPiv	H	51	95
Me	p-Cl	H	35	87
Me	3-Cl-4-OMe	H	47	90
Me	3-Me-4-OMe	H	40	75
Me	3-CF$_3$-4-OMe	H	39	89
Et	H	H	61	72
Pr	H	H	52	76
H	H	H	69	58

Yu also proposed a possible transition state for this enantioselective C–H olefination (Figure 5). A chiral carbon-Pd intermediate could be formed, followed by olefination to give the corresponding chiral product.

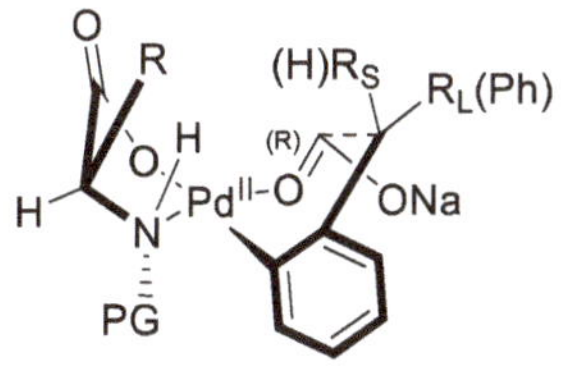

Figure 5. The proposed transition state. Reproduced from Reference [81].

Based on the recent development of enantioselective C–H iodination using Pd^{II}/MPAA catalysts for kinetic resolution through C–H hydroxylation and iodination [82,83], Yu and coworkers developed a kinetic resolution method to achieve enantioselective C–H olefinations of α-hydroxy and α-amino phenylacetic acids utilizing Pd^{II}-catalyzed system in 2016 (Scheme 27) [84].

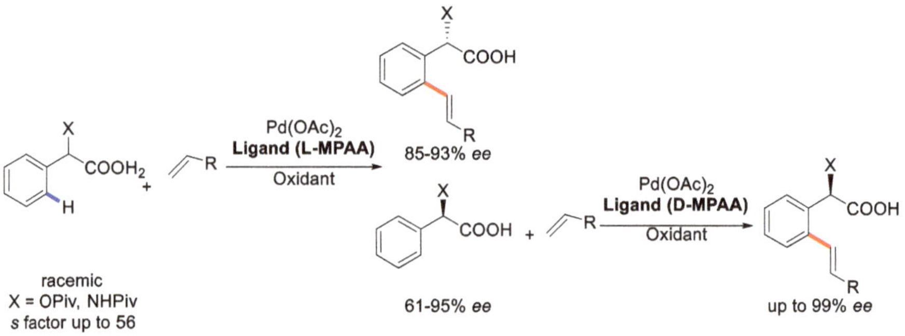

Scheme 27. Enantioselective C–H olefination by kinetic resolution. Reproduced from Reference [84].

In this paper, Yu also employed mono-N-protected amino acid (**MPAA**) as ligands to enable enantioselective C–H bond olefinations. Of different **MPAA** ligands screened, Boc-L-Thr(Bz)-OH (**L38**) gave the best selectivity factor (s) [85] of 54 (90% ee and 45% yield). Notably, the loading of Pd(OAc)$_2$ could be reduced to 5 mol % without a pronounced erosion of the selectivity, and the addition of 0.4 equivalent of olefin increased the enantioselectivity to 93%.

With the optimized reaction condition, a series of olefin coupling partners were subjected to this transformation and a broad range of electron-deficient olefins were well tolerated. Of note, acrylates were a good coupling partner affording s factors ranging from 46 to 54. Vinyl amides and vinyl phosphates also proceeded smoothly. A wide range of different substituted mandelic acid substrates were also successfully olefinated with reasonable s factors (Scheme 28).

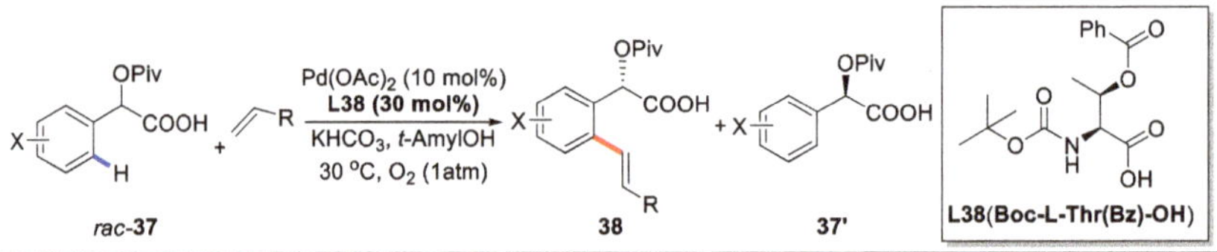

X	R	Yield (%)	ee (%) 38	ee (%) 37'	s
3-Cl	CO$_2$Me	45	90	87	54
3-Cl	CO$_2$Et	46	90	83	48
3-Cl	CO$_2$Bn	41	91	73	46
3-Cl	CONMe$_2$	43	92	77	56
3-Cl	PO(OEt)$_2$	42	91	80	52
3-Cl	COMe	39	92	67	48
3-Cl	Ph	46	90	82	48
3-CF$_3$	CO$_2$Me	35	93	58	50
3-Me	CO$_2$Me	43	89	70	36
3-Ph	CO$_2$Me	41	90	64	37
3-OPiv	CO$_2$Me	40	92	65	47
4-F	CO$_2$Me	37	90	61	22
4-OMe	CO$_2$Me	41	86	67	20
3,4-Cl$_2$	CO$_2$Me	40	89	86	50
3-OMe-4-OPiv	CO$_2$Me	43	85	75	28
H	CO$_2$Me	39	91	64	23

Scheme 28. Enantioselective C–H olefination/kinetic resolution of mandelic acids [84].

Moreover, the scope of different substituted racemic phenylglycine substrates was investigated and most of them furnished the products with synthetically useful *s* factors (Scheme 29).

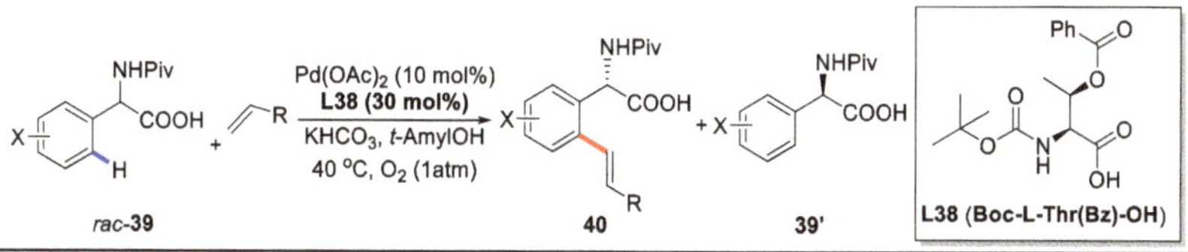

X	R	Yield (%)	*ee* (%)		*s*
			40	39′	
3-Cl	CO₂Me	41	91	79	51
4-Cl	CO₂Me	43	90	83	49
4-F	CO₂Me	38	88	72	34
4-OMe	CO₂Me	40	87	70	30
H	CO₂Me	45	89	79	41
H	Ph	44	88	81	39

Scheme 29. Enantioselective C–H olefination/kinetic resolution of phenylglycines [84].

Furthermore, for the remaining starting materials **37′** with 87% *ee*, a following olefination protocol using the opposite configuration ligand could afford the corresponding chiral product with 99% *ee* (Scheme 30).

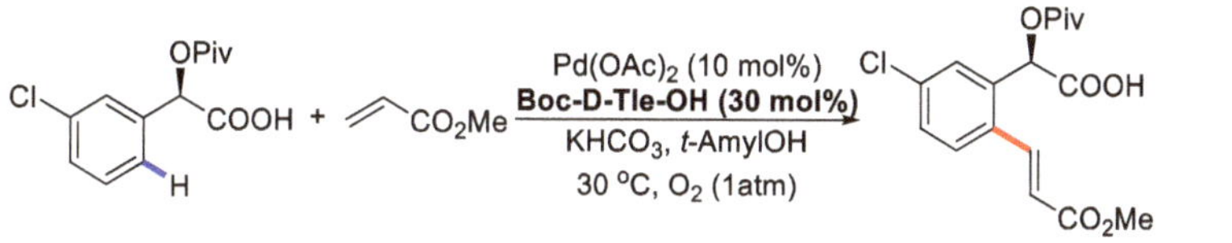

37′, 87% *ee*

38′, 71% yield, 99% *ee*

Scheme 30. Enantioselective C–H olefination of the remaining starting material [84].

In analogy to previously mentioned mechanism, there are two proposed intermediates (Figure 6), **TS$_S$** and **TS$_R$**, respectively, and **TS$_S$** is the favored configuration due to the less steric repulsion between the Boc group and larger OPiv moiety. This stereomodel can well rationalize the origin of the enantioselectivity.

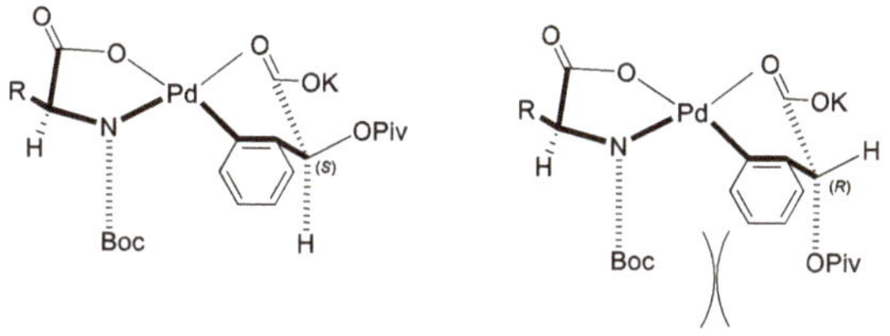

41a
TS$_S$, favored

41b
TS$_R$, disfavored

Figure 6. The proposed intermediate state. Reproduced from Reference [84].

Another interesting finding on enantioselective C–H activation/Heck reaction was provided by Shi and coworkers in 2017 [86]. In this article, atroposelective synthesis of axially chiral biaryls by C–H olefination, utilizing the Pd^II/transient chiral auxiliary (**TCA**) as the catalytic system, was achieved. Of note, in this asymmetric C–H olefination reaction, the chiral free amino acid played as a transient chiral auxiliary (**TCA**), which promoted the C–H activation process instead of only serving as a simple chiral ligand. Of the various **TCA** examined, the L-*tert*-leucine (**T1**) was found to be the optimal. As expected, the other atropisomer was obtained, while the opposite **TCA** (D-*tert*-leucine) was employed (Scheme 31).

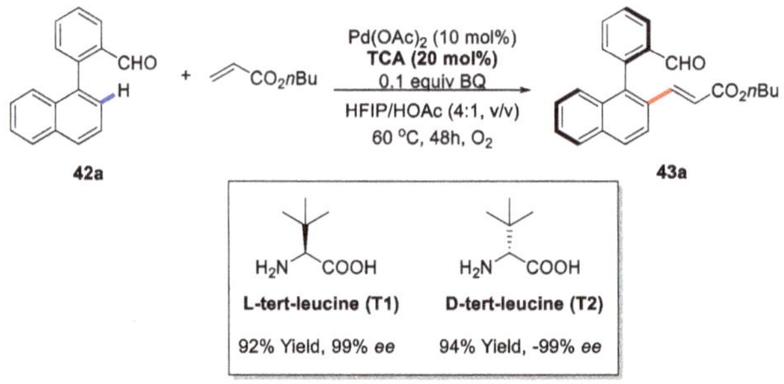

Scheme 31. Optimization of the reaction [86].

With the optimized reaction condition, the substrate generality was inspected. A broad range of biaryls with different substituents were tolerated to this protocol. It ought to be noted that C–H olefination of biaryls with substituents at either 6- or 2′-position or less hindered substituents at both 6- and 2′-position proceeded successfully in a dynamic kinetic resolution (DKR) manner, giving enantioenriched products in good to excellent yields and excellent enantioselectivities (95 to >99% *ee*) (Scheme 32).

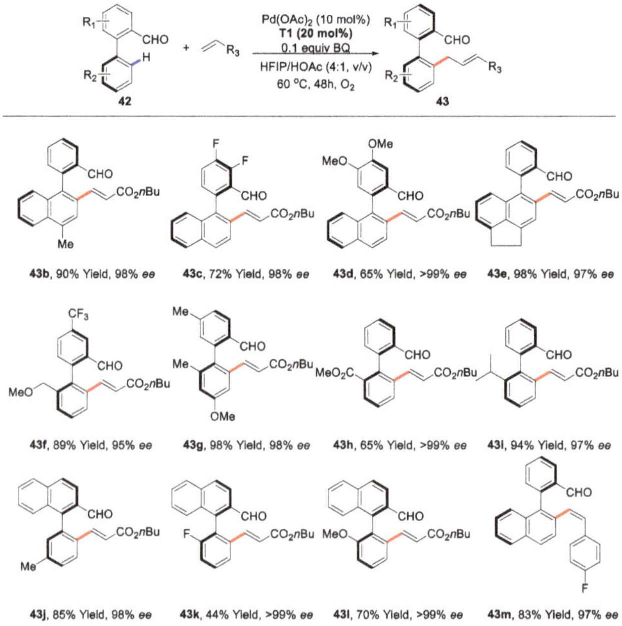

Scheme 32. Substrate scope of biaryls concerning C–H olefination/DKR [86].

Nonetheless, for the biaryls bearing sterically more hindered substituents at both 6- and 2′-position, C–H olefination would occur through a kinetic resolution (KR) manner in excellent selectivities (Scheme 33). Moreover, various acrylates and styrenes were investigated and most of them were compatible to this method, except that electronrich styrenes were determined as inert coupling partners (Schemes 32 and 33).

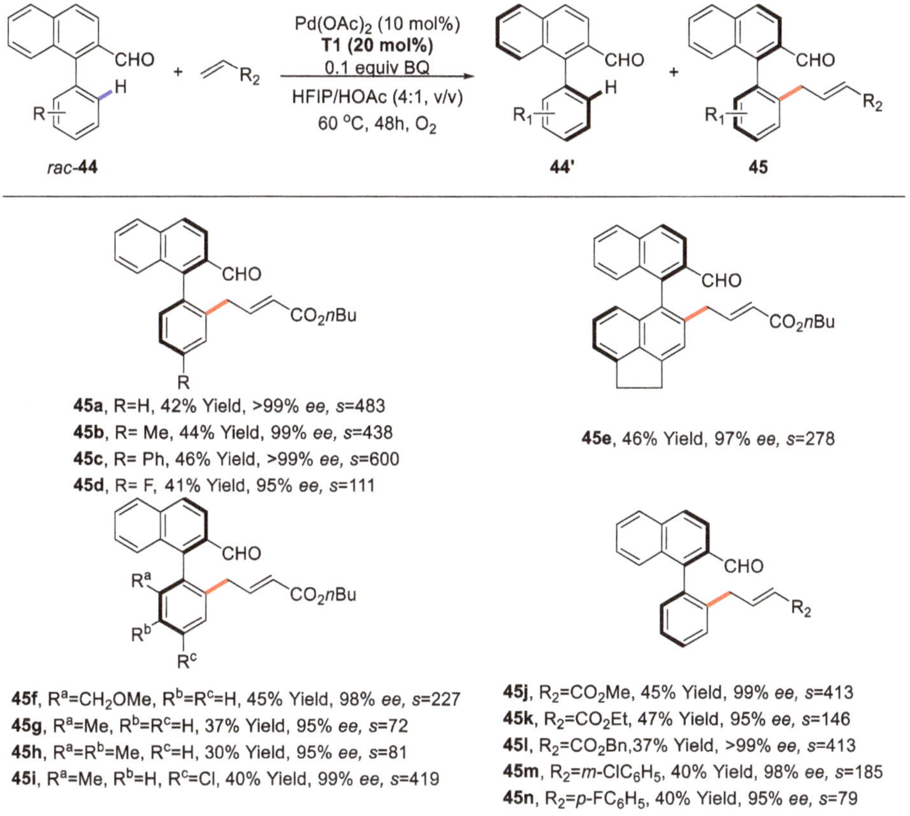
CHO
H
R
rac-44
R2
Pd(OAc)2 (10 mol%)
T1 (20 mol%)
0.1 equiv BQ
HFIP/HOAc (4:1, v/v)
60 °C, 48h, O2
R1
44'
45
CO2nBu
45a, R=H, 42% Yield, >99% ee, s=483
45b, R= Me, 44% Yield, 99% ee, s=438
45c, R= Ph, 46% Yield, >99% ee, s=600
45d, R= F, 41% Yield, 95% ee, s=111
45e, 46% Yield, 97% ee, s=278
Ra
Rb
Rc
45f, Ra=CH2OMe, Rb=Rc=H, 45% Yield, 98% ee, s=227
45g, Ra=Me, Rb=Rc=H, 37% Yield, 95% ee, s=72
45h, Ra=Rb=Me, Rc=H, 30% Yield, 95% ee, s=81
45i, Ra=Me, Rb=H, Rc=Cl, 40% Yield, 99% ee, s=419
45j, R2=CO2Me, 45% Yield, 99% ee, s=413
45k, R2=CO2Et, 47% Yield, 95% ee, s=146
45l, R2=CO2Bn,37% Yield, >99% ee, s=413
45m, R2=m-ClC6H5, 40% Yield, 98% ee, s=185
45n, R2=p-FC6H5, 40% Yield, 95% ee, s=79

Scheme 33. Substrate scope of biaryls concerning C–H olefination/KR [86].

As far as the mechanism was concerned, it was proposed that the chiral amino acid would react with the racemic substrate to give the imine intermediates **A** and **B** reversibly. Then, C–H cleavage of **B** took place selectively due to the minor steric repulsion, resulting in intermediate **C** with axially stereoenriched biaryl palladacycle. Then, intermediate **C** underwent a typical Heck reaction with olefin to afford intermediate **D**, which would be hydrolyzed to furnish the desired chiral biaryls **(*Ra*)-E**. Meanwhile, the Pd^0 was reoxidised into Pd^{II} to close the catalytic cycle (Scheme 34).

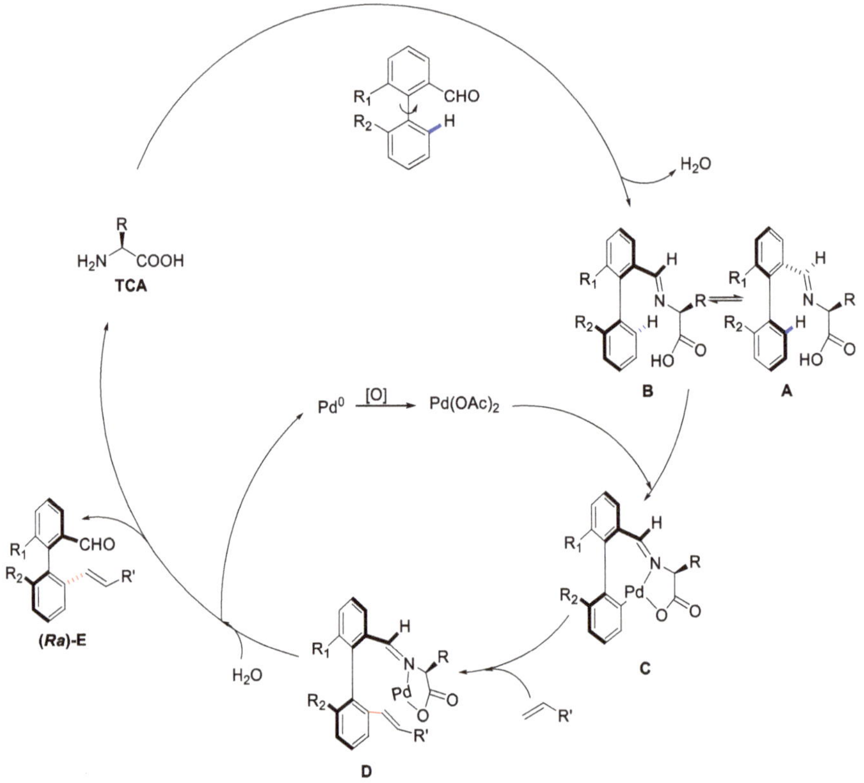

Scheme 34. Hypothesis on the mechanism for the C–H olefination/DKR [86].

The Pd^{II}/MPAA catalysts were also applied in the enantioselective olefinations of N,N-dimethylaminomethylferrocene by Wu group (Scheme 35) [87].

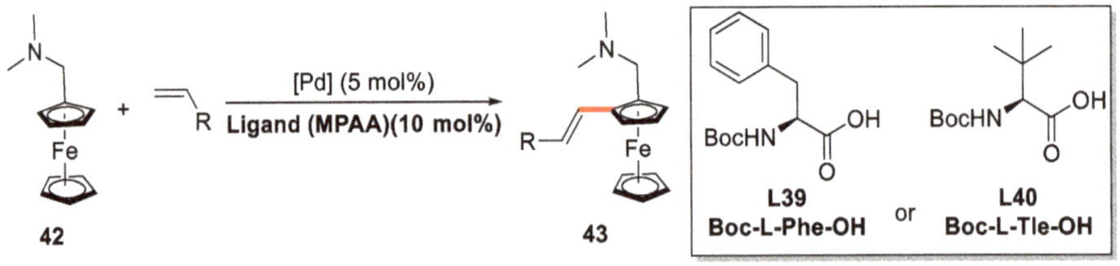

Scheme 35. Pd-catalyzed direct C–H olefination of N,N-dimethylaminomethylferrocene [87].

In the proposed mechanism, the cyclopalladated complex **A** was first generated by palladium species through coordination with the substrate and the ligand. Then, intermediate **A** underwent a typical Heck reaction to furnish the desired product. It is worth noting that in this catalytic system, the *N,N*-dimethylaminomethylferrocenium **C** was generated in situ by air and served as a terminal oxidant to regenerate active Pd^{II} from the reduced Pd^0 species, completing the catalytic cycle. Therefore, no external oxidant was needed (Scheme 36).

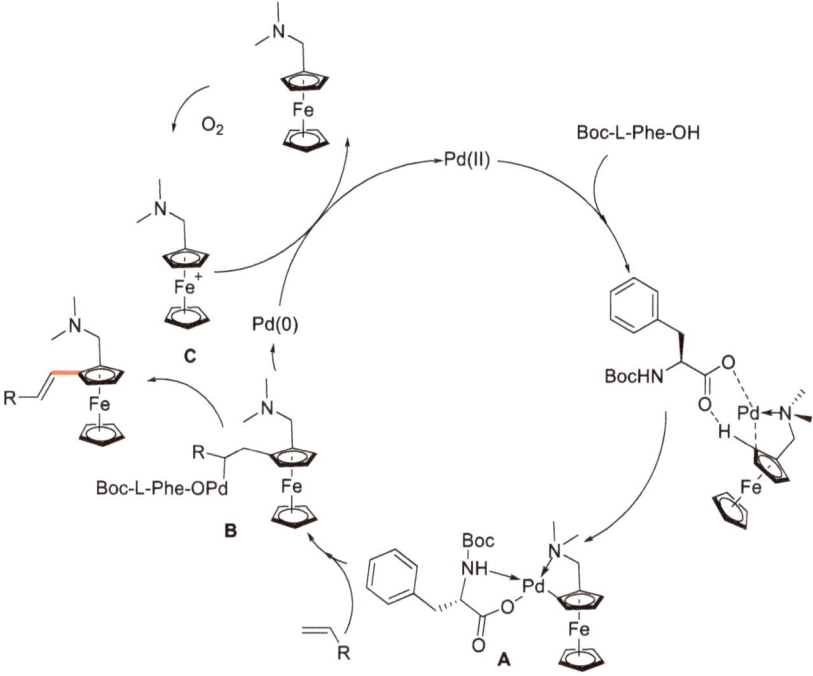

Scheme 36. The proposed mechanism for the reaction. Reproduced from Reference [87].

N,N-dimethylaminomethylferrocene and butyl acrylate were chosen as the model substrates. It was found the introduction of mono-*N*-protected amino acids (**MPAA**) as ligands to Pd^{II}, not just induced the high enantioselectivities, but also dramatically increased the reaction yields. Both Boc-L-Phe-OH (**L39**) and Boc-L-Tle-OH (**L40**) gave excellent yields and enantioselectivities for the model reaction (Scheme 35). Under the optimized condition, various acrylates, styrenes, and even aliphatic olefins were tested (Scheme 37). All of them worked well with excellent enantioselectivities (up to 99%) and yields (up to 98%).

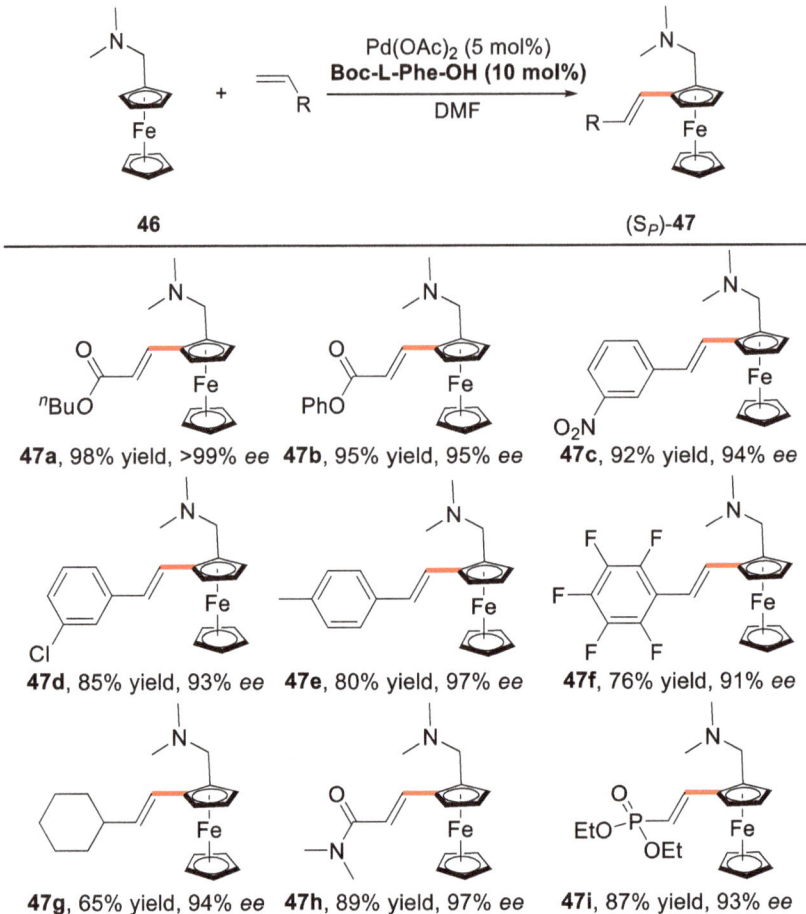

Scheme 37. Substrate scope for the enantioselective C–H olefination reaction [87].

Another impressive development was revealed by You and coworkers [88]. They also employed the Pd^{II}/**MPAA** catalyst system in the enantioselective oxidative C–H/C–H cross-coupling of ferrocenes with heteroarenes to prepare planar chiral ferrocenes.

The Boc-L-Ile-OH proved to be the best ligand, affording the desired product in a C2 regioselective manner with 71% yield and nearly perfect 99% *ee* under the optimal reaction conditions with air as the oxidant. Subsequently, various ferrocene derivatives were evaluated and the reaction exhibited excellent tolerance towards different functional groups (Scheme 38).

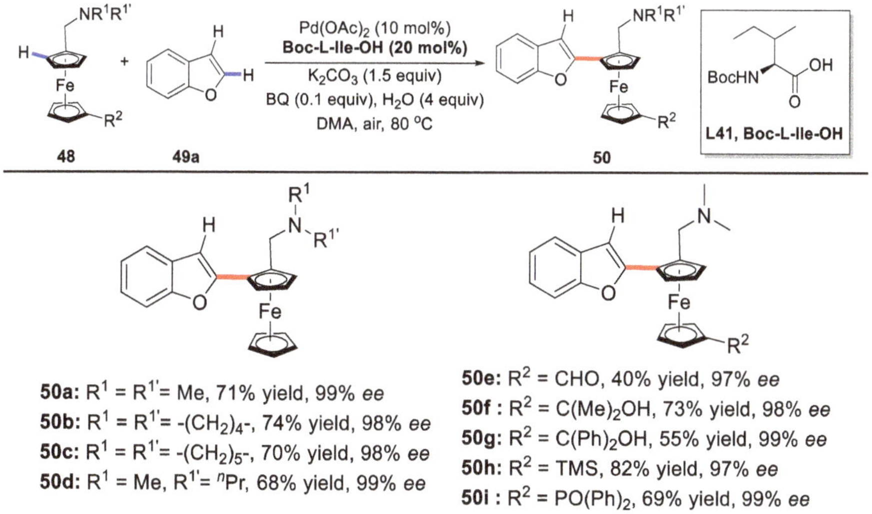

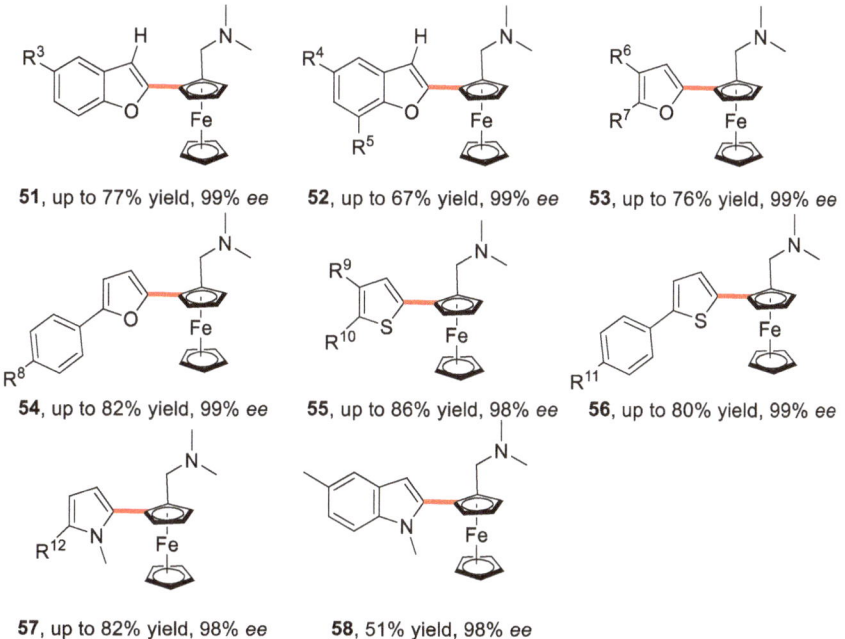

Scheme 38. Enantioselective cross-coupling reactions of benzofurans [88].

Moreover, the scope of heteroarenes was explored as well. Various substituted benzofurans, furans, thiophenes, pyrroles, and indoles worked smoothly under this condition, providing impressive yields and enantioselectivities (up to 99%) (Scheme 39).

51, up to 77% yield, 99% ee
52, up to 67% yield, 99% ee
53, up to 76% yield, 99% ee
54, up to 82% yield, 99% ee
55, up to 86% yield, 98% ee
56, up to 80% yield, 99% ee
57, up to 82% yield, 98% ee
58, 51% yield, 98% ee

Scheme 39. Enantioselective cross-coupling reactions with different heteroarenes [88].

Besides, the newly synthesized planar chiral ferrocenes could be further elaborated into useful ligands for asymmetric transformations (Scheme 40).

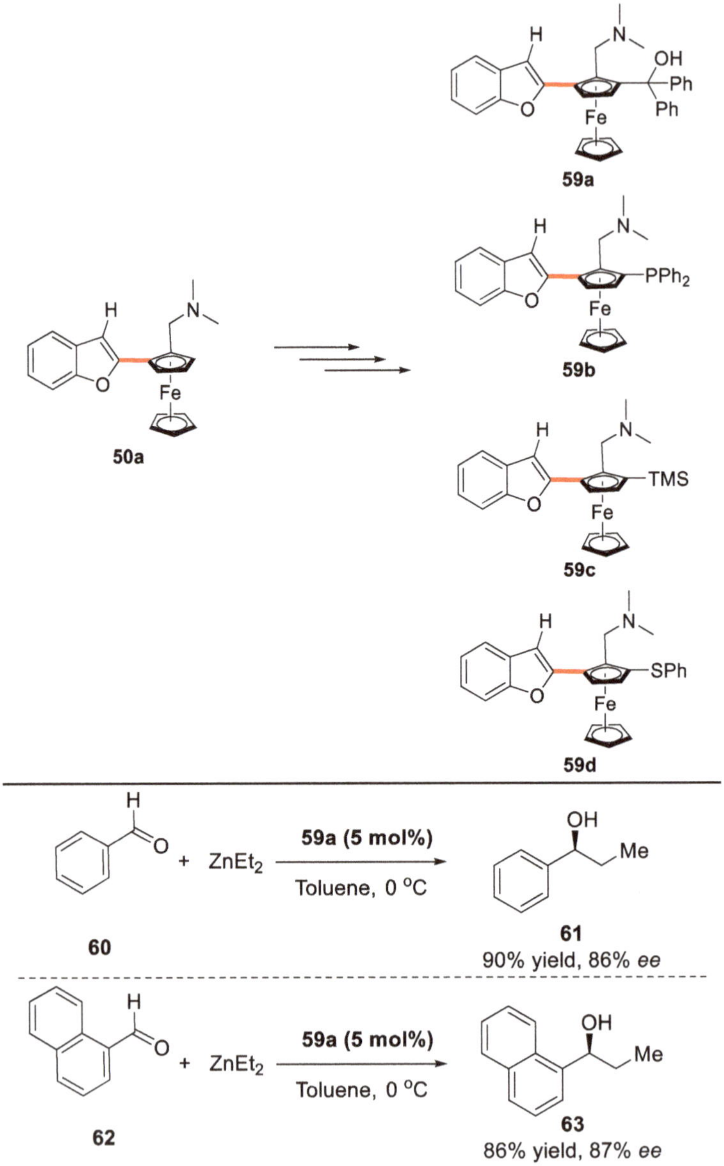

Scheme 40. Applications of the protocol [88].

Another excellent work on synthesizing novel axially chiral biaryls by direct C–H bond olefination was fulfilled by You and coworkers [89]. This dehydrogenative coupling reaction was catalyzed by chiral Cp/rhodium complexes. Further screening revealed that catalysts with bulky substituents usually led to diminished yield and e.r. value (Scheme 41). Finally, the combination of the **Cat.1**, Ag_2CO_3 (1.0 equiv), and $Cu(OAc)_2$ (20 mol %) in methanol was determined to be the optimal condition, affording the desired product with 94% yield and 90:10 e.r.

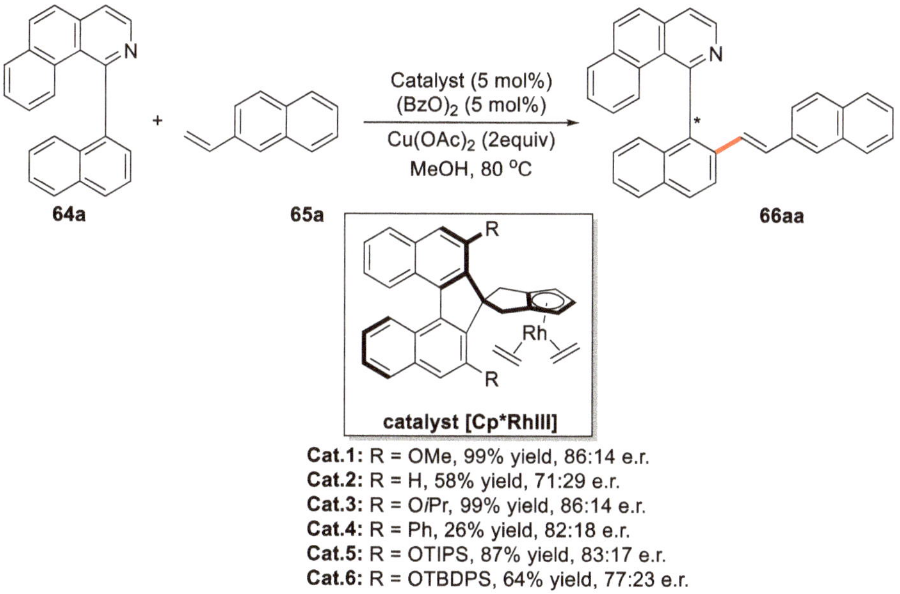
64a
65a
66aa
Catalyst (5 mol%)
(BzO)2 (5 mol%)
Cu(OAc)2 (2equiv)
MeOH, 80 °C
catalyst [Cp*RhIII]
Cat.1: R = OMe, 99% yield, 86:14 e.r.
Cat.2: R = H, 58% yield, 71:29 e.r.
Cat.3: R = O*i*Pr, 99% yield, 86:14 e.r.
Cat.4: R = Ph, 26% yield, 82:18 e.r.
Cat.5: R = OTIPS, 87% yield, 83:17 e.r.
Cat.6: R = OTBDPS, 64% yield, 77:23 e.r.

Scheme 41. Catalyst screening for the enantioselective oxidative Heck coupling reaction [89].

Under the optimized condition, biaryl substrates bearing different EDG or EWG proceeded efficiently to generate the desired alkenylated products in moderate to excellent yields and enantioselectivities, with up to 97% conversion and 93:7 e.r. Additionally, different olefins were well tolerated to this reaction, including styrenes, acrylates, acrylamides, and vinyl phosphonate esters. Notably, ethylene was also introduced, which gave the desired product with 86:14 e.r. Moreover, the gram-scale reaction also worked well (Scheme 42).

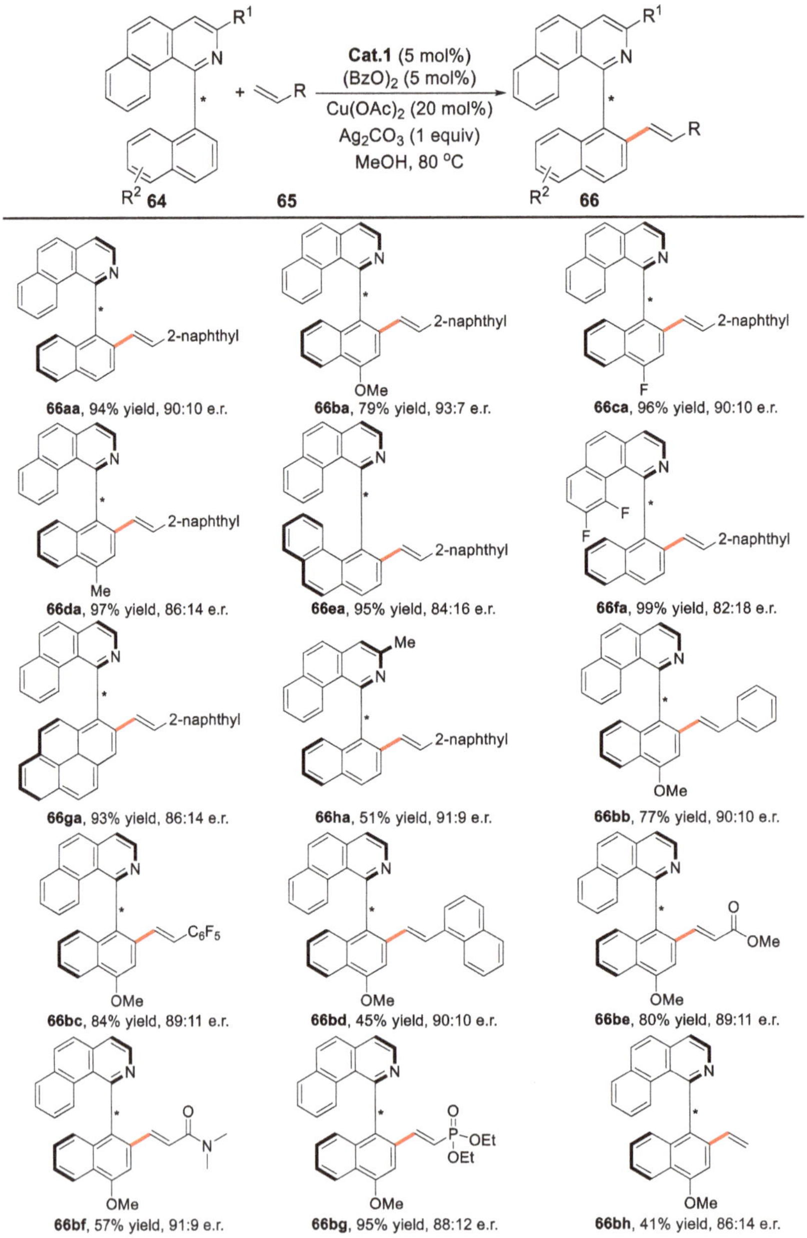

Scheme 42. The investigation of substrate scope [89].

Finally, the product was successfully utilized as ligands in the rhodium-catalyzed conjugate addition of phenylboronic acid to cyclohexenone reaction (Scheme 43), which sufficiently demonstrated the potential of this method.

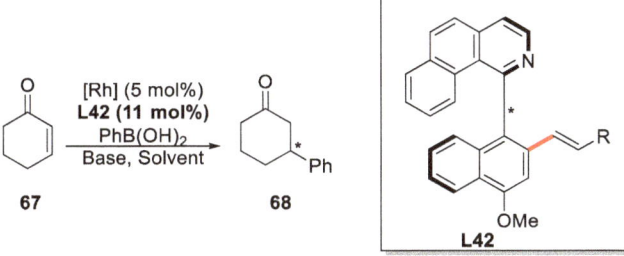

Scheme 43. The application of enantioenriched biaryl products [89].

4. Summary

Although traditional cross-coupling reactions have revolutionized organic chemistry and are widely applied in modern organic synthesis, the need for prefunctionalized starting materials has prompted chemists to investigate more atom and step economic alternatives. Therefore, C–H activation has emerged as a powerful tool to achieve C–C bond formation, allowing for the transformation of otherwise unreactive C–H bonds, thus maximizing the overall operational efficiency and decreasing the amount of stoichiometric metallic waste. The combinations of C–H activation/C–C cross-coupling reactions provide unlimited possibilities for synthetic chemists to access complex molecules.

This review summarizes the recent development on eantioselective C–H activation/Mizoroki-Heck reaction and Suzuki reaction. Due to the low reactivity of the C–H bonds, and the selectivity problem rooted in the abundance of C–H bonds, these transformations are extremely difficult to achieve. However, thanks to the increased mechanistic studies, chemists continually develop a better understanding of the mechanical aspects ruling these transformations. The Pd^{II}/MPAA systems developed by Yu group have been utilized successfully and represents one of the most important progresses.

It should also be pointed out that the concept of toxic heavy metals and benign lighter metals should not be taken for granted. Recently, studies revealed that some palladium, rhodium compounds, which were often considered heavy and toxic, might be less toxic than lighter metals [90]. This may change our traditional views on the toxic effects of metal salts in favor of Pd-catalyzed C–H activation.

Even though apparent advancements in this area have been made, more general protocols are highly demanded. Much research efforts as far as to design new chiral catalysts and chiral ligands, expand the substrate scope, and improve the efficiency of these transformations are still needed before a more general, atom-economical, and more environmentally friendly process become the method of choice for chemists in industrial or academic settings.

Acknowledgments: We thank Shanghai Jiao Tong University, National "1000-Youth Talents Plan", "1000 Talents Plan" of Zhejiang Province for financial support.

Author Contributions: Shuai Shi, Khan Shah Nawaz, Muhammad Kashif Zaman and Zhankui Sun analyzed the data and wrote the paper.

Conflicts of Interest: The author declares no conflicts of interest.

References

1. Mo, J.; Wang, L.; Liu, Y.; Cui, X. Transition-metal-catalyzed direct C–H functionalization under external-oxidant-free conditions. *Synthesis* **2015**, *47*, 439–459. [CrossRef]
2. Li, S.-S.; Qin, L.; Dong, L. Rhodium-catalyzed C–C coupling reactions via double C–H activation. *Org. Biomol. Chem.* **2016**, *14*, 4554–4570. [CrossRef] [PubMed]
3. Xue, X.-S.; Ji, P.; Zhou, B.; Cheng, J.-P. The Essential Role of Bond Energetics in C–H Activation/Functionalization. *Chem. Rev.* **2017**, *117*, 8622–8648. [CrossRef] [PubMed]

4. Gensch, T.; Hopkinson, M.; Glorius, F.; Wencel-Delord, J. Mild metal-catalyzed C–H activation: Examples and concepts. *Chem. Soc. Rev.* **2016**, *45*, 2900–2936. [CrossRef] [PubMed]
5. Musaev, D.G.; Figg, T.M.; Kaledin, A.L. Versatile reactivity of Pd-catalysts: Mechanistic features of the mono-N-protected amino acid ligand and cesium-halide base in Pd-catalyzed C–H bond functionalization. *Chem. Soc. Rev.* **2014**, *43*, 5009–5031. [CrossRef] [PubMed]
6. Yang, Y.; Lan, J.; You, J. Oxidative C–H/C–H Coupling Reactions between Two (Hetero) arenes. *Chem. Rev.* **2017**, *117*, 8787–8863. [CrossRef] [PubMed]
7. Labinger, J.A. Platinum-Catalyzed C–H Functionalization. *Chem. Rev.* **2017**, *117*, 8483–8496. [CrossRef] [PubMed]
8. Wencel-Delord, J.; Glorius, F. CH bond activation enables the rapid construction and late-stage diversification of functional molecules. *Nat. Chem.* **2013**, *5*, 369–375. [CrossRef] [PubMed]
9. Neufeldt, S.R.; Sanford, M.S. Controlling site selectivity in palladium-catalyzed C–H bond functionalization. *Acc. Chem. Res.* **2012**, *45*, 936–946. [CrossRef] [PubMed]
10. Engle, K.M.; Mei, T.-S.; Wasa, M.; Yu, J.-Q. Weak coordination as a powerful means for developing broadly useful C–H functionalization reactions. *Acc. Chem. Res.* **2012**, *45*, 788–802. [CrossRef] [PubMed]
11. Song, G.; Wang, F.; Li, X. C–C, C–O and C–N bond formation via rhodium (iii)-catalyzed oxidative C–H activation. *Chem. Soc. Rev.* **2012**, *41*, 3651–3678. [CrossRef] [PubMed]
12. Hickman, A.J.; Sanford, M.S. High-valent organometallic copper and palladium in catalysis. *Nature* **2012**, *484*, 177–185. [CrossRef] [PubMed]
13. Patureau, F.W.; Wencel-Delord, J.; Glorius, F. Cp* Rh-Catalyzed C—H Activations. Versatile Dehydrogenative Cross-Couplings of Csp^2 C—H Positions with Olefins, Alkynes, and Arenes. *ChemInform* **2013**, *44*. [CrossRef]
14. Arockiam, P.B.; Bruneau, C.; Dixneuf, P.H. Ruthenium (II)-catalyzed C–H bond activation and functionalization. *Chem. Rev.* **2012**, *112*, 5879–5918. [CrossRef] [PubMed]
15. Yeung, C.S.; Dong, V.M. Catalytic dehydrogenative cross-coupling: Forming carbon-carbon bonds by oxidizing two carbon-hydrogen bonds. *Chem. Rev.* **2011**, *111*, 1215–1292. [CrossRef] [PubMed]
16. Wencel-Delord, J.; Dröge, T.; Liu, F.; Glorius, F. Towards mild metal-catalyzed C–H bond activation. *Chem. Soc. Rev.* **2011**, *40*, 4740–4761. [CrossRef] [PubMed]
17. Cho, S.H.; Kim, J.Y.; Kwak, J.; Chang, S. Recent advances in the transition metal-catalyzed twofold oxidative C–H bond activation strategy for C–C and C–N bond formation. *Chem. Soc. Rev.* **2011**, *40*, 5068–5083. [CrossRef] [PubMed]
18. Baudoin, O. Transition metal-catalyzed arylation of unactivated $C(sp^3)$–H bonds. *Chem. Soc. Rev.* **2011**, *40*, 4902–4911. [CrossRef] [PubMed]
19. Rech, J.C.; Yato, M.; Duckett, D.; Ember, B.; LoGrasso, P.V.; Bergman, R.G.; Ellman, J.A. Synthesis of Potent Bicyclic Bisarylimidazole c-Jun N-Terminal Kinase Inhibitors by Catalytic C—H Bond Activation. *J. Am. Chem. Soc.* **2007**, *129*, 490–491. [CrossRef] [PubMed]
20. O'Malley, S.J.; Tan, K.L.; Watzke, A.; Bergman, R.G.; Ellman, J.A. Total synthesis of (+)-Lithospermic acid by asymmetric intramolecular alkylation via catalytic C—H bond activation. *J. Am. Chem. Soc.* **2005**, *127*, 13496–13497. [CrossRef] [PubMed]
21. Hinman, A.; Du Bois, J. A stereoselective synthesis of (−)-tetrodotoxin. *J. Am. Chem. Soc.* **2003**, *125*, 11510–11511. [CrossRef] [PubMed]
22. Dai, X.; Wan, Z.; Kerr, R.G.; Davies, H.M. Synthetic and isolation studies related to the marine natural products (+)-elisabethadione and (+)-elisabethamine. *J. Org. Chem.* **2007**, *72*, 1895–1900. [CrossRef] [PubMed]
23. Trost, B.M. Atom Economy-A Challenge for Organic Synthesis: Homogeneous Catalysis Leads the Way. *Angew. Chem. Int. Ed. Engl.* **1995**, *34*, 259–281. [CrossRef]
24. Wehn, P.M.; Du Bois, J. A Stereoselective Synthesis of the Bromopyrrole Natural Product (−)-Agelastatin A. *Angew. Chem.* **2009**, *121*, 3860–3863. [CrossRef]
25. Chen, K.; Baran, P.S. Total synthesis of eudesmane terpenes by site-selective C–H oxidations. *Nature* **2009**, *459*, 824–828. [CrossRef] [PubMed]
26. Tsai, A.S.; Bergman, R.G.; Ellman, J.A. Asymmetric Synthesis of (−)-Incarvillateine Employing an Intramolecular Alkylation via Rh-Catalyzed Olefinic C—H Bond Activation. *J. Am. Chem. Soc.* **2008**, *130*, 6316–6317. [CrossRef] [PubMed]
27. Bowie, A.L.; Hughes, C.C.; Trauner, D. Concise synthesis of (±)-rhazinilam through direct coupling. *Org. Lett.* **2005**, *7*, 5207–5209. [CrossRef] [PubMed]

28. Baran, P.S.; Corey, E. A short synthetic route to (+)-austamide, (+)-deoxyisoaustamide, and (+)-hydratoaustamide from a common precursor by a novel palladium-mediated indole → dihydroindoloazocine cyclization. *J. Am. Chem. Soc.* **2002**, *124*, 7904–7905. [CrossRef] [PubMed]
29. Miyaura, N.; Yamada, K.; Suzuki, A. A new stereospecific cross-coupling by the palladium-catalyzed reaction of 1-alkenylboranes with 1-alkenyl or 1-alkynyl halides. *Tetrahedron Lett.* **1979**, *20*, 3437–3440. [CrossRef]
30. Jagtap, S. Heck Reaction—State of the Art. *Catalysts* **2017**, *7*, 267. [CrossRef]
31. Sonogashira, K.; Tohda, Y.; Hagihara, N. A convenient synthesis of acetylenes: Catalytic substitutions of acetylenic hydrogen with bromoalkenes, iodoarenes and bromopyridines. *Tetrahedron Lett.* **1975**, *16*, 4467–4470. [CrossRef]
32. King, A.O.; Okukado, N.; Negishi, E.-I. Highly general stereo-, regio-, and chemo-selective synthesis of terminal and internal conjugated enynes by the Pd-catalysed reaction of alkynylzinc reagents with alkenyl halides. *J. Chem. Soc. Chem. Commun.* **1977**, 683–684. [CrossRef]
33. Milstein, D.; Stille, J. A general, selective, and facile method for ketone synthesis from acid chlorides and organotin compounds catalyzed by palladium. *J. Am. Chem. Soc.* **1978**, *100*, 3636–3638. [CrossRef]
34. Tamao, K.; Sumitani, K.; Kumada, M. Selective carbon-carbon bond formation by cross-coupling of Grignard reagents with organic halides. Catalysis by nickel-phosphine complexes. *J. Am. Chem. Soc.* **1972**, *94*, 4374–4376. [CrossRef]
35. Heck, R.F.; Negishi, E.-I.; Suzuki, A. Nobel Prizes 2010. *Angew. Chem. Int. Ed.* **2010**, *49*, 8300.
36. Wencel-Delord, J.; Colobert, F. Asymmetric C(sp^2)–H Activation. *Chem. A Eur. J.* **2013**, *19*, 14010–14017. [CrossRef] [PubMed]
37. Pan, S.C. Organocatalytic C–H activation reactions. *Beilstein J. Org. Chem.* **2012**, *8*, 1374. [CrossRef] [PubMed]
38. Newhouse, T.; Baran, P.S. If C–H Bonds Could Talk: Selective C–H Bond Oxidation. *Angew. Chem. Int. Ed.* **2011**, *50*, 3362–3374. [CrossRef] [PubMed]
39. Peng, H.-M.; Dai, L.-X.; You, S.-L. Enantioselective Palladium-Catalyzed Direct Alkylation and Olefination Reaction of Simple Arenes. *Angew. Chem. Int. Ed.* **2010**, *49*, 5826–5828. [CrossRef] [PubMed]
40. Giri, R.; Shi, B.-F.; Engle, K.M.; Maugel, N.; Yu, J.-Q. Transition metal-catalyzed C–H activation reactions: Diastereoselectivity and enantioselectivity. *Chem. Soc. Rev.* **2009**, *38*, 3242–3272. [CrossRef] [PubMed]
41. Zhao, Y.-L.; Wang, Y.; Luo, Y.-C.; Fu, X.-Z.; Xu, P.-F. Asymmetric C–H functionalization involving organocatalysis. *Tetrahedron Lett.* **2015**, *56*, 3703–3714. [CrossRef]
42. Musaev, D.G.; Kaledin, A.; Shi, B.-F.; Yu, J.-Q. Key mechanistic features of enantioselective C–H bond activation reactions catalyzed by [(chiral mono-N-protected amino acid)–Pd (II)] complexes. *J. Am. Chem. Soc.* **2012**, *134*, 1690–1698. [CrossRef] [PubMed]
43. Zhang, G.; Zhang, Y.; Wang, R. Catalytic Asymmetric Activation of a Csp^3–H Bond Adjacent to a Nitrogen Atom: A Versatile Approach to Optically Active α-Alkyl α-Amino Acids and C1-Alkylated Tetrahydroisoquinoline Derivatives. *Angew. Chem. Int. Ed.* **2011**, *50*, 10429–10432. [CrossRef] [PubMed]
44. Wasa, M.; Engle, K.M.; Lin, D.-W.; Yoo, E.J.; Yu, J.-Q. Pd (II)-catalyzed enantioselective C–H activation of cyclopropanes. *J. Am. Chem. Soc.* **2011**, *133*, 19598–19601. [CrossRef] [PubMed]
45. Li, Q.; Yu, Z.-X. Enantioselective Rhodium-Catalyzed Allylic C–H Activation for the Addition to Conjugated Dienes. *Angew. Chem.* **2011**, *123*, 2192–2195. [CrossRef]
46. Davies, H.M.; Manning, J.R. Catalytic C–H functionalization by metal carbenoid and nitrenoid insertion. *Nature* **2008**, *451*, 417–424. [CrossRef] [PubMed]
47. Haines, B.E.; Yu, J.-Q.; Musaev, D.G. Enantioselectivity Model for Pd-Catalyzed C–H Functionalization Mediated by the Mono-N-protected Amino Acid (MPAA) Family of Ligands. *ACS Catal.* **2017**, *7*, 4344–4354. [CrossRef]
48. Motevalli, S.; Sokeirik, Y.; Ghanem, A. Rhodium-Catalysed Enantioselective C–H Functionalization in Asymmetric Synthesis. *Eur. J. Org. Chem.* **2016**, *2016*, 1459–1475. [CrossRef]
49. Loxq, P.; Manoury, E.; Poli, R.; Deydier, E.; Labande, A. Synthesis of axially chiral biaryl compounds by asymmetric catalytic reactions with transition metals. *Coord. Chem. Rev.* **2016**, *308*, 131–190. [CrossRef]
50. Qin, Y.; Lv, J.; Luo, S. Catalytic asymmetric α-C(sp^3)–H functionalization of amines. *Tetrahedron Lett.* **2014**, *55*, 551–558. [CrossRef]
51. Yang, L.; Huang, H. Asymmetric catalytic carbon–carbon coupling reactions via C–H bond activation. *Catal. Sci. Technol.* **2012**, *2*, 1099. [CrossRef]

52. Wasa, M.; Engle, K.M.; Yu, J.-Q. Cross-Coupling of C(sp^3)–H Bonds with Organometallic Reagents via Pd(II)/Pd(0) Catalysis. *Isr. J. Chem.* **2010**, *50*, 605–616. [CrossRef] [PubMed]
53. Hansen, J.; Davies, H.M. High symmetry dirhodium (II) paddlewheel complexes as chiral catalysts. *Coord. Chem. Rev.* **2008**, *252*, 545–555. [CrossRef] [PubMed]
54. Enders, D.; Balensiefer, T. Nucleophilic carbenes in asymmetric organocatalysis. *Acc. Chem. Res* **2004**, *37*, 534–541. [CrossRef] [PubMed]
55. Singh, B.K.; Kaval, N.; Tomar, S.; Eycken, E.V.; Parmar, V.S. Transition metal-catalyzed carbon-carbon bond formation Suzuki, Heck, and Sonogashira reactions using microwave and microtechnology. *Org. Process Res. Dev.* **2008**, *12*, 468–474. [CrossRef]
56. Cantillo, D.; Kappe, C.O. Immobilized Transition Metals as Catalysts for Cross-Couplings in Continuous Flow—A Critical Assessment of the Reaction Mechanism and Metal Leaching. *ChemCatChem* **2014**, *6*, 3286–3305. [CrossRef]
57. Shu, W.; Pellegatti, L.; Oberli, M.A.; Buchwald, S.L. Continuous-Flow Synthesis of Biaryls Enabled by Multistep Solid-Handling in a Lithiation/Borylation/Suzuki-Miyaura Cross-Coupling Sequence. *Angew. Chem.* **2011**, *123*, 10853–10857. [CrossRef]
58. Basle, O.; Li, C.-J. Copper-Catalyzed Oxidative sp^3 C–H Bond Arylation with Aryl Boronic Acids. *Org. Lett.* **2008**, *10*, 3661–3663. [CrossRef] [PubMed]
59. Kirchberg, S.; Tani, S.; Ueda, K.; Yamaguchi, J.; Studer, A.; Itami, K. Oxidative Biaryl Coupling of Thiophenes and Thiazoles with Arylboronic Acids through Palladium Catalysis: Otherwise Difficult C4-Selective C–H Arylation Enabled by Boronic Acids. *Angew. Chem. Int. Ed.* **2011**, *50*, 2387–2391. [CrossRef] [PubMed]
60. Yamaguchi, K.; Yamaguchi, J.; Studer, A.; Itami, K. Hindered biaryls by C–H coupling: Bisoxazoline-Pd catalysis leading to enantioselective C–H coupling. *Chem. Sci.* **2012**, *3*, 2165. [CrossRef]
61. Yamaguchi, K.; Kondo, H.; Yamaguchi, J.; Itami, K. Aromatic C–H coupling with hindered arylboronic acids by Pd/Fe dual catalysts. *Chem. Sci.* **2013**, *4*, 3753. [CrossRef]
62. Shi, B.-F.; Maugel, N.; Zhang, Y.-H.; Yu, J.-Q. Pd(II)-catalyzed enantioselective activation of C(sp^2)–H and C(sp^3)–H bonds using monoprotected amino acids as chiral ligands. *Angew. Chem. Int. Ed. Engl.* **2008**, *47*, 4882–4886. [CrossRef] [PubMed]
63. Xiao, K.-J.; Lin, D.-W.; Miura, M.; Zhu, R.-Y.; Gong, W.; Wasa, M.; Yu, J.-Q. Palladium(II)-catalyzed enantioselective C(sp(3))–H activation using a chiral hydroxamic acid ligand. *J. Am. Chem. Soc.* **2014**, *136*, 8138–8142. [CrossRef] [PubMed]
64. Dembitsky, V.M. Bioactive cyclobutane-containing alkaloids. *J. Nat. Med.* **2008**, *62*, 1–33. [CrossRef] [PubMed]
65. Kurosawa, K.; Takahashi, K.; Tsuda, E. SNF4435C and D, Novel Immunosuppressants Produced by a Strain of Streptomyces spectabilis. *J. Antibiot.* **2001**, *54*, 541–547. [CrossRef] [PubMed]
66. Fu, G.C. Applications of planar-chiral heterocycles as ligands in asymmetric catalysis. *Acc. Chem. Res.* **2006**, *39*, 853–860. [CrossRef] [PubMed]
67. Arae, S.; Ogasawara, M. Catalytic asymmetric synthesis of planar-chiral transition-metal complexes. *Tetrahedron Lett.* **2015**, *56*, 1751–1761. [CrossRef]
68. Gómez Arrayás, R.; Adrio, J.; Carretero, J.C. Recent applications of chiral ferrocene ligands in asymmetric catalysis. *Angew. Chem. Int. Ed.* **2006**, *45*, 7674–7715. [CrossRef] [PubMed]
69. Fu, G.C. Asymmetric catalysis with "planar-chiral" derivatives of 4-(dimethylamino) pyridine. *Acc. Chem. Res.* **2004**, *37*, 542–547. [CrossRef] [PubMed]
70. Colacot, T.J. A concise update on the applications of chiral ferrocenyl phosphines in homogeneous catalysis leading to organic synthesis. *Chem. Rev.* **2003**, *103*, 3101–3118. [CrossRef] [PubMed]
71. Dai, L.-X.; Tu, T.; You, S.-L.; Deng, W.-P.; Hou, X.-L. Asymmetric catalysis with chiral ferrocene ligands. *Acc. Chem. Res.* **2003**, *36*, 659–667. [CrossRef] [PubMed]
72. Gao, D.-W.; Shi, Y.-C.; Gu, Q.; Zhao, Z.-L.; You, S.-L. Enantioselective synthesis of planar chiral ferrocenes via palladium-catalyzed direct coupling with arylboronic acids. *J. Am. Chem. Soc.* **2013**, *135*, 86–89. [CrossRef] [PubMed]
73. Parmar, D.; Sugiono, E.; Raja, S.; Rueping, M. Complete field guide to asymmetric BINOL-phosphate derived Brønsted acid and metal catalysis: History and classification by mode of activation; Brønsted acidity, hydrogen bonding, ion pairing, and metal phosphates. *Chem. Rev.* **2014**, *114*, 9047–9153. [CrossRef] [PubMed]

74. Brak, K.; Jacobsen, E.N. Asymmetric Ion-Pairing Catalysis. *Angew. Chem. Int. Ed.* **2013**, *52*, 534–561. [CrossRef] [PubMed]
75. Mahlau, M.; List, B. Asymmetric Counteranion-Directed Catalysis: Concept, Definition, and Applications. *Angew. Chem. Int. Ed.* **2013**, *52*, 518–533. [CrossRef] [PubMed]
76. You, S.-L.; Cai, Q.; Zeng, M. Chiral Brønsted acid catalyzed Friedel–Crafts alkylation reactions. *Chem. Soc. Rev.* **2009**, *38*, 2190–2201. [CrossRef] [PubMed]
77. Du, Z.-J.; Guan, J.; Wu, G.-J.; Xu, P.; Gao, L.-X.; Han, F.-S. Pd(II)-catalyzed enantioselective synthesis of P-stereogenic phosphinamides via desymmetric C–H arylation. *J. Am. Chem. Soc.* **2015**, *137*, 632–635. [CrossRef] [PubMed]
78. Laforteza, B.N.; Chan, K.S.; Yu, J.-Q. Enantioselective ortho-C–H cross-coupling of diarylmethylamines with organoborons. *Angew Chem Int Ed Engl* **2015**, *54*, 11143–11146. [CrossRef] [PubMed]
79. Mizoroki, T.; Mori, K.; Ozaki, A. Arylation of olefin with aryl iodide catalyzed by palladium. *Bull. Chem. Soc. Jpn.* **1971**, *44*, 581. [CrossRef]
80. Heck, R.F.; Nolley, J., Jr. Palladium-catalyzed vinylic hydrogen substitution reactions with aryl, benzyl, and styryl halides. *J. Org. Chem.* **1972**, *37*, 2320–2322. [CrossRef]
81. Shi, B.-F.; Zhang, Y.-H.; Lam, J.K.; Wang, D.-H.; Yu, J.-Q. Pd(II)-catalyzed enantioselective C–H olefination of diphenylacetic acids. *J. Am. Chem. Soc.* **2010**, *132*, 460–461. [CrossRef] [PubMed]
82. Chu, L.; Xiao, K.-J.; Yu, J.-Q. Room-temperature enantioselective C–H iodination via kinetic resolution. *Science* **2014**, *346*, 451–455. [CrossRef] [PubMed]
83. Gao, D.-W.; Gu, Q.; You, S.-L. Pd (II)-catalyzed intermolecular direct C–H bond Iodination: An efficient approach toward the synthesis of axially chiral compounds via kinetic resolution. *ACS Catal.* **2014**, *4*, 2741–2745. [CrossRef]
84. Xiao, K.-J.; Chu, L.; Yu, J.-Q. Enantioselective C–H Olefination of alpha-Hydroxy and alpha-Amino Phenylacetic Acids by Kinetic Resolution. *Angew. Chem. Int. Ed. Engl.* **2016**, *55*, 2856–2860. [CrossRef] [PubMed]
85. Kagan, H.; Fiaud, J. Kinetic resolution. *Top. Stereochem* **1988**, *18*, 21.
86. Yao, Q.-J.; Zhang, S.; Zhan, B.-B.; Shi, B.-F. Atroposelective Synthesis of Axially Chiral Biaryls by Palladium-Catalyzed Asymmetric C–H Olefination Enabled by a Transient Chiral Auxiliary. *Angew. Chem. Int. Ed. Engl.* **2017**, *56*, 6617–6621. [CrossRef] [PubMed]
87. Pi, C.; Li, Y.; Cui, X.; Zhang, H.; Han, Y.; Wu, Y. Redox of ferrocene controlled asymmetric dehydrogenative Heck reaction via palladium-catalyzed dual C–H bond activation. *Chem. Sci.* **2013**, *4*, 2675. [CrossRef]
88. Gao, D.-W.; Gu, Q.; You, S.-L. An Enantioselective Oxidative C–H/C–H Cross-Coupling Reaction: Highly Efficient Method to Prepare Planar Chiral Ferrocenes. *J. Am. Chem. Soc.* **2016**, *138*, 2544–2547. [CrossRef] [PubMed]
89. Zheng, J.; You, S.-L. Construction of axial chirality by rhodium-catalyzed asymmetric dehydrogenative Heck coupling of biaryl compounds with alkenes. *Angew. Chem. Int. Ed. Engl.* **2014**, *53*, 13244–13247. [CrossRef] [PubMed]
90. Egorova, K.S.; Ananikov, V.P. Which Metals are Green for Catalysis? Comparison of the Toxicities of Ni, Cu, Fe, Pd, Pt, Rh, and Au Salts. *Angew. Chem. Int. Ed. Engl.* **2016**, *55*, 12150–12162. [CrossRef] [PubMed]

© 2018 by the authors. Licensee MDPI, Basel, Switzerland. This article is an open access article distributed under the terms and conditions of the Creative Commons Attribution (CC BY) license (http://creativecommons.org/licenses/by/4.0/).

Solvent-Free Mizoroki-Heck Reaction Applied to the Synthesis of Abscisic Acid and Some Derivatives

Geoffrey Dumonteil, Marie-Aude Hiebel and Sabine Berteina-Raboin *

Institut de Chimie Organique et Analytique (ICOA), Université d'Orléans UMR CNRS 7311, BP 6759, rue de Chartres, 45067 Orleans CEDEX 2, France; Geoffrey.Dumonteil@sanofi.com (G.D.); marie-aude.hiebel@univ-orleans.fr (M.-A.H.)
* Correspondence: sabine.berteina-raboin@univ-orleans.fr; Tel.: +33-238-494-856

Received: 28 February 2018; Accepted: 13 March 2018; Published: 15 March 2018

Abstract: Abscisic acid (ABA) is a natural product, which is a well-known phytohormone. However, this molecule has recently exhibited interesting biological activities, emphasizing the need for a simple and direct access to new analogues based on the ABA framework. Our strategy relies on a pallado-catalyzed Mizoroki-Heck cross-coupling as key reaction performed in solvent and ligand free conditions. After a careful optimization, we succeeded in accessing various (E/Z)-dienes and (E/E/Z)-trienes in moderate to good yields without isomerization and applied the same approach to the synthesis of ABA in an environmentally sound manner.

Keywords: Mizoroki-Heck; abscisic acid; solvent-free

1. Introduction

Polyenic scaffolds constitute an important functionality among organic compounds and have a high synthetic interest since medicinally relevant molecules and natural products exhibit diene fragments (Figure 1) [1].

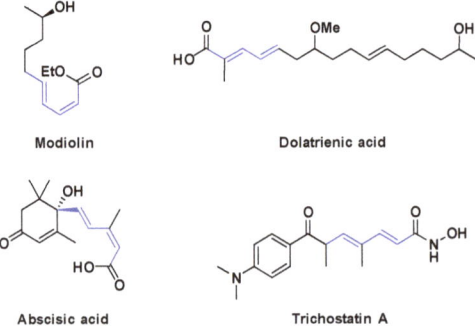

Figure 1. Examples of biologically active natural products containing diene moieties.

Several synthetic methods have therefore been developed to obtain these carotenoid moieties in iterative process or in convergent methods [2]. Traditionally, the olefination reaction was extensively used. However, it is often associated with the uncontrolled production of E and Z isomers which may require careful purification [3]. Then transition metal-catalyzed cross-coupling reactions galvanized the synthesis of these complex conjugated molecules. Catalysts such as ruthenium [4], zirconium [5–7], zinc [8] and nickel [9] were successfully used. The use of palladium has been widely reported with

Negishi [10,11], Stille [12–15], Suzuki-Miyaura [16–18], Sonogashira [19], Kumada-Tamao [20] and Mizoroki-Heck [21–25] cross-coupling reactions [26–28]. Lately, the use of single unsaturated units as building blocks was promoted to respond to the challenging but crucial control of the configuration of the double bond generated [29–33]. In our effort to develop environmentally benign tools [34–36], we herein report the use of the Mizoroki-Heck reaction, which requires simple and directly accessible starting materials to build stereocontrolled dienes and trienes. Unlike other cross-coupling approaches, which may require several steps to install the pre-activated partners, the Mizoroki-Heck reaction enables the direct formation of dienes from terminal olefin substrates. The efficiency of our method was then evaluated in the synthesis of abscisic acid, an important phytohormone [37–43] which has been recently reported to have interesting biological effects [44,45].

2. Results and Discussion

2.1. Optimization of the Mizoroki-Heck Reaction

To achieve the optimized conditions, the cross-coupling reaction of 1-ethenyl-3-methylcyclohex-2-en-1-ol **1** with methyl (2Z)-3-iodobut-2-enoate **2** was selected as the model reaction under the standard conditions previously described by Cossy and co-workers (Table 1, entry 1) [46]. Surprisingly, only degradation was observed. Since conjugated products are prompted to make versatile rearrangement [47,48], the reaction was next performed in a flask protected from natural light. A small amount of the expected product **3a** was isolated along with the side product **4** (entry 2). The formation of **4** can be explained by a 1,3-rearrangement of the allylic alcohol, a transformation previously described by Qu and co-workers under a thermal activation in water [48]. The use of acetonitrile moreover of an additional ligand induced no positive change (entries 3 and 4) [49]. However, a significant improvement was obtained by replacing silver acetate by silver carbonate and 1.5 equiv. of silver carbonate proved to be the optimized amount (entries 5–7). A moderate heating is recommended since the formation of **3a** was significantly reduced at 80 °C.

Table 1. Optimization of the Mizoroki-Heck reaction between **1** and **2**.

Entry	1 (equiv.)	2 (equiv.)	Additive (equiv.)	Solvent	Time (h)	Yield 3:4 (%)
1	1.2	1	AgOAc (1.1)	DMF	17	0:0 [a]
2	1.2	1	AgOAc (1.1)	DMF	17	5:4 [b]
3	1.2	1	AgOAc (1.1)	MeCN	17	9:0 [b]
4	1.2	1	AgOAc (1.1) P(oTol)$_3$ (0.1)	MeCN	17	9:20 [b]
5	1.2	1	Ag$_2$CO$_3$ (1.5)	MeCN	17	63:0 [b,c]
6	1.2	1	Ag$_2$CO$_3$ (2)	MeCN	17	63:0 [b,c]
7	1.2	1	Ag$_2$CO$_3$ (1.1)	MeCN	17	54:0 [b,c]
8	1.2	1	Ag$_2$CO$_3$ (1.5)	MeCN	17	10:0 [b,c,d]
9	1.2	1	Ag$_2$CO$_3$ (1.5)	none	1	63:0 [b,c]
10	1	2	Ag$_2$CO$_3$ (1.5)	none	1	60:0 [b,c]
11	1	1 + 1	Ag$_2$CO$_3$ (1.5)	none	1	50:0 [b,c]
12	1.2	1	Ag$_2$CO$_3$ (1.5)	none	1	62:0 [b,d]

[a] The reaction was performed under natural light. [b] Reaction performed protected from natural light. [c] Reaction performed under air. [d] The reaction temperature was set at 80 °C. [e] Reaction performed under inert atmosphere.

Next, neat conditions were tried, and even if the reaction mixture was a thick paste, the expected product **3a** was isolated in similar yield (63%) in a considerably shorter time (1 h vs. 17 h). To the best of our knowledge, this is the first example of neat Mizoroki-Heck reaction for the formation of dienes. The vinylation of acrylic substrates has been already reported in solvent-free conditions but it usually requires the use of a ligand, palladium supported catalyst, palladium nanocatalyst or microwave activation [50–57]. Finally, **2** was introduced in excess, in one portion or in sequential addition, without improving the yield (entries 10 and 11). Hence the optimized reaction conditions were as follows: in a flask protected from light, **1** (1.2 equiv.) and **2** (1 equiv.) in presence of Pd(OAc)$_2$ (5 mol %) and Ag$_2$CO$_3$ (1.5 equiv.) at 50 °C for 1 h. It should be noted that working under an inert atmosphere did not improve the yield of the products **3a** and **4** (entry 12).

2.2. Substrate Scope

A diversity of terminal olefin substrates was tested in the coupling reaction with methyl (2Z)-3-iodobut-2-enoate **2**. The results are reported in Scheme 1. Various allylic cyclohexenols and cyclohexanols were examined and the expected dienes were obtained in moderate to good yields **3a–3f**. The presence of an unsaturation and/or different methyl substituents on the ring had little influence on the efficiency of the reaction. The variation of the yield observed was more substrate-dependent since **3c** and **3d** appeared to be very unstable and degraded spontaneously if not stored at low temperature in dark conditions. For these compounds, the reaction was tried at room temperature; however, the cross-coupling reaction failed completely. The stability issue was even more pronounced for 1-ethenylcyclopentan-1-ol, since its formation from cyclopentanone was difficult. The cross-coupling reaction was performed on the crude starting material, which could explain the low yield observed. Surprisingly, the resulting product **3g** was completely bench stable. Satisfyingly, sterically hindered secondary alcohols **3h** and sensitive tertiary alcohols (**3a–3g**) were well tolerated under our optimized conditions. Even the volatile ethenylcyclohexane, which required working in a sealed tube, and the unstable styrene led to the corresponding dienes **3i** and **3j** in 42% and 78% yields. It is worth noting that all the coupling products were obtained as pure (E,Z)-dienes. The configuration of the diene was confirmed by ^{1}H NMR (Supplementary Materials: Figure S1). The chemical shift of the hydrogen alpha to the ester moiety is in accordance with the values reported in the literature for a (E/Z)-diene, around 5.5 ppm (vs. 6.0 ppm for a E/E fragment) [58].

Finally, different vinylic iodides were submitted to our solvent-free Mizoroki-Heck conditions. (2Z)-3-iodobut-2- enenitrile **2a**, Methyl (2Z)-3-iodoacrylate **2b**, and 4-nitrophenyl(2Z)-3-iodobut-2-enoate **2c** were successfully introduced on 1-ethenyl-3-methylcyclohex-2-en-1-ol **1** leading to expected compounds (**3k–3m**). Compound **3k** appeared to be more sensitive to degradation than **3a**, but the (E,Z) configuration remained unchanged. The reaction conditions were then extended to the formation of (E,E,Z) tertiary trienol **3n**, using Methyl (2Z,4E)-5-iodo-3-methylpenta-2,4-dienoate (**2e**), which was isolated in 56% yield.

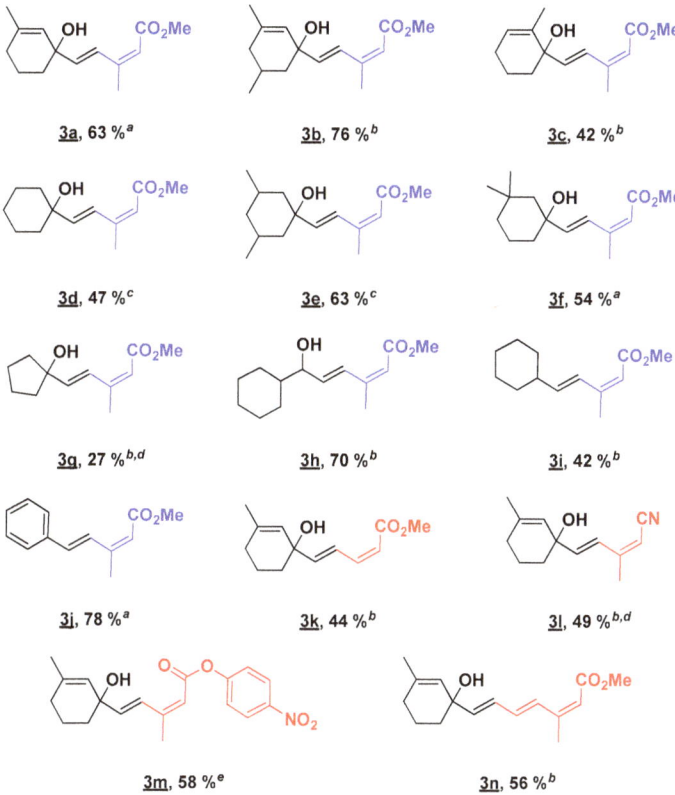

Scheme 1. Scope of substrates. [a] 1 h of reaction time. [b] 2 h of reaction time. [c] 1 h 40 min was required for the reaction time. [d] sealed tube. [e] 40 min of reaction time.

2.3. Synthesis of ABA

Having a method to obtain the diene scaffold in hand, we focused on the synthesis of abscisic acid (ABA). Several syntheses have already been reported in the literature [59–61]. Most of them are based on the introduction of the carbon skeleton of the side chain via the corresponding alkyne in one step [62–65] or with trimethylsilylacetylene [66,67], which is then functionalized by a Sonogashira reaction (Scheme 2). Our strategy requires a reduction step to obtain the final (E,Z) diene fragment. Our approach is based on the introduction of the side chain with our solvent-free Mizoroki-Heck reaction between methyl (2Z)-3-iodobut-2-enoate **2** and **7**. Our key cross-coupling precursor was straightforwardly obtained from the commercially available 2,6,6-trimethylcyclohex-2-ene-1,4-dione **5**.

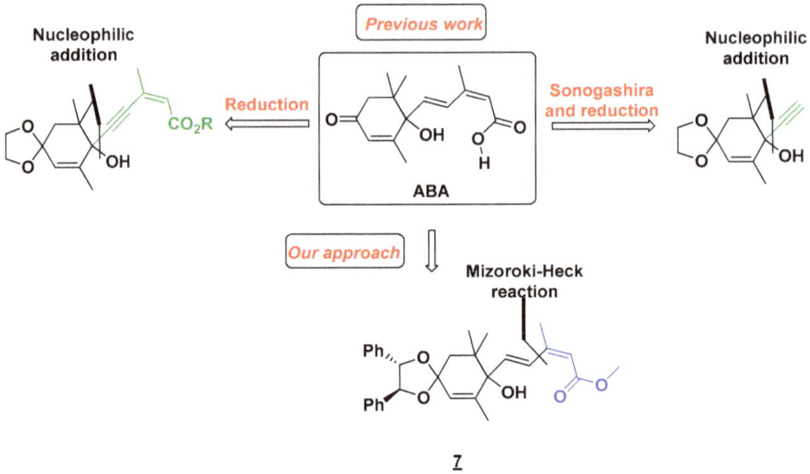

Scheme 2. Retrosynthetic approaches for the synthesis of ABA.

First, the mono protection of the diketone **5** was tried following the conditions described by Ferrer and co-workers (Scheme 3) [68]. (*S,S*)-hydrobenzoin, our chiral auxiliary, was heated in presence of a catalytic amount of *p*TsOH in benzene. However only the degradation of the hydrobenzoin into benzaldehyde was observed. We then decided to use pyridinium *p*-toluenesulfonate (PPTS), which is a milder acidic catalyst [69].

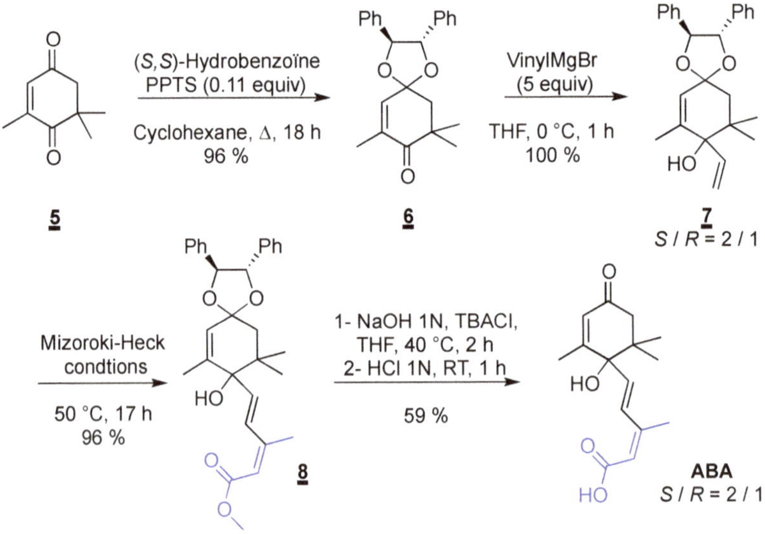

Scheme 3. Synthesis of ABA.

This time, in benzene the desired product **6** was isolated after 5 days in a promising 24% yield. A survey of different solvents was made and fortunately, the environmentally sound cyclohexane significantly improved the yield and the reaction time since **6** was obtained in 96% yield after 18 h. Next, the Grignard reaction with an excess of vinylmagnesium bromide gave quantitatively **7** as a mixture of inseparable diastereoisomers. The R/S ratio of the newly formed center was determined by HPLC. Disappointingly, working at a lower temperature ($-20\,^\circ$C or $-78\,^\circ$C) did not improve this ratio much while it considerably reduced the formation of **7**. Nevertheless, our key step was performed on the mixture of both isomers and the expected (E/Z)-diene **8** was isolated in 96% yield without racemization, the R/S ratio remained unchanged during the formation of the diene **8**. A saponification followed by an acidic treatment enabled the formation of abscisic acid with the same R/S ration and therefore enantiomerically enriched in its S isomer. The final product was synthetized in four steps from **5** in 54% global yields.

3. Conclusions

To conclude, we managed to develop an efficient, environmentally sound method to synthetize delicate dienes and trienes via a Mizoroki-Heck reaction. The configuration of the double bonds was controlled, and no isomerization was observed. The salient features of our approach are the association of simple terminal olefins with various vinylic iodides, palladium acetate under air without any ligand or solvent. Our optimized solvent-free Mizoroki-Heck reaction was next successfully applied to the synthesis of ABA. This method offers a short new pathway where solvents and reagents were chosen to give an environmentally friendlier alternative to the synthesis already available in the literature.

Supplementary Materials: The following are available online at http://www.mdpi.com/2073-4344/8/3/115/s1, Figure S1: ^{1}H NMR and ^{13}C NMR Spectra of all compounds.

Acknowledgments: We acknowledge the Region Centre for financial support.

Author Contributions: Sabine Berteina-Raboin conceived and designed the experiments. Geoffrey Dumonteil performed the experiments. Geoffrey Dumonteil, Marie-Aude Hiebel and Sabine Berteina-Raboin analyzed the data and wrote the paper.

Conflicts of Interest: The authors declare no conflict of interest.

References

1. Hopf, H.; Maas, G. Synthesis and Transformation of Radialenes. In *The Chemistry of Dienes and Polyenes, Vol. 1*; Rappoport, Z., Ed.; Wiley: Chichester, UK, 1997; ISBN 978-0-471-96512-1.
2. De Paolis, M.; Chataigner, I.; Maddaluno, J. Recent advances in stereoselective synthesis of 1,3-dienes. *Top. Curr. Chem.* **2012**, *327*, 87–146. [CrossRef]
3. Takeda, T. (Ed.) *Modern Carbonyl Olefination: Methods and Applications*; Wiley-VCH: Weinheim, Germany, 2004; ISBN 978-3-527-60538-5.
4. Grubbs, R.H. (Ed.) *Handbook of Metathesis, Vol. 1*; Wiley-VCH: Weinheim, Germany, 2003; ISBN 978-3-527-30616-9.
5. Van Horn, D.E.; Negishi, E. Selective carbon-carbon bond formation via transition metal catalysts. 8. Controlled carbometalation. Reaction of acetylenes with organoalane-zirconocene dichloride complexes as a route to stereo- and regio-defined trisubstituted olefins. *J. Am. Chem. Soc.* **1978**, *100*, 2252–2254. [CrossRef]
6. Negishi, E. Bimetallic catalytic systems containing Ti, Zr, Ni, and Pd. Their applications to selective organic syntheses. *Pure Appl. Chem.* **1981**, *53*, 2333–2356. [CrossRef]
7. Negishi, E.; Van Horn, D.E.; Yoshida, T. Controlled carbometalation. 20. Carbometalation reaction of alkynes with organoalene-zirconocene derivatives as a route to stereo-and regiodefined trisubstituted alkenes. *J. Am. Chem. Soc.* **1985**, *107*, 6639–6647. [CrossRef]
8. Robinson, C.Y.; Brouilllette, W.J.; Muccio, D.D. Reactions of vinylogous phosphonate carbanions and Reformatskii reagents with a sterically hindered ketone and enone. *J. Org. Chem.* **1989**, *54*, 1992–1997. [CrossRef]
9. Horie, H.; Kurahashi, T.; Matsubara, S. Selective synthesis of trienes and dienes via nickel-catalyzed intermolecular cotrimerization of acrylates and alkynes. *Chem. Commun.* **2010**, *46*, 7229–7231. [CrossRef] [PubMed]

10. Negishi, E.; Okukado, N.; King, A.O.; Van Horn, D.E.; Spiegel, B.I. Selective carbon-carbon bond formation via transition metal catalysts. 9. Double metal catalysis in the cross-coupling reaction and its application to the stereo- and regioselective synthesis of trisubstituted olefins. *J. Am. Chem. Soc.* **1978**, *100*, 2254–2256. [CrossRef]
11. Negishi, E.; Hu, Q.; Wang, Z.; Yin, N. *The Chemistry of Organozinc Compounds*; Rappoport, Z., Marek, I., Eds.; John Wiley & Sons, Ltd.: Chichester, UK, 2006; Chapter 11; p. 453, ISBN 978-0-470-09339-9.
12. Stille, J.K.; Milstein, D. A general, selective, and facile method for ketone synthesis from acid chlorides and organotin compounds catalyzed by palladium. *J. Am. Chem. Soc.* **1978**, *100*, 3636–6638. [CrossRef]
13. Stille, J.K. The Palladium-Catalyzed Cross-Coupling Reactions of Organotin Reagents with Organic Electrophiles. *Angew. Chem. Int. Ed. Engl.* **1986**, *25*, 508–524. [CrossRef]
14. Farina, V. New perspectives in the cross-coupling reactions of organostannanes. *Pure Appl. Chem.* **1996**, *68*, 73–78. [CrossRef]
15. Farina, V.; Krishnamurthy, V.; Scott, W.J. The Stille Reaction. *Org. React.* **1997**, *50*. [CrossRef]
16. Miyaura, N.; Suzuki, A. Palladium-Catalyzed Cross-Coupling Reactions of Organoboron Compounds. *Chem. Rev.* **1995**, *95*, 2457–2483. [CrossRef]
17. Stanforth, S.P. Catalytic cross-coupling reactions in biaryl synthesis. *Tetrahedron* **1998**, *54*, 263–303. [CrossRef]
18. Suzuki, A. Recent advances in the cross-coupling reactions of organoboron derivatives with organic electrophiles 1995–1998. *J. Organomet. Chem.* **1999**, *576*, 147–168. [CrossRef]
19. Kotovshchikov, Y.N.; Latyshev, G.V.; Lukashev, N.V.; Beletskaya, I.P. Alkynylation of steroids via Pd-free Sonogashira coupling. *Org. Biomol. Chem.* **2015**, *13*, 5542–5555. [CrossRef] [PubMed]
20. Murahashi, S.-I. Palladium-catalyzed cross-coupling reaction of organic halides with Grignard reagents, organolithium compounds and heteroatom nucleophiles. *J. Organomet. Chem.* **2002**, *653*, 27–33. [CrossRef]
21. Dounay, A.B.; Overman, L.E. The Asymmetric Intramolecular Heck Reaction in Natural Product Total Synthesis. *Chem. Rev.* **2003**, *103*, 2945–2964. [CrossRef] [PubMed]
22. Bräse, S.; De Meijeren, F. *Metal-Catalyzed Cross-Coupling Reactions*; De Meijere, A., Diederich, F., Eds.; Wiley-VCH: New York, NY, USA, 2004; Chapter 5, ISBN 978-3-527-61953-5.
23. Fayol, A.; Fang, Y.-Q.; Lautens, M. Synthesis of 2-Vinylic Indoles and Derivatives via a Pd-Catalyzed Tandem Coupling Reaction. *Org. Lett.* **2006**, *8*, 4203–4206. [CrossRef] [PubMed]
24. Battace, A.; Zair, T.; Doucet, H.; Santelli, M. Heck Vinylations Using Vinyl Sulfide, Vinyl Sulfoxide, Vinyl Sulfone, or Vinyl Sulfonate Derivatives and Aryl Bromides Catalyzed by a Palladium Complex Derived from a Tetraphosphine. *Synthesis* **2006**, *20*, 3495–3505. [CrossRef]
25. McConville, M.; Saidi, O.; Blacker, J.; Xiao, J. Regioselective Heck Vinylation of Electron-Rich Olefins with Vinyl Halides: Is the Neutral Pathway in Operation? *J. Org. Chem.* **2009**, *74*, 2692–2698. [CrossRef] [PubMed]
26. Hartwig, J.F. *Handbook of Organopalladium Chemistry for Organic Synthesis*; Negishi, E., Ed.; Wiley: Chichester, UK, 2002; ISBN 978-0-47-121246-1.
27. Beletskaya, I.P.; Cheprakov, A.V. *Comprehensive Coordination Chemistry II*; McCleverty, J.A., Meyer, T.J., Eds.; Wiley-VCH: Weinheim, UK, 2004; ISBN 978-0-080-91316-2.
28. Wang, K.; Chen, S.; Zhang, H.; Xu, S.; Ye, F.; Zhanga, Y.; Wang, J. Pd(0)-catalyzed cross-coupling of allyl halides with α-diazocarbonyl compounds or N-mesylhydrazones: Synthesis of 1,3-diene compounds. *Org. Biomol. Chem.* **2016**, *14*, 3809–3820. [CrossRef] [PubMed]
29. O'Nei, G.W.; Phillips, A.J. Total Synthesis of (−)-Dictyostatin. *J. Am. Chem. Soc.* **2006**, *128*, 5340–5341. [CrossRef] [PubMed]
30. Negishi, E.; Wang, G.; Rao, H.; Xu, Z. Alkyne Elementometalation-Pd-Catalyzed Cross-Coupling. Toward Synthesis of All Conceivable Types of Acyclic Alkenes in High Yields, Efficiently, Selectively, Economically, and Safely: "Green" Way. *J. Org. Chem.* **2010**, *75*, 3151–3182. [CrossRef] [PubMed]
31. Eto, K.; Yoshino, M.; Takahashi, K.; Ishihara, J.; Hatakeyama, S. Total Synthesis of Oxazolomycin A. *Org. Lett.* **2011**, *13*, 5398–5401. [CrossRef] [PubMed]
32. Schmidt, B.; Kunz, O. One-Flask Tethered Ring Closing Metathesis–Electrocyclic Ring Opening for the Highly Stereoselective Synthesis of Conjugated Z/E-Dienes. *Eur. J. Org. Chem.* **2012**, *5*, 1008–1018. [CrossRef]
33. Souris, C.; Frébault, F.; Patel, A.; Audisio, D.; Houk, K.N.; Maulide, N. Stereoselective Synthesis of Dienyl-Carboxylate Building Blocks: Formal Synthesis of Inthomycin C. *Org. Lett.* **2013**, *15*, 3242–3245. [CrossRef] [PubMed]

34. Hiebel, M.-A.; Fall, Y.; Scherrmann, M.-C.; Berteina-Raboin, S. Straightforward Synthesis of Various 2,3-Diarylimidazo[1,2-a]pyridines in PEG400 Medium through One-Pot Condensation and C–H Arylation. *Eur. J. Org. Chem.* **2014**, *21*, 4643–4650. [CrossRef]
35. Hiebel, M.-A.; Berteina-Raboin, S. Iodine-catalyzed regioselective sulfenylation of imidazoheterocycles in PEG400. *Green Chem.* **2015**, *17*, 937–944. [CrossRef]
36. Tber, Z.; Hiebel, M.-A.; Akssira, M.; Guillaumet, G.; Berteina-Raboin, S. Use of Ligand-Free Iron/Copper Cocatalyst for Nitrogen and Sulfur Cross-Coupling Reaction with 6-Iodoimidazo[1,2-a]pyridine. *Synthesis* **2015**, *47*, 1780–1790. [CrossRef]
37. Addicott, F.T.; Carns, H.R. *Abscisic Acids*; Addicott, F.T., Ed.; Praeger: New York, NY, USA, 1983; Chapter 1.
38. Finkelstein, R.F.; Tenbarge, K.M.; Shumway, J.E.; Crunch, M.L. Role of ABA in Maturation of Rapeseed Embryos. *Plant Physiol.* **1985**, *78*, 630–636. [CrossRef] [PubMed]
39. Léon-Kloosterziel, K.M.; Van de Bunt, G.A.; Zeevaart, J.A.D.; Koornneef, M. Arabidopsis Mutants with a Reduced Seed Dormancy. *Plant Physiol.* **1996**, *110*, 233–240. [CrossRef] [PubMed]
40. Shinozaki, K.; Yamaguchi-Shinozaki, K. Molecular responses to dehydration and low temperature: Differences and cross-talk between two stress signaling pathways. *Curr. Opin. Plant. Biol.* **2000**, *3*, 217–223. [CrossRef]
41. Himmelbach, A.; Yang, Y.; Grill, E. Relay and control of abscisic acid signaling. *Curr. Opin. Plant. Biol.* **2003**, *6*, 470–479. [CrossRef]
42. Nambara, E.; Marion-Poll, A. Abscisic acid biosynthesis and catabolism. *Annu. Rev. Plant. Biol.* **2005**, *56*, 165–186. [CrossRef] [PubMed]
43. Todoroki, Y.; Narita, K.; Muramatsu, T.; Shimamura, H.; Ohnishi, T.; Mizutani, M.; Ueno, K.; Hirai, N. Synthesis and biological activity of amino acid conjugates of abscisic acid. *Bioorg. Med. Chem.* **2011**, *19*, 1743–1750. [CrossRef] [PubMed]
44. Guri, A.J.; Hontecillas, R.; Si, H.; Liu, D.; Bassaganya-Riera, J. Dietary abscisic acid ameliorates glucose tolerance and obesity-related inflammation in db/db mice fed high-fat diets. *Clin. Nutr.* **2007**, *26*, 107–116. [CrossRef] [PubMed]
45. Bellotti, M.; Salis, A.; Grozio, A.; Damonte, G.; Vigliarolo, T.; Galatini, A.; Zocchi, E.; Benatti, U.; Millo, E. Synthesis, structural characterization and effect on human granulocyte intracellular cAMP levels of abscisic acid analogs. *Bioorg. Med. Chem.* **2015**, *23*, 22–32. [CrossRef] [PubMed]
46. Brandt, D.; Bellosta, V.; Cossy, J. Stereoselective Synthesis of Conjugated Trienols from Allylic Alcohols and 1-Iodo-1,3-dienes. *Org. Lett.* **2012**, *14*, 5594–5597. [CrossRef] [PubMed]
47. McCubbin, J.A.; Voth, S.; Krokhin, O.V. Mild and Tunable Benzoic Acid Catalysts for Rearrangement Reactions of Allylic Alcohols. *J. Org. Chem.* **2011**, *76*, 8537–8542. [CrossRef] [PubMed]
48. Li, P.-F.; Wang, H.-L.; Qu, J. 1,n-Rearrangement of Allylic Alcohols Promoted by Hot Water: Application to the Synthesis of Navenone B, a Polyene Natural Product. *J. Org. Chem.* **2014**, *79*, 3955–3962. [CrossRef] [PubMed]
49. Knowles, J.P.; O'Connor, V.E.; Whiting, A. Studies towards the synthesis of the northern polyene of viridenomycin and synthesis of Z-double bond analogues. *Org. Biomol. Chem.* **2011**, *9*, 1876–1886. [CrossRef] [PubMed]
50. Diaz-Ortiz, A.; Prieto, P.; Vazquez, E. Heck Reactions under Microwave Irradiation in Solvent-Free Conditions. *Synlett* **1997**, *3*, 269–270. [CrossRef]
51. Leadbeater, N.E.; Williams, V.A.; Barnard, T.M.; Collins, M.J., Jr. Solvent-Free, Open-Vessel Microwave-Promoted Heck Couplings: From the mmol to the mol Scale. *Synlett* **2006**, *18*, 2953–2958. [CrossRef]
52. Du, L.-H.; Wang, Y.-G. Microwave-Promoted Heck Reaction Using Pd(OAc)$_2$ as Catalyst under Ligand-Free and Solvent-Free Conditions. *Synth. Commun.* **2007**, *37*, 217–222. [CrossRef]
53. Liu, G.; Hou, M.; Song, J.; Jiang, T.; Fan, H.; Zhang, Z.; Han, B. Immobilization of Pd nanoparticles with functional ionic liquid grafted onto cross-linked polymer for solvent-free Heck reaction. *Green Chem.* **2010**, *12*, 65–69. [CrossRef]
54. Firouzabadi, H.; Iranpoor, N.; Kazemi, F.; Gholinejad, M. Palladium nano-particles supported on agarose as efficient catalyst and bioorganic ligand for CC bond formation via solventless Mizoroki–Heck reaction and Sonogashira–Hagihara reaction in polyethylene glycol (PEG 400). *J. Mol. Catal. A Chem.* **2012**, *357*, 154–161. [CrossRef]

55. Khazaei, A.; Rahmati, S.; Hekmatian, Z.; Saeednia, S. A green approach for the synthesis of palladium nanoparticles supported on pectin: Application as a catalyst for solvent-free Mizoroki–Heck reaction. *J. Mol. Catal. A Chem.* **2013**, *372*, 160–166. [CrossRef]
56. Zolfigol, M.A.; Azadbakht, T.; Khakyzadeh, V.; Nejatyami, R.; Perrin, D.M. C(sp2)-C(sp2) cross coupling reactions catalyzed by an active and highly stable magnetically separable Pd-nanocatalyst in aqueous media. *RSC Adv.* **2014**, *4*, 40036–40042. [CrossRef]
57. Khazaei, A.; Khazaei, M.; Rahmati, S. A green method for the synthesis of gelatin/pectin stabilized palladium nano-particles as efficient heterogeneous catalyst for solvent-free Mizoroki–Heck reaction. *J. Mol. Catal. A Chem.* **2015**, *398*, 241–247. [CrossRef]
58. Trost, B.M.; Conway, W.P.; Strege, P.E.; Dietsche, T.J. New Synthetic reactions. Alkylative elimination. *J. Am. Chem. Soc.* **1974**, *96*, 7165–7167. [CrossRef]
59. Roberts, D.L.; Heckmann, R.A.; Hege, B.P.; Bellin, S.A. Synthesis of (RS)-abscisic acid. *J. Org. Chem.* **1968**, *33*, 3566–3569. [CrossRef]
60. Constantino, M.G.; Losco, P. A novel synthesis of (±)-abscisic acid. *J. Org. Chem.* **1989**, *54*, 681–683. [CrossRef]
61. Cornforth, J.; Hawes, J.E.; Mallaby, R. A Stereospecific Synthesis of (±)-Abscisic Acid. *Aust. J. Chem.* **1992**, *45*, 179–185. [CrossRef]
62. Mayer, H.J.; Rigassi, N.; Schwieter, U.; Weedon, B.C.L. Synthesis of Abscisic Acid. *Helv. Chim. Acta* **1976**, *59*, 1424–1427. [CrossRef]
63. Kienzle, F.; Mayer, H.; Minder, R.E.; Thommen, H. Synthese von optisch aktiven, natürlichen Carotinoiden und strukturell verwandten Verbindungen. III. Synthese von (+)-Abscisinsäure, (−)-Xanthoxin, (−)-Loliolid, (−)-Actinidiolid und (−)-Dihydroactinidiolid. *Helv. Chim. Acta* **1978**, *61*, 2616–2627. [CrossRef]
64. Constantino, M.G.; Donate, P.M.; Petragnani, N. An efficient synthesis of (±)-abscisic acid. *J. Org. Chem.* **1986**, *51*, 253–254. [CrossRef]
65. Rose, P.A.; Abrams, S.R.; Shaw, A.C. Synthesis of chiral acetylenic analogs of the plant hormone abscisic acid. *Tetrahedron Asymmetry* **1992**, *3*, 443–450. [CrossRef]
66. Hanson, J.R.; Uyanik, C. An efficient synthesis of the plant hormone abscisic acid. *J. Chem. Res.* **2003**, *7*, 426–427. [CrossRef]
67. Smith, T.R.; Clark, A.J.; Clarkson, G.J.; Taylor, P.C.; Marsh, A. Concise enantioselective synthesis of abscisic acid and a new analogue. *Org. Biomol. Chem.* **2006**, *4*, 4186–4192. [CrossRef] [PubMed]
68. Ferrer, E.; Alibés, R.; Busqué, F.; Figueredo, M.; Font, J.; De March, P. Enantiodivergent Synthesis of Cyclohexenyl Nucleosides. *J. Org. Chem.* **2009**, *74*, 2425–2432. [CrossRef] [PubMed]
69. Mash, E.A.; Torok, D.S. Homochiral ketals in organic synthesis. Diastereoselective cyclopropanation of alpha beta-unsaturated ketals derived from (S,S)-(−)-hydrobenzoin. *J. Org. Chem.* **1989**, *54*, 250–253. [CrossRef]

© 2018 by the authors. Licensee MDPI, Basel, Switzerland. This article is an open access article distributed under the terms and conditions of the Creative Commons Attribution (CC BY) license (http://creativecommons.org/licenses/by/4.0/).

Article

MgAl-Layered Double Hydroxide Solid Base Catalysts for Henry Reaction: A Green Protocol

Magda H. Abdellattif [1],* and Mohamed Mokhtar [2,3],*

[1] Pharmaceutical Chemistry Department, Deanship of Scientific Research Taif University, Taif 21974, Saudi Arabia
[2] Chemistry Department, Faculty of Science, King Abdulaziz University, Jeddah 21589, Saudi Arabia
[3] Physical Chemistry Department, National Research Centre, El Buhouth St., Dokki, Cairo 12622, Egypt
* Correspondence: magdah11uk@hotmail.com or m.hasan@tu.edu.sa (M.H.A.); mmokhtar2000@yahoo.com or mmoustafa@kau.edu.sa (M.M.)

Received: 28 February 2018; Accepted: 27 March 2018; Published: 29 March 2018

Abstract: A series of MgAl-layered double hydroxide (MgAl-HT), the calcined form at 500 °C (MgAlO$_x$), and the rehydrated one at 25 °C (MgAl-HT-RH) were synthesized. Physicochemical properties of the catalysts were characterized by X-ray diffraction (XRD) and scanning electron microscopy (SEM). Surface area of the as-synthesized, calcined, and rehydrated catalysts was determined by N$_2$ physisorption at −196 °C. CO$_2$ temperature-programmed desorption (CO$_2$-TPD) was applied to determine the basic sites of catalysts. The catalytic test reaction was carried out using benzaldehyde and their derivatives with nitromethane and their derivatives. The Henry products (1–15) were obtained in a very good yield using MgAl-HT-RH catalyst either by conventional method at 90 °C in liquid phase or under microwave irradiation method. The mesoporous structure and basic nature of the rehydrated solid catalyst were responsible for its superior catalytic efficiency. The robust nature was determined by using the same catalyst five times, where the product % yield was almost unchanged significantly.

Keywords: C–C bond formation; Henry reaction; solid base catalyst; layered double hydroxide

---

1. Introduction

The fine chemical industry has experienced remarkable interest over the past few years due to the high requirements for products like pharmaceuticals, pesticides, fragrances, flavorings, and food additives [1].

The classical methods for the C–C coupling in the Henry reaction using soluble bases such as alkali metal hydroxides, carbonates, bicarbonates, alkoxides, alkaline earth metal hydroxide, aluminium ethoxides, complexes, and also organic bases such as primary, secondary, and tertiary amines, usually resulted in dehydrated products [2]. Therefore, careful control of the basic properties of the reaction medium is vital to obtain better yields of β-nitroalcohols. However, the efforts done by the researchers in the literature required longer reaction times and produced moderate yields [3,4]. The stoichiometric organic synthesis that largely applied so far resulted in large quantities of inorganic salts as byproducts; the disposal of such material causes a serious problem due to the important environmental issues [5]. The homogenous catalytic methodologies reported in the literature have many disadvantages, such as disposal of waste and difficulty recovering the catalyst from the products. In the last decade, there were notable improvements in the development of heterogeneous catalysts for the Henry reaction [6].

The extraordinary growth in the industry's struggle has pushed researchers to advance more effective catalytic processes in the synthesis of fine chemicals. The products of the Henry reaction, those representing C–C bond development, are substantial materials widely used in frequent organic syntheses [7]. The supreme challenge in the selective synthesis of 2-nitroalkanols in the multiple product options such as aldol olefin and its polymer and Cannizzaro products is the selection of an accurate kind of base [8]. Noteworthy developments to the Henry reaction have been realized by means of silyl nitronates in the presence of fluoride ions or instead α-α doubly deprotonated primary nitroalkanes [9]. Both of these processes have shown to be valuable for the stereo selective preparation of vicinal amino alcohols under drastic conditions, which reduces diastereoselectivity with aromatic aldehydes. Hereafter, to find better yields and diastereoselectivity of 2-nitroalcohols, it is essential to advance novel procedures employing heterogeneous catalysts with basic appeal [10].

Heterogeneous catalysis encouraged by solid catalysts such as basic alumina [11] and alumina–KF [12] and homogeneous phase transfer catalysis with surfactants [13] in bi-phase systems are two opposing tactics that were discovered and are intended to attain higher atom selectivity. The solid base catalysts provide an alternative to the classical soluble bases with emphasis on avoiding the environmental problems caused by salt formation and hazardous conditions [14]. Previous work in the synthesis of fine chemicals using layered double hydroxides revealed the importance of such materials and discovered its environmentally favorable routes in comparison to the other catalysts [15–20].

Layered double hydroxide materials (LDHs) have unique features as they represent the basis for new environment-benign technologies concerning inexpensive and highly efficient pathways to catalyze chemical reactions. The general formula of LDHs is $[M^{(II)}{}_{1-n} M^{(III)}{}_n(OH)_2]^{n+} [(A^{m-}_{n/m}) \cdot xH_2O]^{n-}$ where A is the interlayer anion with valence m- and negatively charged anions such as NO_3^-, SO_4^{2-}, and CO_3^{2-} encounter the positively charged cationic sheet and the valence "n" is equal to the molar ratio of $M^{III}/(M^{II} + M^{III})$ [18]. The heat treatment of LDHs carried out in the temperature range 400–500 °C leads to a breakdown of the layered structure forming metal oxides mixture. However, the collapsed metal oxide mixture could reform the layered double hydroxide structure after water/anion treatment. LDHs, calcined metal oxide mixtures, or reformed LDH-like structures represent solid base catalysts for different organic reactions in fine chemical production [16–18].

In the present study, MgAl-layered double hydroxide, its calcined form at 500 °C (MgAlO$_x$), and the rehydrated form (MgAl-HT-RH) were synthesized and tested for the Henry reaction between nitroalkanes and different aldehydes. To the author's knowledge, this is the second trial after pioneering work by V. J. Bulbule et al. [21]. However, the present study should be the first extensive study to understand the effect of the mesoporous and basic nature of such catalysts in Henry reactions under the reaction conditions. The obtained promising results could open the gate towards a robust catalyst and a benign process in the Henry reaction.

2. Results and Discussion

2.1. Elemental Chemical Analysis (ICP)

ICP analysis of MgAl-HT was achieved to govern its chemical composition. The analysis discovered that the Mg/Al molar ratio in the solid was 2.8, which is very near to the minimal molar composition of the as-synthesized Mg/Al molar ratio of 3 in the precipitate. This result confirmed the efficacy of the preparation procedure.

2.2. X-ray Diffraction (XRD)

X-ray powder diffraction patterns of the as-synthesized MgAl-HT, thermally treated at 500 °C (MgAlO$_x$), and rehydrated MgAl-HT-RH catalysts are shown in Figure 1. A typical crystalline carbonate containing hydrotalcite phase structure (Ref. Pattern 22-0700, JCPDS) with strong (003), (006), (009), (110), (113) and broadened (015), (018) reflections was observed for the MgAl-HT sample. A crystalline periclase MgO phase was obtained upon thermal treatment of as-synthesized catalyst at 500 °C (MgAlO$_x$) (Ref. Pattern 45-0946, JCPDS) [22]. Thanks to the memory effect, the hydration of the calcined materials using an aqueous alkaline solution of NaOH led to the formation of a layered double hydroxide-like structure of lower intensity than the original MgAl-HT material. In the present study, we intended to hydrate the calcined material in aqueous alkaline solution to maintain the structure of the layered material and improve its basic nature by introduction of some terminal hydroxyl ions (Brönsted basic sites) [23]. The crystallite size derived from the Scherrer equation [24] showed that MgAl-HT-RH is much lower in size (20 nm) than MgAl-HT (160 nm). The pronounced decrease in the crystallite size of the rehydrated layered double hydroxide structure could improve the catalytic performance towards the Henry reaction.

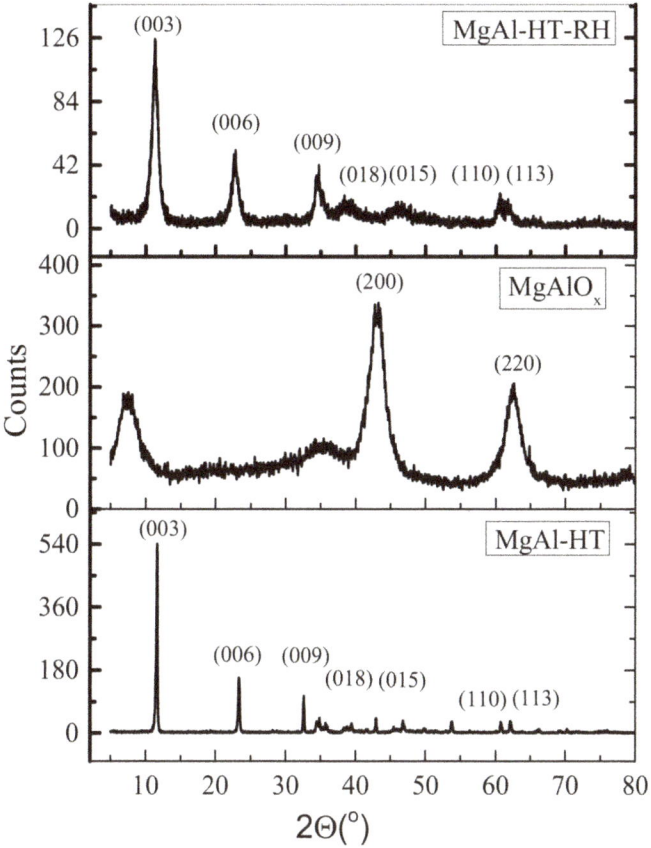

Figure 1. XRD patterns of all the investigated catalysts.

2.3. Scanning Electron Microscopy (SEM)

SEM images of all the investigated catalysts are given in Figure 2. The hydrothermal treatment under autogenous pressure at 170 °C for the coprecipitated MgAl-HT sample (Figure 2A) resulted in the formation of uniform hexagonal platelets of the layered material with 180 nm particle size. The calcination of MgAl-HT led to a pronounced collapse in the layered structure due to the removal of the interlayer anions and thermal decomposition of the hydroxide carbonate into the corresponding metal oxides (MgAlO$_x$) (Figure 2C) [22]. Alkaline treatment of the mixed oxide led to the building of the interlayer gallery between hexagonal platelets of relatively small particle size (20 nm) for MgAl-HT-RH (Figure 2B).

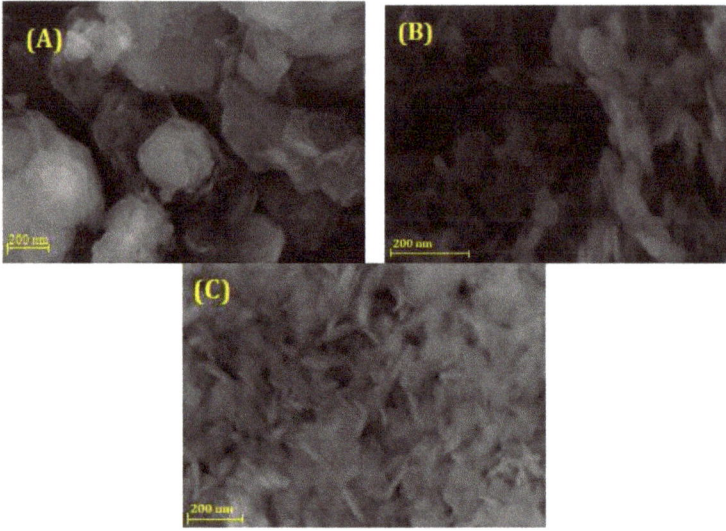

Figure 2. SEM images of: (**A**) MgAl-HT; (**B**) MgAl-HT-RH; (**C**) MgAlO$_x$.

2.4. N$_2$ Physisorption

N$_2$ adsorption/desorption isotherms of all the synthesized materials are given in Figure 3. Mesoporous isotherms of Type IV were detected. H3-hysteresis was recorded, which is characteristic of the occurrence of open, relatively large pores that could facilitate reactant/product diffusion through the catalysts [15]. BET surface area of all the investigated solid materials was calculated and is depicted in Table 1. MgAl-HT showed the smallest surface area of all samples (84 m^2/g), while the calcined sample (MgAlO$_x$) recorded the biggest BET surface area (167 m^2/g). The obvious rise in surface area was allocated to the creation of craters through the layers due to development of CO$_2$ and H$_2$O [25]. Rehydration of the calcined layered double hydroxide in the alkaline solution by mechanical stirring at room temperature led to an increase in surface area as a result of the breaking of particles and a noticeable exfoliation of the crystals [26].

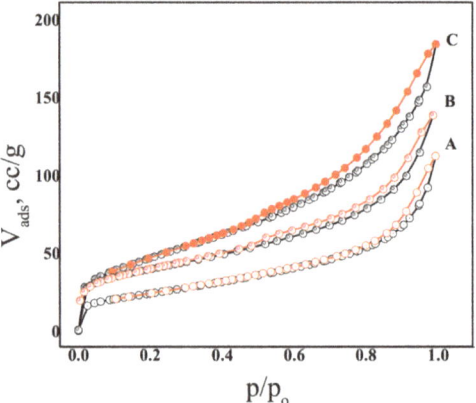

Figure 3. Adsorption-desorption isotherms of all the investigated catalysts; A: MgAl-HT; B: MgAlOx; C: MgAl-HT-RH.

Table 1. BET-surface area, total pore volume and pore radii of all the investigated samples obtained from N_2 adsorption/desorption isotherms.

Sample	S_{BET} (m^2/g)	V_p (cm^3/g)	r_p (Å)	C-Constant
MgAl-HT	84	0.1642	39	177
MgAlO$_x$	167	0.2645	34	86
MgAl-HT-RH	134	0.2014	30	567

2.5. CO_2 Temperature-Programmed Desorption (CO_2-TPD)

The measure of the basicity of the diverse solids was attained by TPD of CO_2. It is well known that types of basic sites can be detected by CO_2 uptake, which is related to diverse types of carbonate coordination in the lamellar interplanetary of layered double hydroxide. The monodentate, bidentate, and bicarbonate anions are often fashioned through the saturation of CO_2 to the basic materials [27]. Monodentate and bidentate carbonate creation contains low-coordination oxygen anions and are then considered as strong basic sites and the creation of bicarbonate anions involves surface hydroxyl groups [28]. The MgAl-HT sample displayed four desorption peaks of relatively low intensity (Figure 4). The low temperature peaks in the range 150–400 °C could be attributed to the desorption of weakly confined CO_2 and breakdown of remaining carbonate ions existing in the MgAl-HT sample [29]. The other two desorption peaks at 450 °C and 665 °C were mostly credited to bicarbonate groups fashioned by the contact of CO_2 with hydroxyl groups in the MgAl-HT and the progress of powerfully attached surface metal carbonate species, respectively. The calcined catalyst (MgAlO$_x$) presented three desorption peaks at 150 °C, 420 °C, and between 510 and 520 °C. The desorption peak at about 420 °C can be accredited to the influence of mostly bidentate carbonate species, together with bicarbonate species. The attendance of a peak at 540 °C is owed to the occurrence of monodentate species. The presence of the two peaks signified the occurrence of OH$^-$ groups with diverse strengths. The differences in CO_2 uptake values between as-synthesized and calcined catalysts can be credited to the presence of misdeeds or linear imperfections in the platelets of the calcined sample. SEM images showed the collapse in the layered structure of hydrotalcite upon thermal treatment at 500 °C. The high intensities of peaks at high temperature could be explained by the formation of Lewis basic sites due to the MgO formation as complemented by XRD data. As we observed, the hydration of the calcined metal oxides (MgAl-HT-RH) resulted in a material with high peak intensity at 420 °C, 500 °C, and a small peak at 620 °C. The pronounced increase in the peak intensity at 420 °C could be attributed to the formation of terminal Brönsted OH$^-$ basic sites together with some Lewis basic sites at high desorption temperature [18]. The pronounced increase in low- and high-temperature basic sites of MgAl-HT-RH could provide superior catalytic activity for this particular catalyst towards solid base catalyzed Henry reactions.

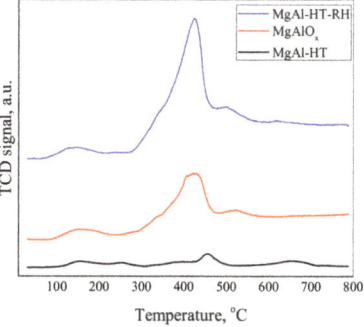

Figure 4. CO_2-TPD profile of all the investigated materials.

2.6. Catalytic Activity Study

The Henry reaction is a base-catalyzed, C–C bond-forming reaction between nitroalkanes and aldehydes or ketones. The catalytic efficacy of MgAl-HT, MgAlO$_x$, and MgAl-HT-RH was tested for the Henry reaction (Scheme 1). The reaction between nitroalkanes (**1a–c**) with different aromatic aldehydes (**2a–e**) in the presencew of all the catalysts was carried out utilizing conventional methods and solvent-free, microwave-assisted reactions to attain only a single isolable product in each case (as investigated by TLC). The identified products were 2-methyl-2-nitro-1-arylpropan-1-ol derivatives (**3a–e**) in the case of 2-nitropropane as a reactant, or 2-nitro-vinylbenzene and 2-nitroprop-1-enylbenzene derivatives (**4a–j**) in the cases of nitromethane and nitroethane as reactants, respectively.

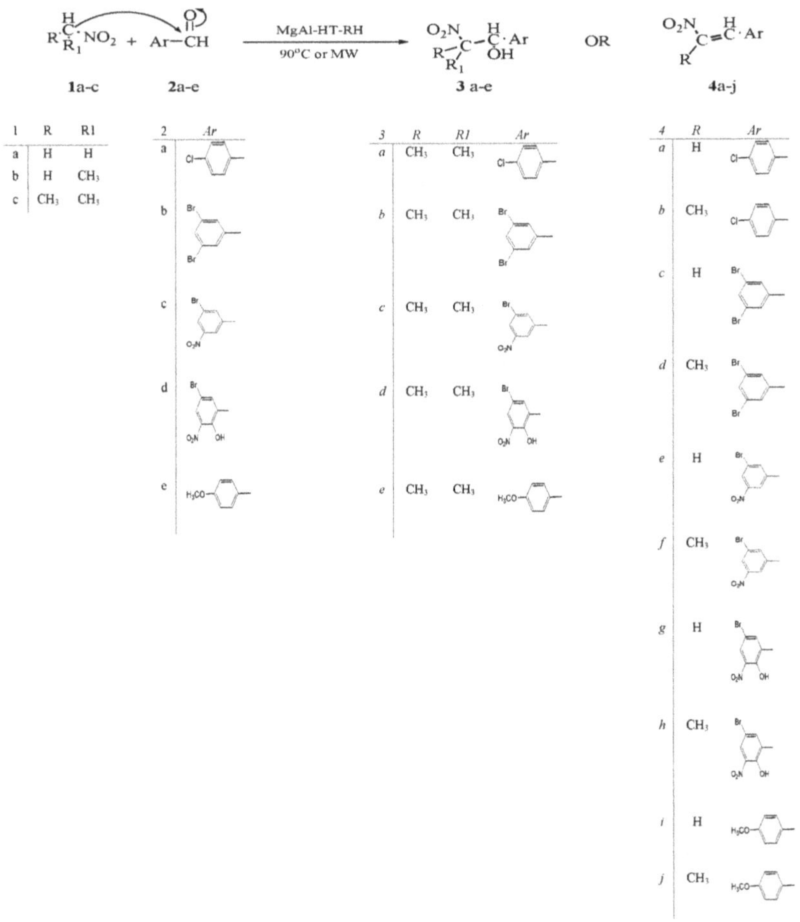

Scheme 1. Reaction of different aldehydes with nitroalkanes utilizing catalyst under different reaction conditions.

All the structures of the isolated products **3a–e** and **4a–j** were elucidated using IR, ^{1}H NMR, ^{13}C NMR, and MS analyses. The IR showed a characteristic band for OH groups for **3a–e** products, which was not recorded for products **4a–j**. The mass bands of the isolated products displayed peaks matching their molecular ions.

The catalytic reaction using different investigated catalysts were carried out between **1a** and **2a** under conventional and microwave irradiation and the obtained results are given in Table 2. It is revealed from this table that the MgAl-HT-RH catalyst resulted in satisfactory yields of the products. The reaction carried out under microwave irradiation exhibited improved yield (99%) in a very short reaction time (14 min) in comparison to the conventional conditions (90%, 5 h). Therefore, this specific catalyst was nominated as the greatest catalyst to assess the other reactions and the obtained results are briefed in Figure 5.

Table 2. Catalytic reaction of **1a** with **2a** under conventional and microwave irradiation conditions.

Catalyst	Conventional Method *		Microwave Method **		Product Structure (4a)
	Time (h)	Yield (%)	Time (min.)	Yield (%)	
MgAl-HT	8	61	20	90	
MgAlO$_x$	6	77	18	96	
MgAl-HT-RH	5	90	14	98	

* Reaction conditions: nitromethane (10 mmol), 4-cholorobenzaldehyd (10 mmol), catalyst (0.2 g), 90 °C. ** Reaction conditions: solvent-free conditions, nitromethane (10 mmol), 4-cholorobenzaldehyd (10 mmol), catalyst (0.2 g), MW irradiation (300 W).

It is seen from Figure 5 that MgAl-HT-RH displays an effectual activity and higher % yield in assessment to the reported data [21]. The comparison between this particular catalyst (MgAl-HT-RH) and other heterogeneous catalysts used to catalyze the Henry reaction reported in literature is presented in Table 3. The data obtained from this table revealed that the MgAl-HT-RH catalyst is the best heterogenous catalyst ever used until now for the synthesized Henry products presented in Table 3. The advanced activity of this catalyst is owed to the high surface area (134 m^2/g) and strong Lewis and Brönsted basic sites associated with this catalyst. The hydration of MgAlO$_x$ resulted in the formation of terminal OH$^-$ that increased the basicity of the catalyst [22], accordingly enhancing the catalytic activity of MgAl-HT-RH towards the Henry reaction.

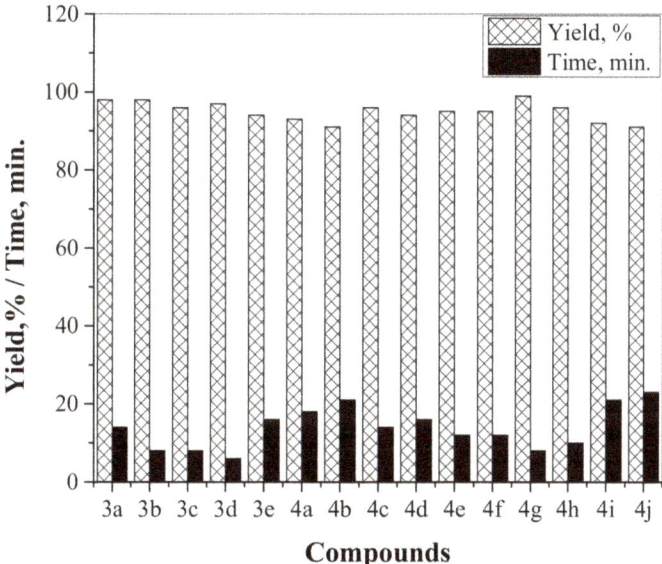

Figure 5. %Yield and reaction time of all the synthesized products utilizing MgAl-HT-RH under microwave irradiation.

Table 3. Henry products over MgAl-HT-RH catalyst and different synthesis routes for the products in the literature.

Compound	Reactants	Henry Product Structure	Current Work		Literature Data		
			Yield %	Time min	Yield %	Time h	Ref.
3a	1c,2a		95	12	71	11	[30]
3b	1c,2b		99	8	-	-	-
3c	1c,2c		92	21	-	-	-
3d	1c,2d		96	10	-	-	-
3e	1c,2e		91	23	68	9	[31]
4a	1a,2a		98	14	78	6	[32]
4b	1b,2a		93	18	80	6	[33]
4c	1a,2b		98	8	75	6	[34]

Table 3. Cont.

		Structure					
4d	1b,2b	(Br, Br substituted)	91	21	-	-	-
4e	1a,2c	(Br, NO₂ substituted)	97	6	76	6	[35]
4f	1b,2c	(Br, NO₂ substituted)	94	16	79	7	[36]
4g	1a,2d	(HO, NO₂, Br substituted)	96	8	-	-	-
4h	1b,2d	(HO, NO₂, Br substituted)	96	14	-	-	-
4i	1a,2e	(OMe substituted)	94	16	68	8	[32]
4j	1b,2e	(OMe substituted)	95	12	78	12	[37]

It was vital to study the stability of MgAl-HT-RH catalysts under the microwave operation. Therefore, a particular reaction between **1a** and **2a** was repeated five times with a redeveloped catalyst. The solid catalyst was subsequently filtered and washed by ethanol for each catalytic cycle, then dried in vacuo. The recovered catalyst was tested numerous times (five periods). The catalytic activity was recorded for the time of the achievement of the reaction and the results attained are given in Figure 6.

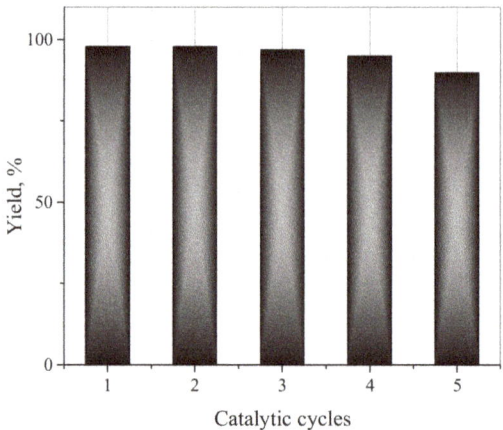

Figure 6. %Yield of **3a** utilizing MgAl-HT-RH for five cycles.

Figure 6 shows that the redeveloped catalyst achieved the reaction capably underneath constant reaction conditions even being subsequently utilized for five times. The observed decay in the catalytic activity was recorded after recycling the catalyst for the fifth time. Our previous studies using XRD technique for the reused catalysts in Aza–Michael addition reactions showed that the reused catalyst was contaminated by organic moieties being adsorbed on the catalyst active sites and caused the temporary poisoning of the regenerated catalyst [15]. The minor deterioration detected in the catalytic action of the MgAl-HT-RH catalyst after reuse for five times could be assigned to the temporary poisoning from organic poisons and/or to the loss in weight by the filtration/washing process.

3. Experimental Section

3.1. Reagents

All chemicals used were of analytical grade purchased from Sigma Alderich (Dorset, UK).

3.2. Catalyst Synthesis

Catalysts were prepared by coprecipitation methods as in the literature [38,39]. Mixing of magnesium nitrate Mg $(NO_3)_2 \cdot 6H_2O$ (0.2213 mol) and aluminium nitrate Al $(NO_3)_3$ (0.0885 mol) in a 0.2213 L of dist. H_2O produced mix (A). Mixing of 0.7162 mol of NaOH and 0. 2084 mol of Na_2CO_3 in a 0.221 L produced mix (B). The steps were as follows: Drop, using a burette, drops from Mix (A) and Mix (B) to a round-bottomed flask 1 L containing 0.5 L distilled water under vigorous stirring and heating at 60 °C and measure pH during precipitation to be 10–11. Keep the temperature at 60 °C overnight (16 h) in water bath. Filter using Whatmann1 filter paper and wash the cake with hot distilled water until pH = 7. Dry the filtrate at 80 °C in an oven for 16 h. The as-synthesized solid was nominated as MgAl-HT. Upon heat treatment, a certain weight of as-synthesized hydrotalcite at 450 °C for 6 h under N_2 atmosphere, a mixture of metal oxide catalyst named $MgAlO_x$ was prepared. Dissolution of $MgAlO_x$ in alkaline solution of 1 M NaOH under vigorous stirring at the room temperature resulted in the creation of layered rehydrated hydrotalcite form nominated as MgAl-HT-RH.

3.3. Catalyst Characterization

The chemical analysis of the prepared solid catalysts was evaluated by using ICP-AES, Optima 7300DV (Perkin Elmmer Corporation, Waltham, MA, USA) apparatus.

X-ray diffraction (XRD) analysis using a Bruker diffractometer (Bruker D8 advance target, Bruker, Karlsruhe, Germany) was utilized using Cu Kα1 and a monochromator (λ = 1.5405 Å) at 40 kV and 40 mA. Particle size of the solid was calculated by means of Scherrer Equation: $D = B\lambda/\beta_{1/2}\cos\theta$, where D is the average particle size of the solid phase under examination, B is the Scherrer constant (0.89), λ is wavelength of the X-ray beam used (1.5405 Å), $\beta_{1/2}$ is the full width at half maximum (FWHM) of the peak, and θ is the angle of diffraction.

A JEOL JSM840A instrument (JEOL, Tokyo, Japan) Scanning Electron Microscope (SEM) was utilized to investigate the morphology of solid samples. Prior to each measurement, the sample was placed on an aluminum block utilizing carbon tape. Physisorption of N_2 at −196 °C using a NOVA 3200e automated gas sorption system (Quantachrome, Boynton Beach, FL, USA) was applied to investigate the pore structure of the solids. Before every measurement, adsorbent was pretreated at 150 °C for 6 h. The Brunauer–Emmett–Teller (BET) equation was applied to determine the specific surface area, while the average pore radius was deduced from the equation: $2V_p/S_{BET}$, as V_p is the total pore volume (at $P/P_0 = 0.98$).

CO_2 TPD analysis was performed using CHEMBET 3000 (Quantachrome, FL, USA). Outgassing of the sample at 100 °C (1 h) was performed while passing helium to detach physisorbed water. Then the saturation of the sample with the CO_2 at 120 °C occurred. The temperature-programmed desorption was simply achieved by ramping the adsorbent temperature at 10 °C/min to 800 °C.

3.4. Characterization of Reaction Products

All melting points of the reaction products were measured on a Gallenkamp melting point apparatus and were uncorrected. ^1H NMR spectra, ^{13}C NMR spectra were recorded on a Bruker AM250 NMR spectrometer (Bruker, Karlsruhe, Germany) using $CDCl_3$ as solvent for the samples. Mass spectra were recorded on Shimadzu LCMS-QP 800 LC-MS (Kyoto, Kyoto Prefecture, Japan), IR for the synthesized compounds were recorded in potassium bromide discs on a Shimadzu FTIR 8101 PC infrared spectrophotometer. Elemental analysis was obtained using a PerkinElmer 2400 II series CHN Analyser (Waltham, MA, USA). Thin-layer chromatography (TLC) was carried out on precoated Merck silica gel F254 plates. Microwave experiments were carried out using CEM Discover Labmate™ microwave apparatus (300 W with ChemDriver™ Software).

3.5. Typical Procedure for the Catalytic Test Reaction

3.5.1. Method A: Conventional Method

A mixture of aldehydes **2a–e** (10 mmol), nitroalkanes **1a–c** (10 mmol), and 0.2 g of catalyst were heated together in a three-necked, round-bottomed flask at 90 °C. The progress of the reaction was monitored by TLC. Upon completion of the reaction, the mixture was cooled and the product was extracted by dissolution in hot alcohol. The catalyst was filtered off and washed by alcohol prior to drying and reuse. After evaporation of volatile materials under vacuum, compounds **3a–e** and **4a–j** were recrystallized from the EtOH/DMF mixture.

3.5.2. Method B: Microwave Irradiation

A mixture of aldehydes **2a–e** (10 mmol), nitroalkanes **1a–c** (10 mmol), and 0.2 g of catalyst were added in a Teflon vial and irradiated by microwave (300 W) for a required time to complete the reaction (Table 3) in a 2-min interval. The reaction progress was monitored using TLC (eluent; Diethyl ether: chloroform). Then the product mixture was cooled and extracted using ethanol. The catalyst was filtered off and the product compounds were purified by crystallization using an EtOH/DMF solvent mixture to afford the pure crude β-nitroalcoohls **3a–e** and the nitrolakenes **4a–j** an excellent yield.

Physical and spectral data of the titled compounds 3a–e and 4a–j are listed below (Supplementary Materials).

3a: 1-(4-Chlorophenyl)-2-methyl-2-nitropropan-1-ol.

mp. 192 °C, IR (KBr) υ max/cm^{-1}: 1530 (NO$_2$), ^{1}H NMR (DMSO): δ 1.1 (s, 3H, CH$_3$), δ 5.1 (s, H, CH), δ 7.27–7.4 (dd, 4H, ArH's), ^{13}C NMR (CDCl$_3$): δ 25.5, 39.5, 77.82, 81.1 126.66, 130.8, 134.5, MS (m/z): 230.01 (M$^+$), Anal. calcd. for C$_{10}$H$_{12}$ClNO$_3$ (229), C-52.30; Cl-15.44; H-5.27; N-6.10% Found: C-52.2; Cl-15.44; H-4.01; N-6.8%.

3b: 1-(3,5-Dibromophenyl)-2-methyl-2-nitropropan-1-ol.

mp. above 300 °C, IR (KBr) υ max/cm^{-1}: 3550 (OH), 1523–1550 (NO$_2$), 670 (C-Br), ^{1}H NMR (DMSO): δ 3.15 (d, 2H, CH$_2$), 7.27–7.4 (dd, 4H, ArH's), 5.2 (t, 1H, –CH), ^{13}C NMR (CDCl3): δ 25.5, 39.8, 77.8, 88.7, 124.4, 129.8, 131.1, 135.1, MS (m/z): 353.3 (M$^+$), Anal. Calcd. for C$_{10}$H$_{11}$Br$_2$NO$_3$ (353), Br-45.27; C-34.02; H-3.14; N-3.97% Found: Br-45.3, C-34.01; H-3.15; N-3.95%.

3c: 1-(3-Bromo-5-nitronitrophenyl)-2-methyl-2-nitropropan-1-ol.

mp. above 300 °C, IR (KBr) υ max/cm^{-1}: 3650 (OH), 1520–1570 (2NO$_2$), ^{1}H NMR (DMSO): δ 1.66 (s, 6H, 2CH$_3$), 4.50 (s, H, CH), 7.75–8.29 (m, 3H, ArH's), ^{13}C NMR (CDCl$_3$): δ 16.80, 81.8, 93.0, 122.8, 123.1, 123.7, 139.4, 144.0,149.7 MS (m/z): 208. (M$^+$), Anal. calcd. for C$_{10}$H$_{11}$N$_2$O$_5$Br (328), C-50.00; H-5.04; N-11.66 % Found: C-49.90; H-5; N-11.66%.

3d: 4-Bromo-2-(1-hydroxy-2-methyl-2-nitropropyl)-6-nitrophenol.

mp. above 300 °C, IR (KBr) υ max/cm^{-1}: 3500–3700 (2OH), 1520–1580 (2NO$_2$), 690 (C-Br), ^{1}H NMR (DMSO): δ 1.09, (s, 3H,CH3), 5.2 (s, 1H, –CH), 6.5–7.2 (m, 2H, ArH's), ^{13}C NMR (CDCl$_3$): δ 25.1, 39.48, 88.7, 110.86, 112.36, 117.5, 118.3, 145.3, 147.9, MS (m/z): 334.9 (M$^+$), Anal. calcd. for C$_{10}$H$_{11}$BrN$_2$O$_6$ (335), Br-23.84; C-35.9; H-3.3; N-8.4% Found: Br-Br-23.84; C-35.75; H-3.35; N-8.36%.

3e: 1-(4-Methoxyphenyl)-2-methyl-2-nitropropan-1-ol.

mp. above 300 °C, IR (KBr) υ max/cm^{-1}: 3650 (OH), 1535 (NO$_2$), 1050 (C–O), ^{1}H NMR (DMSO), δ 1.04 (s, 3H, CH$_3$), 3.14 (s, 3H, CH$_3$), 5.2 (s, H, CH), 6.9–7.3 (m, 4H, ArH's), ^{13}C NMR (CDCl$_3$): δ 18.02, 44.95, 65.01, 77.8, 84.7, 114.16, 120.86, 129.6, 159.99, MS (m/z): 225.3 (M$^+$), Anal. calcd. for C$_{11}$H$_{15}$NO$_4$ (225), C-58.66; H-6.71; N-6.22% Found: C-58.66; H-6.71; N-6.22%.

4a: p-2-Nitroethenylchlorobenzene.

mp. above 300 °C, IR (KBr) υ max/cm^{-1}: 1515–1560 (NO$_2$), ^{1}H NMR (DMSO): δ 5.1 (d, 1H, –C=C), 5.6 (d, 1H, –C=C) δ 7.27–7.45 (dd, 4H, Ar), ^{13}C NMR (CDCl$_3$): δ 77.8, 111.22, 127.4, 128.2, 131.8, 134.3, 134.7, MS (m/z): 183.01 (M$^+$), Anal. calcd. for C$_8$H$_6$ClNO$_2$ (183.6), C-52.61; Cl-19.31; H-3.3; N-7.65% Found: C-52.66; Cl-19.3; H-3.3; N-7.63%.

4b: 1-(p-Chlorophenyl)-2-nitro-1-propene.

mp. above 300 °C, IR (KBr) υ max/cm^{-1}: 1525–1570 (NO$_2$), ^{1}H NMR (DMSO): δ 1.7 (d, 3H, CH$_3$), 4.9 (q, 1H, –C=CH), 7.2–7.5 (dd, 4H, Ar), ^{13}C NMR (CDCl$_3$): δ 24.5, 77.8, 84.66, 119.8, 128.1, 130, 131.5, 134.7, MS (m/z): 197 (M$^+$), Anal. calcd. for C$_9$H$_8$ClNO$_2$ (197), C-54.13; Cl-18; H-4.1; N-7% Found: C-50.2; Cl-17.5; H-4. 1; N-6.8%.

4c: 2-Nitroethenyl-3,5-dibromobenzene.

mp. above 300 °C, IR (KBr) υ max/cm^{-1}: 1480 (NO$_2$), 670 (C-Br), ^{1}H NMR (DMSO): δ δ 5.2 (d, 1H, –C=CH), 6.2 (d, 1H, –C=CH), 7.7–7.9 (m, 3H, ArH's), ^{13}C NMR (CDCl$_3$): δ 77.8, 112.33, 123, 128, 131.1, 135.4, 139.5, MS (m/z): 307 (M$^+$), Anal. calcd. for C$_8$H$_5$Br$_2$NO$_2$ (307), Br-57.98; C-31.27; H-1.62; N-4.56% Found: Br-57.7; C-30.57; H-1.6; N-4.46%.

4d: 1-(3,5-Dibromophenyl)-2-nitro-1-propene.

mp. above 300 °C, IR (KBr) υ max/cm^{-1}: 1545 (NO$_2$), 672 (C-Br), ^{1}H NMR (DMSO): δ 1.7 (d, 3H,CH$_3$) 5.9 (q, 1H, –C=CH), 7.7–7.9 (m, 3H, ArH's), ^{13}C NMR (CDCl$_3$): δ 24.9, 77.8, 119.7, 123, 128.77, 130, 131.7, 139.4 MS (m/z): 321.9 (M$^+$), Anal. calcd. for C$_9$H$_9$Br$_2$NO$_3$ (322), Br-49.8; C-34; H-2.2; N-4.4% Found: Br-49.79; C-33.68; H-2.20; N-4.36%.

4e: 2-(2-Nitroethenyl)-4-bromo-6-nitrophenol.

mp. above 300 °C, IR (KBr) υ max/cm^{-1}: 3600 (OH), 1520–1580 (2NO$_2$), 675 (C-Br), ^{1}H NMR (DMSO): δ 5.1 (d, 1H, C=CH), 5.63 (H, OH), 6.1(d, 1H, C=CH), 6.9–7.2 (dd, 2H, ArH's), ^{13}C NMR (CDCl$_3$): δ 77.8, 112.3, 113, 115, 120, 126, 130, 135.8, 138.5, MS (m/z): 289 (M$^+$), Anal. calcd. for C$_8$H$_5$Br$_2$NO$_2$ (289), Br-27.64; C-33.24; H-1.74; N-9.69% Found: Br-27.64; C-33.24; H-1.74; N-9.69%.

4f: 2-[-2-Nitro-1-propenyl]-4-bromo-6-nitrophenol.

mp. above 300 °C, IR (KBr) υ max/cm^{-1}: 3650 (OH), 1540 (2NO$_2$), 675 (C-Br), ^{1}H NMR (DMSO): δ 1.75(d, 3H, CH$_3$), 4.95 (q, H, C=CH), 5.6 (s, H, OH), 6.9–7.2 (dd, 2H, ArH's), ^{13}C NMR (CDCl$_3$): δ 24.82, 77.82, 114, 115.86, 119, 120, 126, 13.5, 136.1, 138.9, MS (m/z): 302 (M$^+$), Anal. calcd. for C$_9$H$_7$BrN$_2$O$_5$ (303), Br-26.37; C-35.67; H-2.33; N-9.24% Found: Br-26.37; C-35.67; H-2.33; N-9.24%.

4g: 4-bromo-2-Nitro-6-(2-nitovinyl) phenol.

mp. 232 °C, IR (KBr) υ max/cm^{-1}: 1510–1555 (2NO$_2$), ^{1}H NMR (DMSO): δ 7.29 (d, H, C=H), 7.36 (d, H, C=H), 7.68,8.07 (2, 2H, ArH's), ^{13}C NMR (CDCl$_3$): δ 117.5, 121.7, 122.2, 131.6, 133.9, 134.9, 135.5, 150.1, MS (m/z): 298 (M$^+$), Anal. calcd. for C$_8$H$_5$N$_2$O$_5$Br (298), C-32.21; H-1.67; N-9.39% Found: C-32.42; H-1.53; N-8.94%.

4h: 4-bromo-2-Nitro-6-(2-nitroprop-1-enyl) phenol.

mp. 229 °C, IR (KBr) υ max/cm^{-1}: 1515–1560 (2NO$_2$), ^{1}H NMR (DMSO): δ 1.75, (s, 3H, CH3), 7.14 (s, 1H, C=CH), 7.69,2.07 (2s, 2H, ArH's), ^{13}C NMR (CDCl$_3$): δ 16.9, 116.5, 119.7, 127.5, 127.6, 137.4, 137.7, 141.0, 149 MS (m/z): 307.5 (M$^+$), Anal. calcd. for C$_9$H$_7$N$_2$O$_5$Br (312), C-34.62; H-2.24; N-8.97% Found: C-34.93; H-2.12; N-8.46%.

4i: p-[-2-Nitroethenyl]methoxybenzene.

mp. 266 °C, IR (KBr) υ max/cm^{-1}: 1545 (NO$_2$), 1150 (C–O), ^{1}H NMR (DMSO): δ 3.15 (s, H, CH$_3$), 5.2 (d, 1H, –C=CH), 6.1 (d, 1H, C=CH), 7–7.4 (m, 4H, ArH's), ^{13}C NMR (CDCl$_3$): δ 65.5, 77.8, 113, 127.7, 128.66, 135.4, 156.5, MS (m/z): 179 (M$^+$), Anal. calcd. for C$_9$H$_9$NO$_3$ (197), C-60.33; H-5.06; N-7.82% Found: C-60.33; H-5.06; N-7.82%.

4j: 1-(p-Methoxyphenyl)-2-nitro-1-propene.

mp. above 300 °C, IR (KBr) υ max/cm^{-1}: 1565 (NO$_2$), 1210 (C–O), ^{1}H NMR (DMSO): δ 1.75 (d, 3H, CH$_3$), 3.14 (s, 3H, CH$_3$), 4.9 (m, 1H, –C=CH), 7–7.4 (m, 4H, ArH's), ^{13}C NMR (CDCl$_3$): δ 24, 77.8, 84.7, 112.86, 119.83, 127.7, 128.5, 130.33, 156.79, MS (m/z): 193.1 (M$^+$), Anal. calcd. for C$_{10}$H$_{11}$NO$_3$ (193), C-56.86; H-6.20; N-6.63 % Found: C-56.86; H-6.20; N-6.63%.

4. Conclusions

In brief, solid base catalysts from the layered double hydroxide family, MgAl-HT, its calcined mixture (MgAlO$_x$), and activated alkali-treated oxide mixture (MgAl-HT-RH) were successfully synthesized and fully characterized using different techniques. The Henry reaction between benzaldehyde and nitromethane over solid base catalysts was attained. MgAl-HT-RH catalyst gave a precious advantage over all other solid base catalysts utilizing conventional or microwave-assisted reaction conditions. The microwave irradiation technique introduced high yields of β-alcohol derivatives using the layered double hydroxide catalysts in a very short time. Novel Henry products were synthesized in good yield (96%) in the current work for the first time. The relatively large surface area, mesoporous nature, and strong basic sites of rehydrated catalyst (MgAl-HT-RH) were responsible for the power of catalytic activity. The catalyst was reusable and its activity could be sustained after five catalytic cycles. The catalysts' superior efficiency and sustainability for carbon–carbon coupling via the Henry reaction makes them a promising candidate for further coupling reactions.

Supplementary Materials: The following are available online at http://www.mdpi.com/2073-4344/8/4/133/s1, Figure S1: Spectral data of compounds 3a–e and 4a–j.

Acknowledgments: The authors are very grateful to Taif University, Taif, KSA, because this work was financially supported by Taif University, Taif, KSA, under project number 1/437/4938. Authors would like to acknowledge Nesreen Said I. Ahmed at NRC, Cairo, Egypt for her valuable discussion and data interpretation.

Author Contributions: Magda H. Abdellattif and Mohamed Mokhtar conceived and designed the experiments; Magda H. Abdellattif performed the experiments; Magda H. Abdellattif and Mohamed Mokhtar analyzed the data; Magda H. Abdellattif contributed reagents/materials/analysis tools; Mohamed Mokhtar wrote the paper.

Conflicts of Interest: The authors declare no conflict of interest.

References

1. Olah, G.A.; Krishnamurti, R.; Prakash, G.K.S. *Comprehensive Organic Synthesis*; Trost, B.M., Fleming, I., Eds.; Pergamon Press: Oxford, UK, 1991; Volume III, p. 293.
2. Rosini, G. *Comprehensive Organic Synthesis*; Heathcock, C.H., Trost, B.M., Fleming, I., Eds.; Pergamon Press: Oxford, UK, 1991; Volume 2, p. 321.
3. Ballini, R.; Bosica, G.; Forconi, P. Nitroaldol (Henry) reaction catalyzed by amberlyst A-21 as a far superior heterogeneous catalyst. *Tetrahedron* **1996**, *52*, 1677–1684. [CrossRef]
4. Contantino, U.; Curini, M.; Marmottini, F.; Rosati, O.; Pisani, E. Potassium Exchanged Layered Zirconium Phosphate as Base Catalyst in the Synthesis of 2-Nitroalkanols. *Chem. Lett.* **1994**, *23*, 2215–2218. [CrossRef]
5. Sheldon, R.A. Catalysis: The key to waste minimization. *J. Chem. Technol. Biotechnol.* **1997**, *68*, 381–388. [CrossRef]
6. De Vries, A.H.M.; de Vries, J.G.; van Assema, F.B.J.; de Lange, B.; Mink, D.; Hyett, D.J. Asymmetric Synthesis of (S)-2-Indolinecarboxylic Acid by Combining Biocatalysis and Homogeneous Catalysis. *ChemCatChem* **2011**, *3*, 289–292. [CrossRef]
7. Choudary, B.M.; Kantam, M.L.; Reddy, C.V.; Rao, K.K.; Figueras, F. Henry reactions catalysed by modified Mg–Al hydrotalcite: An efficient reusable solid base for selective synthesis of β-nitroalkanols. *Green Chem.* **1999**, *1*, 187–189. [CrossRef]
8. Choudary, B.M.; Kantam, M.L.; Kavita, B. Synthesis of 2-nitroalkanols by Mg_3Al_2O-t-Bu hydrotalcite. *J. Mol. Catal. A Chem.* **2001**, *169*, 193–197. [CrossRef]
9. Seebach, D.; Beck, A.K.; Mukhopdyay, T.; Thomas, E.H. Diastereoselective Synthesis of Nitroaldol Derivatives. *Helv. Chim. Acta* **1982**, *65*, 1101–1133. [CrossRef]
10. Rosini, G.; Ballini, R.; Sorrenti, P. Synthesis of 2-Nitroalkanols on Alumina Surfaces without Solvent: A Simple, Mild and Convenient Method. *Synthesis* **1983**, *1983*, 1014–1016. [CrossRef]
11. Melot, J.M.; Texier-Boullet, F.; Foucaud, A. Preparation and oxidation of α-nitro alcohols with supported reagents. *Tetrahedron Lett.* **1986**, *27*, 493–496. [CrossRef]
12. Ballini, R.; Bosica, G. Nitroaldol Reaction in Aqueous Media: An Important Improvement of the Henry Reaction. *J. Org. Chem.* **1997**, *62*, 425–427. [CrossRef] [PubMed]
13. Kloetstra, K.R.; van Bekkum, H. Base and acid catalysis by the alkali-containing MCM-41 mesoporous molecular sieve. *J. Chem. Soc. Chem. Commun.* **1995**, 1005–1006. [CrossRef]
14. Mokhtar, M.; Saleh, T.S.; Ahmed, N.S.; Al-Thabaiti, S.A.; Al-Shareef, R.A. An eco-friendly N-sulfonylation of amines using stable and reusable Zn–Al–hydrotalcite solid base catalyst under ultrasound irradiation. *Ultrason. Sonochem.* **2011**, *18*, 172–176. [CrossRef] [PubMed]
15. Mokhtar, M.; Saleh, T.S.; Basahel, S.N. Mg–Al hydrotalcites as efficient catalysts for Aza-Michael addition reaction: A green protocol. *J. Mol. Catal. A Chem.* **2012**, *353–354*, 122–131. [CrossRef]

16. Saleh, T.S.; Narasimharao, K.; Ahmed, N.S.; Basahel, S.N.; Al-Thabaiti, S.A.; Mokhtar, M. Mg–Al hydrotalcite as an efficient catalyst for microwave assisted regioselective 1,3-dipolar cycloaddition of nitrilimines with the enaminone derivatives: A green protocol. *J. Mol. Catal. A Chem.* **2013**, *367*, 12–22. [CrossRef]
17. Narasimharao, K.; Al-Sabban, E.; Saleh, T.; Gallastegu, A.G.; Sanfiz, A.C.; Basahel, S.; Al-Thabaiti, S.; Alyoubi, A.; Obaid, A.; Mokhtar, M. Microwave assisted efficient protocol for the classic Ullman homocoupling reaction using Cu-MG-Al hydrotalcite catalysts. *J. Mol. Catal. A Chem.* **2013**, *379*, 152–162. [CrossRef]
18. Basahel, S.N.; Al-Thabaiti, S.A.; Narasimharao, K.; Ahmed, N.S.; Mokhtar, M. Nanostructured Mg–Al Hydrotalcite as Catalyst for Fine Chemical Synthesis. *J. Nanosci. Nanotechnol.* **2014**, *14*, 1931–1946. [CrossRef] [PubMed]
19. Sanfiz, A.C.; Vega, N.M.; de Marco, M.; Mokhtar, D.I.M.; Bawaked, S.M.; Basahel, S.N.; Al-Thabaiti, S.A.; Alyoubi, A.O.; Shaffer, M.S.P. Self-condensation of acetone over Mg–Al layered double hydroxide supported on multi-walled carbon nanotube catalysts. *J. Mol. Catal. A Chem.* **2015**, *398*, 50–57. [CrossRef]
20. Phukan, M.; Borah, K.J.; Borah, R. Henry reaction in environmentally benign methods using imidazole as catalyst. *Green Chem. Lett. Rev.* **2009**, *2*, 249–253. [CrossRef]
21. Bulbule, V.J.; Deshpande, V.H.; Velu, S.; Sudalai, A.; Sivasankar, S.; Sathe, V.T. Heterogeneous Henry reaction of aldehydes: Diastereoselective synthesis of nitroalcohol derivatives over Mg–Al hydrotalcites. *Tetrahedron* **1999**, *55*, 9325–9332. [CrossRef]
22. Mokhtar, M.; Inayat, A.; Ofili, J.; Schwieger, W. Thermal decomposition, gas phase hydration and liquid phase reconstruction in the system Mg/Al hydrotalcite/mixed oxide: A comparative study. *Appl. Clay Sci.* **2010**, *50*, 176–181. [CrossRef]
23. Tichit, D.; Fajula, F. Layered double hydroxides as solid base catalysts and catalyst precursors. *Stud. Surf. Sci. Catal.* **1999**, *125*, 329–340.
24. Scherrer, P. *Göttinger Nachrichten Gesell*; Springer: Berlin, Germany, 1918; Volume 2, p. 98.
25. Reichle, W.T.; Kang, S.Y.; Everhardt, D.S. The nature of the thermal decomposition of a catalytically active anionic clay mineral. *J. Catal.* **1986**, *101*, 352–359. [CrossRef]
26. Abelló, S.; Medina, F.; Tichit, D.; Pérez-Ramírez, J.; Sueiras, J.E.; Salagre, P.; Cesteros, Y. Aldol condensation of campholenic aldehyde and MEK over activated hydrotalcite. *Appl. Catal. B Environ.* **2007**, *70*, 577–584. [CrossRef]
27. Chimentao, R.J.; Abello, S.; Medina, F.; Llorca, J.; Sueiras, J.E.; Cesteros, Y.; Salagre, P. Defect-induced strategies for the creation of highly active hydrotalcites in base-catalyzed reactions. *J. Catal.* **2007**, *252*, 249–257. [CrossRef]
28. Prinetto, F.; Ghiotti, G.; Durand, R.; Tichit, D. Investigation of Acid−Base Properties of Catalysts Obtained from Layered Double Hydroxides. *J. Phys. Chem. B* **2000**, *104*, 1117–11126. [CrossRef]
29. Abello, S.; Medina, F.; Tichit, D.; Perez-Ramirez, J.; Rodriguez, X.; Sueiras, J.E.; Salagre, P.; Cesteros, Y. Study of alkaline-doping agents on the performance of reconstructed Mg–Al hydrotalcites in aldol condensations. *Appl. Catal. A Gen.* **2005**, *281*, 191–198. [CrossRef]
30. Klein, T.A.; Schkeryantz, J.M. Tandem Hass-Bender/Henry reaction for the synthesis of dimethylnitro alcohols from benzylic halides. *Tetrahedron Lett.* **2005**, *46*, 4535–4538. [CrossRef]
31. Soengas, R.G.; Silva, A.M.S. Indium-catalyzed Henry-type reaction of aldehydes with bromonitroalkanes. *Synlett* **2012**, *23*, 873–876. [CrossRef]
32. Moustakim, M.; Clark, P.G.K.; Trulli, L.; de Arriba, A.L.F.; Ehebauer, M.T.; Chaikuad, A.; Murphy, E.J.; Mendez-Johnson, J.; Daniels, D.; Hou, C.-D.; et al. Discovery of a PCAF Bromodomain Chemical Probe. *Angew. Chem.* **2017**, *56*, 827–831. [CrossRef] [PubMed]
33. Zhu, F.-X.; Zhao, P.-S.; Sun, X.-J.; An, L.-T.; Deng, Y.; Wu, J.-M. Direct synthesis and application of bridged diamino-functionalized periodic mesoporous organosilicas with high nitrogen contents. *J. Solid State Chem.* **2017**, *255*, 70–75. [CrossRef]
34. Pagano, M.A.; Poletto, G.; di Maira, G.; Cozza, G.; Ruzzene, M.; Sarno, S.; Bain, J.; Elliott, M.; Moro, S.; Zagotto, G.; et al. Tetrabromocinnamic acid (TBCA) and related compounds represent a new class of specific protein kinase CK2 inhibitors. *ChemBioChem* **2007**, *8*, 129–139. [CrossRef] [PubMed]
35. Chen, B.-C.; Hynes, J., Jr.; Pandit, C.R.; Zhao, R.; Skoumbourdis, A.P.; Wu, H.; Sundeen, J.E.; Leftheris, K. A general largescale synthesis of 2-alkyl-7-methoxyindoles. *Heterocycles* **2001**, *55*, 951–960. [CrossRef]
36. Bandgarand, B.P.; Uppalla, L.S. Gel entrapped base catalyzed (GEBC) Henry reaction: Synthesis of conjugated nitroalkenes. *Synth. Commun.* **2000**, *30*, 2071–2075. [CrossRef]

37. Tanemura, K.; Suzuki, T. Base-catalyzed reactions enhanced by solid acids: Amine-catalyzed nitroaldol (Henry) reactions enhanced by silica gel or mesoporous silica SBA-15. *Tetrahedron Lett.* **2018**, *59*, 392–396. [CrossRef]
38. Li, T.; Miras, H.N.; Song, Y. Polyoxometalate (POM)–Layred Double Hydroxides (LDH) Composite Materials: Design and Catalytic Applications. *Catalysts* **2017**, *7*, 260. [CrossRef]
39. Nakhate, A.V.; Rasal, K.B.; Deshmukh, G.P.; Gupta, S.S.R.; Mannepalli, L.K. Synthesis of quinoxaline derivatives from terminal alkynes and o-phenylenediamines by using copper alumina catalyst. *J. Chem. Sci.* **2017**, *129*, 1761–1769. [CrossRef]

© 2018 by the authors. Licensee MDPI, Basel, Switzerland. This article is an open access article distributed under the terms and conditions of the Creative Commons Attribution (CC BY) license (http://creativecommons.org/licenses/by/4.0/).

MDPI
St. Alban-Anlage 66
4052 Basel
Switzerland
Tel. +41 61 683 77 34
Fax +41 61 302 89 18
www.mdpi.com

Catalysts Editorial Office
E-mail: catalysts@mdpi.com
www.mdpi.com/journal/catalysts

www.ingramcontent.com/pod-product-compliance
Lightning Source LLC
LaVergne TN
LVHW071951080526
838202LV00064B/6718